Topics in Applied Physics   Volume 25

W0235165

Topics in Applied Physics Founded by Helmut K. V. Lotsch

Laser Beam Propagation in the Atmosphere

Edited by J. W. Strohbehn

With Contributions by

S.F.Clifford M.E.Gracheva A.S.Gurvich A.Ishimaru
S.S.Kashkarov V.V.Pokasov J.H.Shapiro
J.W.Strohbehn P.B.Ulrich J.L.Walsh

With 78 Figures

Springer-Verlag Berlin Heidelberg GmbH 1978

Professor *John W. Strohbehn*

Thayer School of Engineering, Dartmouth College, Hanover, NH 03755, USA

ISBN 978-3-662-31162-2   ISBN 978-3-540-35826-8 (eBook)
DOI 10.1007/978-3-540-35826-8

Library of Congress Cataloging in Publication Data. Main entry under title: Laser beam propagation in the atmosphere. (Topics in applied physics; v. 25). Bibliography: p. Includes index. 1. Laser beams-Atmospheric effects. 2. Atmosphere – Laser observations. 3. Atmospheric turbulence. I. Strohbehn, J. W., 1936-. II. Clifford, Steven Francis, 1943-. QC976.L36L37   551.5'27   78-19100

This work is subject to copyright. All rights are reserved, whether the whole or part of the material is concerned, specifically those of translation, reprinting, re-use of illustrations, broadcasting, reproduction by photocopying machine or similar means, and storage in data banks. Under §54 of the German Copyright Law, where copies are made for other than private use, a fee is payable to the publisher, the amount of the fee to be determined by agreement with the publisher.

© by Springer-Verlag Berlin Heidelberg 1978
Originally published by Springer-Verlag Berlin Heidelberg New York in 1978
Softcover reprint of the hardcover 1st edition 1978

The use of registered names, trademarks, etc. in this publication does not imply, even in the absence of a specific statement, that such names are exempt from the relevant protective laws and regulations and therefore free for general use.

2153/3130-543210

Preface

When optical waves, whether they originate from starlight, incandescent sources, or lasers, propagate through the clear atmosphere of the earth, they experience distortions due to small temperature variations. The two most common manifestations of this phenomena are the twinkling of stars and the shimmering of "heat" over hot pavement or surfaces. The small temperature variations, on the order of $1°$ C or less, are related to the sun's heating of the atmosphere and turbulent motion of the air due to winds and convection. The small temperature perturbations cause changes in the velocity of the light as it passes through the atmosphere, which in turn causes distortions in the intensity (twinkling), phase, angle-of-arrival, and displacement (quivering) of the light beam.

With the invention of the laser in 1962 and the myriad of possible practical applications, from lidar (laser radars) to communications, interest in atmospheric effects on optical waves rapidly expanded. In this volume we attempt to cover the latest developments in this field. Earlier monographs on this subject include those of Chernov (*Wave Propagation in a Random Medium*) and Tatarskii (*Wave Propagation in a Turbulent Medium*, and *the Effects of the Turbulent Atmosphere on Wave Propagation*). The emphasis in this volume is on situations where temperature variations in the atmosphere dominate the laser propagation characteristics. Most of the volume is concerned with problems dominated by the earth's atmospheric turbulence, except for the last chapter which considers the case where the laser beam itself is so powerful that it produces the heating of the atmosphere. This volume does not consider effects on laser beams caused by absorption of the light energy by molecules or particles nor does it consider scattering of the light by aerosols or rain or fog.

The first chapter is an introduction and attempts to provide an overview of the subject and the material included in this volume. The second chapter, by S. F. Clifford, presents the classical theory of optical propagation through a turbulent atmosphere including a description of the atmospheric model. This chapter represents the state of the art of the theory until the middle 1960's and is applicable for a wide range of practical problems. The next chapter, by J. W. Strohbehn, attempts to summarize and compare the modern approaches to laser propagation through the turbulent atmosphere, with particular attention to the region where the classical theory has been shown not to be valid. This theory is necessary for laser propagation over long paths and when the strength of the refractive index turbulence is high. Chapter 4 describes the results of a very thorough set of measurements, performed by M. E. Gracheva et al. in 1972, on laser signals in the region where the classical theory breaks down. While the theory in the chapters by Clifford and Strohbehn assumes the initial wave is plane or spherical, Chapter 5, by A. Ishimaru, concentrates on the subtleties of the theory presented by finite-size laser beams which may or may not be focused. This chapter also introduces the problem of using measurements of laser signals to remotely probe atmospheric characteristics such as the strength of the turbulence or the velocity of the wind. J. Shapiro, in Chapter 6, addresses the communications and imaging problems when an optical signal passes through the turbulent atmosphere. Based on the results of the preceding chapters he defines the optimum receiver for communications and analyzes the receiver's performance. Finally, in the last chapter, J. Walsh and P. Ulrich consider the problem known as thermal blooming. In this situation, the intensity of the laser beam is so great that it heats the atmosphere through which it passes. The heated atmosphere distorts the original laser beam so that the beam size is expanded (blooming) and, in the presence of a wind, displaces the position of the beam. These authors give one of the first thorough reviews of this interesting topic.

It is my pleasure to acknowledge some of the debts I have incurred and the inspiration I have received. In particular, I would like to thank A. Waterman, Jr. who got me started on radio-wave propagation, V. I. Tatarskii with whom I had many interesting discussions on the modern theories of laser propagation, and M. G. Morgan who provided an ideal environment in which to work. Besides the contributors, discussions with R. Lawrence, W. P. Brown, D. Fried, R. Fante, and H. Yura materially influenced the direction of this volume. The authors who wrote most of the text have worked hard and patiently to produce high quality manuscripts. Dean C. F. Long generously provided the time and support necessary for the writing and editing of this volume. Janet Brown quickly and accurately typed my contributions as well as Chapter 7, and Cutting Johnson did a superb job with the figures in Chapter 3, 4, and 7. Finally, a special thanks to my children, Jo, Kris, and Carolyn, for cheerfully enduring the many extra hours and late meals while this project was underway—it is to them that my efforts are dedicated.

Hanover, New Hampshire
July, 1978

J. W. Strohbehn

Contents

Contributors

Clifford, Steven F.
  Wave Propagation Laboratory, Environmental Research Laboratories,
  National Oceanic and Atmospheric Administration,
  Boulder, CO 80302, USA

Gracheva, M. E.
  Institute of Atmospheric Physics, Pyzhevskii Per 3, Moscow ZH-17, USSR

Gurvich, A. S.
  Institute of Atmospheric Physics, Pyzhevskii Per 3, Moscow ZH-17, USSR

Ishimaru, Akira
  Department of Electrical Engineering, University of Washington,
  Seattle, WA 98195, USA

Kashkarov, S. S.
  Institute of Atmospheric Physics, Pyzhevskii Per 3, Moscow ZH-17, USSR

Pokasov, Vl. V.
  Institute of Atmospheric Physics, Pyzhevskii Per 3, Moscow ZH-17, USSR

Shapiro, Jeffrey H.
  Department of Electrical Engineering and Computer Science, and
  Research Laboratory of Electronics, Massachusetts Institute of Technology,
  Cambridge, MA 02139, USA

Strohbehn, John W.
  Thayer School of Engineering, Dartmouth College,
  Hanover, NH 03755, USA

Ulrich, Peter B.
  Naval Research Laboratories, Washington, DC 20375, USA

Walsh, John L.
  Naval Research Laboratories, Washington, DC 20375, USA

1. Introduction

J. W. Strohbehn

This book focuses on the problem of the propagation of laser beams through the atmosphere. In particular it is concerned with line-of-sight propagation problems, i.e., situations where if there were no atmosphere and the waves were propagating in a vacuum the receiver would observe a steady signal from the transmitter. (An example of a non-line-of-sight path would be the case where the transmitter and receiver are separated physically by a mountain, but use scattering from clouds to propagate the signal.) This book is primarily devoted to the question of how the intervening medium affects the properties of the received signal. When categorizing the effects of the medium, four broad categories can be defined: 1) propagation through turbulent media, 2) propagation through turbid media, 3) absorption effects, and 4) high power laser effects. This volume is devoted to the first and last of these categories.

Propagation through the turbulent atmosphere refers to situations where a laser beam is propagating through the clear atmosphere but very small changes in the refractive index are present. These small changes in refractive index, which are typically on the order of 10^{-6}, are related primarily to the small variations in temperature (on the order of $0.1-1°$ C), which are produced by the turbulent motion of the atmosphere. Discussion of the refractive index fluctuations and their relation to turbulence in the atmosphere is given in detail in Chapter 2. While the refractive index variation from the mean value is very small, in a typical situation of practical interest a laser beam propagates through a large number of refractive index inhomogeneities, and hence the cumulative effect can be very significant. Significant enough in fact that five of the following six chapters are devoted to the understanding and ramifications of this topic. These small inhomogeneities initially produce optical phase effects which in turn lead to angle-of-arrival fluctuations or beam wander, intensity fluctuations or scintillations, and beam broadening.

Propagation through a turbid medium refers to waves propagating through a medium composed of large numbers of discrete scatters or aerosol particles (e.g., rain, fog, or dust), which give rise to strong scattering effects. While it is possible to envision a unified theoretical treatment that describes propagation through turbulent and turbid media, the major effects are so different that such a unified treatment is of little value. Note, however that both topics are discussed by *Ishimaru* [1.1, 2]. Two related topics are a discussion of the subject of aerosol particles [1.3] and the use of lasers in monitoring the atmosphere [1.4].

The major distinctions between the two cases are: 1) in turbid media the refractive index variations are large, on the order of unity, while in turbulent media they are small, on the order of 10^{-6}; and 2) in turbid media the particles are discrete, giving sharp variations in refractive index compared to a wavelength while in turbulent media the refractive index variations are smooth, continuous functions of space. Because of these distinctions, in turbid media a single particle gives strong scattering of an optical wave, the scattering tends to be in all directions (including backscattering), depolarization effects are usually significant, and the average beam intensity is strongly attenuated. In contrast, in turbulent media the scattering of a wave by a single inhomogeneity is weak, backscattering and depolarization effects can be ignored, the forward scattering is localized in a fairly small angle (on the order of milliradians), and the average intensity of the beam is not greatly attenuated. Another distinction is that in propagation through turbid media the time variations are usually unimportant but in turbulent media time variations of the signal are very pronounced.

In this volume we have not discussed the problem of propagation through turbid media. The interested reader is referred to the thorough discussion of this topic in the books by *Zuev* [1.5, 6] as well as the publications by *Ishimaru* [1.1, 2].

The third category of effects is absorption, which refers to the direct absorption of energy from the optical wave by atmospheric gases. While laser propagation is clearly a topic of critical importance, it is discussed in detail by *Zuev* [1.5, 6] and also by *Goody* [1.7] and is not included here.

The fourth category is high power effects. By high power effects is meant situations where the energy in the laser beam is great enough so that it can significantly change the properties of the medium. This category can be broken down into three subcategories. First, there is the problem of passage of a laser beam through a clear atmosphere which may be weakly absorbing, i.e., part of the energy of the optical radiation is absorbed by the atmosphere gases. This heating of the atmosphere by the laser beam causes a refractive index change in the medium and hence alters the propagation conditions. In the most obvious case where the intensity of the beam is greatest at its center, the heating is also greatest near the center of the beam and weaker near the edges. This nonuniform heating pattern produces a nonuniform variation in the refractive index which causes the beam to diverge or spread over a larger area than it would in free space. This phenomenon is usually called thermal blooming, and is covered in some detail in Chapter 7, providing the first thorough overview of this topic in the open literature.

There are other high power laser effects which are of interest in laser propagation but which are not included in this volume. One of the interesting problems when considering high power laser propagation through a turbid medium is the effect of the laser on the aerosol particles. In particular, when propagating through rain or fog, the laser energy may cause the water droplets to evaporate or explode, thus leading to changes in the propagation conditions [1.8, 9]. The drop evaporation or explosion, depending on the situation, can lead to less or greater absorption by an optical wave. Finally, if the laser beam is intense enough it can cause ionization of the medium and breakdown effects, again a topic that is not included in this volume but is covered elsewhere [1.10].

In summary, this book is concerned with the effects of turbulence [1.11, 12] and thermal blooming on laser propagation.

1.1 A Brief History

One of the interesting facts about the history of work related to laser propagation through turbulent media is that even though the first working laser was not announced until 1960 [1.13], much of the necessary theoretical work and some of the experimental work had been done before that date, primarily in the Soviet Union. The two initial monographs in this field by *Chernov* [1.14] and *Tatarskii* [1.15] were both published in the Soviet Union before 1960 and published in translation in the United States in 1960 and 1961, respectively. The reason for this early work is that the topic of laser propagation in the atmosphere is really a subtopic of a more general problem, i.e., the propagation of waves in a turbulent medium. This general topic includes a number of other practical applications, such as propagation of starlight through the atmosphere, propagation of sound waves through the atmosphere and ocean, propagation of microwaves through planetary atmospheres, and propagation of radio waves through the ionosphere and interplanetary space. The monographs by *Chernov* and *Tatarskii* were largely motivated by the first two topics. While our book concentrates primarily on topics of importance in laser propagation, a large portion of the material will be of interest to workers in the other areas noted above. Because of the tremendous overlapping of fields, it is difficult to trace a coherent thread through the developments in this field.

Clearly, the most well-known effect of turbulence in the atmosphere is the twinkling (or scintillation) of stars, which refers to the irregular change in brightness of the image. In the early 1950s, scientific interest in the twinkling and quivering of stellar images in telescopes became quite active and a number of papers appeared on this topic. Early experimental work on scintillations was reported by *Mikesell* [1.16] and many others [1.17–21]. Quivering refers to irregular displacements of the image of a star in random directions and is related to the fluctuations in angle of arrival of the incoming light wave. The experimental work on this problem was reviewed by *Kolchinski* [1.22, 23]. More recent directions relating to speckle phenomena are covered by *Dainty* [1.24].

Interest in scattering of sound waves by turbulence was initiated as early as 1941 [1.25] and the topic has been studied extensively since. *Krasilnikov* [1.26, 27] used a geometrical optics approach, while *Mintzer* [1.28–30] applied diffraction theory to the acoustic propagation problem, as did *Obukhov* [1.31] and *Tatarskii* [1.32]. More recent work in this field is discussed in [1.14, 15].

As expected, early work in this field centered on geometrical optics techniques [1.26, 27, 33, 34], and the limitations on this technique were discussed by *Ellison* [1.35] and *Tatarskii* [1.36]. A different approach to the geometrical optics problem for random medium, suggested by *Chernov* [1.37], led to an angular distribution function of the rays that obeys the Einstein-Fokker-Kolmogorov equation. In this approach the propagation of the ray is described as a continuous Markov process.

Given that the limitations in the geometrical optics approximations restrict the path lengths quite severely, for example for optical waves the solutions are valid only for path lengths on the order of tens of meters, attention soon focused on wave optics or diffraction theory techniques [1.28–32]. Both the geometrical optics and the wave optics methods were based on perturbation theories and assumed that the scintillations introduced in the wave were small compared to the mean value. The first attempt to circumvent this restriction was by *Obukhov* [1.31] who used a method suggested by *Rytov* [1.38], which involves taking the logarithm of the field before applying the perturbation technique. When applying the perturbation method to the usual wave equation (Helmholtz equation), the perturbation method leads to the Born approximation. By writing the field in terms of its logarithm, the wave equation is transformed into the Ricatti equation. Application of the perturbation technique at this point tends to be called the method of smooth perturbation (MSP) in the Soviet literature and Rytov's method in the western literature. The respective merits of these two methods generated a large discussion in the literature, probably more than it was worth, and the topic is discussed in some extent in Chapter 2.

The major new development in this field was in the 1960s when *Gracheva* and *Gurvich* [1.39, 40] showed experimentally that the basic theoretical results for the variance of the logarithm of the irradiance were completely inapplicable for long path lengths. Chapter 3 is devoted to theoretical attempts to extend the theory to path lengths where the weak perturbation theory breaks down. The more recent history of the development of these fields is given in the respective chapters.

The historical developments related to the thermal blooming problem are much more difficult to trace since much of the early work in this area was not reported in the open literature. *Walsh* and *Ulrich* describe some of this development in Chapter 7.

1.2 Organization

The body of the material in this book is divided into six chapters. In Chapter 2, *Clifford* describes the classical theory relating to wave propagation through the atmosphere. The first section discusses the properties of the atmosphere important for laser beam propagation. In particular, a turbulence model for the refractive index fluctuations is developed and some details on the applicability of that model in the earth's atmosphere are given. In Section 2.2, expressions for the covariance functions and spectra of the amplitude and phase of a plane wave propagating through a turbulent medium are derived, based on the classical technique of weak perturbations. In Section 2.3, comparisons between the experimental work and the classical theoretical predictions are discussed.

In Chapter 3 *Strohbehn* discusses and compares the recent theories developed to explain phenomena in the region where the wave fluctuations are strong and the weak perturbation theory is not applicable. In practical applications, the classical results are not valid for long path lengths or strong atmospheric turbulence. These conditions exist most often for propagation paths close to the ground on warm summer days. None of the presently existing theories completely solves the strong fluctuation problem, but significant progress has been made. In this chapter, four different methods are discussed: 1) the diagram method, 2) the Markov approximation, 3) the local method of smooth perturbations, and 4) heuristic theories. The first three methods are essentially equivalent and lead to the same set of partial differential equations to describe the various moments of the field, e.g., the mean field, the mutual coherence function, or the intensity covariance function. Section 3.5 describes a less mathematical but perhaps more physically understandable method of deriving some of the results in the strong fluctuation region. Section 3.6 presents the results that have been derived by solving the differential equations for the mean field, the mutual coherence function, and the spectrum of the irradiance. Results for the irradiance probability distribution are also mentioned.

Chapter 4 is a translation of an article published in 1972 by *Gracheva* and *Gurvich* and some of their co-workers adapted to fit the context of the present book. It is definitely the most thorough piece of experimental work performed to date on laser beam propagation through the turbulent atmosphere at long path lengths. Some extremely interesting results are given in this chapter on the important scaling parameters, the form of the irradiance probability distribution, and the behavior of the variance of irradiance fluctuations for long path lengths or strong turbulence. In addition, measurements on the irradiance covariance function and spectrum are reported.

Most of the work mentioned in Chapters 2–4 assumes that the transmitted wave is plane or approximately plane. However, for most situations of interest, the wave is a beam of limited width and may be collimated or focused at the receiver. *Ishimaru* in Chapter 4 discusses the ramifications of beam wave propagation for both the weak and strong fluctuation cases. In the second half of Chapter 4 *Ishimaru* investigates the remote sensing problem. For most communication or transmission applications the atmospheric turbulence distorts the desired signal and causes a degradation in system performance. However, for remote sensing applications, the distorted signal provides a means by which various atmospheric parameters can be measured. For example *Ishimaru* describes how the average wind speed along a beam transmission path can be deduced from the received signal. A more general view of this topic is also covered in [1.4, 41].

In Chapter 6 *Shapiro* looks specifically at the effect of the atmospheric channel on imaging and optical communication systems. In Section 6.1 the propagation model is reviewed. This model is based in large part on the results of the preceding chapters, but some new concepts, such as normal-mode decomposition of the transmitter-atmosphere-receiver system are introduced. Section 6.2 concentrates on imaging systems including incoherent sources; thin lens imaging, interferometric imaging, and phase-compensated imaging. The final subsection deals with the important topic of optimum imaging. Section 6.3 reviews communication applications including the earth-space propagation channel, statistical models for optical detection, diffraction-limited reception, and diversity reception. The chapter concludes with a review of the topic of reciprocity pointing. Extensions of these topics as applied to the detection problem are covered by [1.42–44].

Chapter 7 is devoted to an entirely new topic, thermal blooming in the atmosphere. This is an effect due solely to the high power of the laser beam, which heats the medium and causes its refractive index to change. This chapter is one of the first thorough reviews of this important topic. Following Section 7.1, which gives a general overview of the problem, Section 7.2 covers the electromagnetic theory pertinent to the problem, while Section 7.3 derives the relevant fluid mechanics equations. Section 7.4 discusses the approximate analytical solutions of the electromagnetic and fluid mechanics equations including geometrical optics and wave optics approaches as well as the pasted phase approximation. The chapter concludes (Sect. 7.5) with a review of the pertinent numerical methodologies suitable for computer use and a presentation of some of the computer results.

References

1.1   A. Ishimaru: *Wave Propagation and Scattering in Random Media* (Academic Press, New York 1977)
1.2   A. Ishimaru: Proc. IEEE **65**, 1030 (1977)
1.3   W. H. Marlow (ed.): *Aerosol Dynamics*, Topics in Current Physics (Springer, Berlin, Heidelberg, New York) (in preparation)
1.4   E. D. Hinkley (ed.): *Laser Monitoring of the Atmosphere*, Topics in Applied Physics, Vol. 14 (Springer, Berlin, Heidelberg, New York 1976)
1.5   V. E. Zuev: *Atmospheric Transparency in the Visible and the Infrared* (Israel Program for Scientific Translation 1970)
1.6   V. E. Zuev: *Propagation of Visible and Infrared Radiation in the Atmosphere* (Wiley and Sons, New York 1974)
1.7   R. M. Goody: *Atmospheric Radiation, Vol. 1, Theoretical Basis* (Oxford University Press, Oxford, England 1964)
1.8   G. L. Mullaney, W. H. Christiansen, D. A. Russell: Appl. Phys. Lett. **13**, 143 (1968)
1.9   J. E. Lowder, H. Kleiman, R. W. O'Neil: J. Appl. Phys. **45**, 221 (1974)
1.10 Yu. P. Rayzer: *Laser Spark and Propagation of Discharges* (Izdatelstvo "Nauka", Moscow 1974)

1.11 J.L.Lumley, H.A.Panofsky: *The Structure of Atmospheric Turbulence* (Interscience, New York 1964)
1.12 P.Bradshaw (ed.): *Turbulence*, Topics in Applied Physics, Vol. 12, 2nd ed. (Springer, Berlin Heidelberg, New York 1978)
1.13 T.H.Maiman: Nature **187**, 493 (1960)
1.14 L.A.Chernov: *Wave Propagation in a Random Medium* (McGraw-Hill, New York 1960)
1.15 V.I.Tatarskii: *Wave Propagation in a Turbulent Medium* (McGraw-Hill, New York 1961)
1.16 A.H.Mikesell, A.A.Hoag, J.S.Hall: J. Opt. Soc. Am. **41**, 689 (1951)
1.17 F.Nettlebald: Observatory **71**, 111 (1951)
1.18 H.E.Butler: Proc. Roy. Inst. Acad. A**54**, 321 (1954)
1.19 M.A.Ellison, H.Seddon: Monthly Notices Roy. Astron. Soc. **112**, 73 (1952)
1.20 H.E.Butler: Quart. J. Roy. Meteorol. Soc. **80**, 241 (1954)
1.21 W.M.Protheroe: "Preliminary report on stellar scintillation", Contrib. Perkins Observ., Ser. **2**, 127 (1955)
1.22 I.G.Kolchinski: Astron. Zh. **29**, 350 (1952)
1.23 I.G.Kolchinski: Astron. Zh. **34**, 638 (1957)
1.24 J.C.Dainty: *Laser Speckle and Related Phenomena*, Topics in Applied Physics, Vol. 9 (Springer, Berlin, Heidelberg, New York 1975)
1.25 A.M.Obukhov: Doklady Akad. Nauk. SSSR **30**, 611 (1941)
1.26 V.A.Krasilnikov: Doklady Akad. Nauk. SSSR **47**, 486 (1945)
1.27 V.A.Krasilnikov: Doklady Akad. Nauk. SSSR **58**, 1353 (1947)
1.28 D.Mintzer: J. Acoust. Soc. Am. **25**, 922 (1953)
1.29 D.Mintzer: J. Acoust. Soc. Am. **25**, 1107 (1953)
1.30 D.Mintzer: J. Acoust. Soc. Am. **26**, 186 (1954)
1.31 A.M.Obukhov: Izv. Akad. Nauk SSSR, Ser. Geograf. Geofiz. **2**, 155 (1953)
1.32 V.I.Tatarskii: Zh. Eksp. Teor. Fiz. **25**, 74 (1953)
1.33 P.G.Bergmann: Phys. Rev. **70**, 486 (1946)
1.43 V.A.Krasilnikov: Izv. Akad. Nauk SSSR, Ser. Geograf. Geofiz. **13**, 33 (1949)
1.35 T.H.Ellison: J. Atmos. Terr. Phys. **2**, 14 (1951)
1.36 V.I.Tatarskii: Zh. Eksp. Teor. Fiz. **25**, 84 (1953)
1.37 L.A.Chernov: Zh. Eksp. Teor. Fiz. **24**, 210 (1953)
1.38 S.M.Rytov: Izv. Akad. Nauk. SSSR (Ser. Fiz.) **2**, 223 (1937)
1.39 M.E.Gracheva, A.S.Gurvich: Izv. VUZ, Radiofiz. **8**, 717 (1965)
1.40 M.E.Gracheva: Izv. VUZ, Radiofiz. **10**, 775 (1967)
1.41 H.P.Baltes (ed.): *Inverse Source Problems in Optics*, Topics in Current Physics (Springer, Berlin, Heidelberg, New York) (in preparation)
1.42 R.J.Keyes (ed.): *Optical and Infrared Detectors*, Topics in Applied Physics, Vol. 19 (Springer, Berlin, Heidelberg, New York 1977)
1.43 B.Saleh: *Photoelectron Statistics*, Springer Series in Optical Sciences, Vol. 6 (Springer, Berlin, Heidelberg, New York 1977)
1.44 R.H.Kingston: *Detection of Optical and Infrared Radiation*, Springer Series in Optical Sciences, Vol. 10 (Springer, Berlin, Heidelberg, New York 1978)

2. The Classical Theory of Wave Propagation in a Turbulent Medium

S. F. Clifford

With 9 Figures

An important consideration in any description of laser propagation in the atmosphere is the geometry of the optical path. The influence of various kinds of optical scatterers is often quite sensitive to the transmitter-receiver configuration. For example, the type of atmospheric refractive index irregularities that we consider important in this work, where forward scatter dominates, has a negligible effect when we calculate their contribution to backscatter. We restrict our attention in this book to the problem most commonly known as line-of-sight optical propagation. The receiver is in full view of the transmitter such that, in the absence of the atmosphere, the irradiance would be constant in time, with a value determined by the transmitter geometry plus vacuum diffraction effects. The introduction of the atmosphere between source and receiver produces spatial and temporal fluctuations in the received irradiance and other wave parameters such as phase, angle of arrival, and frequency that are due to random variations in atmospheric refractive index along the optical path.

This problem should be differentiated from wave scattering where the scattering volume is remote from both the transmitter and receiver and, in the absence of refractivity fluctuations to deflect some of the incident wave energy, there would be no received signal. In the line-of-sight case, the transmitter or receiver, or both, are usually imbedded in the turbulent medium, and the intervening refractive index fluctuations merely perturb the vacuum parameters of the wave.

There are two areas where analysis of this model has direct practical relevance. In communications, the motivation for studying the problem arises from the desire to take advantage of the potentially wide bandwidth of an optical system. In this context, the atmospherically induced noise degrades system performance and is something to be overcome. In contrast, this model is also being studied for its potential for remotely sensing atmospheric turbulence parameters, in which context the random fluctuations imposed by the atmosphere are the signal to be studied and the information is contained in them.

In this chapter, we are concerned with the effect on an optical wave of refractive index fluctuations caused by turbulent effects in the atmosphere. As we shall see, these refractive index fluctuations are primarily the result of small temperature fluctuations transported by the turbulent motion of the atmosphere. Changes in the optical signal due to absorption or scattering by molecules or aerosols are not discussed. The changes in refractive index caused by temperature fluctuations are usually smooth random functions of both space and time. As an aid in understanding the effects of the refractive index fluctuations, a region of high or low refractive index is often thought of as an eddy, which may behave very much as a lens. In this model the atmosphere may be thought of as a large number of random lenses, having different shapes and scale sizes, that move randomly through space. It should be kept in mind, however, that the refractive index is varying within an eddy, and no discontinuities in the refractive index are present.

In the earth's atmosphere, the interaction of an electromagnetic wave with a single refractive index fluctuation or eddy is extremely weak. To first order, there is no amplitude change at the position of the eddy, but merely a phase change caused by the change in velocity of the small segment of the wave that traverses the eddy. Because induced phase fluctuations are not the same at different points perpendicular to the direction of propagation, they may cause focusing or defocusing effects, local deviations in the direction of wave propagation, and eventually, through interference, irradiance fluctuations at the receiver. Because these eddies do not have sharp edges, we may ignore reflections. Similarily, the bounding surfaces of the fluctuating region are assumed not to have any sharp discontinuities. The process we consider here is merely diffraction of the wave by random inhomogeneities. As we shall see in Chapter 3, although the effect of an individual eddy on the wave is quite weak, the cumulative effect of many eddies can be extremely strong.

2.1 Description of the Atmosphere

2.1.1 Atmospheric Refractive Index

The refractive index of air at optical frequencies is given approximately by

$$n-1 = 77.6(1 + 7.52 \times 10^{-3}\lambda^{-2})(P/T)10^{-6}, \tag{2.1}$$

where P is the atmospheric pressure in millibars, T is the temperature in Kelvin, and λ is the wavelength of light in μm. Therefore, $n_1 = n-1$ is a measure of the deviation of the refractive index from its free space value, and at sea level it has a typical value of 3×10^{-4}. In writing (2.1), we have neglected a term that depends on the water vapor pressure. Over land, this humidity-dependent term contributes significantly less than 1 %. (For a more extensive analysis of optical refractive index with formulas valid to one part in 10^7, see *Owens* [2.1].) *Wesely* and *Alcaraz* [2.2] and *Friehe* et al. [2.3] indicated that a third contribution to (2.1) arises from the product of the temperature and humidity and that this term could significantly affect the refractive index on over-water paths. We shall be content here to use (2.1) and to ignore the refinements in the model described in [2.2].

2.1.2 The Role of Atmospheric Turbulence

The refractive index fluctuations that we are considering are those that result from the naturally occurring random fluctuations in wind velocity called turbulence. The description and phenomenology of turbulence constitute a vast field, perhaps best described in [2.4–7]. The following arguments will merely sketch the derivation outlined in these texts.

Because turbulence is a random process, it must be described in terms of statistical quantities. It is hopeless to attempt to predict the velocity of a given parcel of air as it is buffeted about in the open atmosphere. In fact, it is nearly impossible to do a complete statistical analysis; therefore, we settle for a more meager description. *Kolmogorov* [2.8], when considering the time variation of the difference of two velocities at two points in space separated by a displacement vector r, found that the mean-square velocity difference could be described by a universal form over a rather broad range of spatial scale sizes of motion. His so-called structure tensor is defined in general by

$$D_{ij}(r) = \langle [v_i(r_1 + r) - v_i(r_1)] [v_j(r_1 + r) - v_j(r_1)] \rangle, \tag{2.2}$$

where v_i and v_j refer to the different components of the velocity, and the angle brackets indicate an ensemble average. Two further assumptions greatly simplify (2.2). The first is the assumption of local homogeneity, which implies that the velocity difference statistics depend only on the displacement vector r. The second is the assumption of local isotropy, which implies that only the magnitude of r is important. The adjective "local" applies because we are considering differences between the velocities at two points and hence only scale sizes of the order of the separation are important (velocity components with scale sizes larger than the separation will tend to be filtered out by the subtraction process).

The impact of these two assumptions on (2.2) allows D_{ij} to be written in the form

$$D_{ij}(r) = [D_{rr}(r) - D_{tt}(r)] n_i n_j + D_{tt}(r) \delta_{ij}, \tag{2.3}$$

where $\delta_{ij} = 1$ for $i = j$, $\delta_{ij} = 0$ for $i \neq j$, and the n_i are components of the unit vector directed along r. The quantities D_{rr} and D_{tt} are, respectively, the structure functions of the wind velocity component parallel and transverse to the displacement r. By making one further assumption of incompressible turbulence, that is, $V \cdot v = 0$ (*Batchelor* [2.4]), we can write D_{tt} in terms of D_{rr}, that is,

$$D_{tt} = \frac{1}{2r} \frac{d}{dr} (r^2 D_{rr}). \tag{2.4}$$

The important result of the analysis leading to (2.3) and (2.4) is that the statistical structure of turbulence can now be described by the single structure function D_{rr} where

$$D_{rr} = \langle [v_r(r_1 + r) - v_r(r_1)]^2 \rangle . \tag{2.5}$$

Kolmogorov [2.8] found that as long as the separation r lay in the so-called inertial subrange of turbulence, D_{rr} has a universal form

$$D_{rr} = C_v^2 r^{2/3}, \tag{2.6}$$

where C_v^2 is the velocity structure constant, a measure of the total amount of energy in the turbulence. This form should persist for values of r between the microscale of turbulence l_0 up to the outer scale of turbulence L_0. The microscale corresponds to the eddy size below which dissipation of energy in the eddy through viscous effects becomes important. The outer scale corresponds to the largest scale size for which the eddies may be considered to be isotropic. (Near the ground l_0 is of the order of millimeters and L_0 of the order of meters. Both quantities appear to increase with height above ground.) If this result is true, and indeed the amount of supporting experimental evidence is amazing (*Myrup* [2.9] and *Kaimal* et al. [2.10]), then (2.6) is all we need to know about turbulence to model most optical propagation through the atmosphere.

2.1.3 The Turbulence Model of Refractivity Fluctuations

We have yet to relate our turbulence model to the refractivity formula (2.1). A critical intermediate step involves the concept of a conservative passive additive (*Tatarskii* [2.11, 12]). Conservative quantities have the property that, if we can identify an air parcel with a certain characteristic number, it will then retain that number no matter how that parcel moves about in space. Strictly speaking, temperature is not a conservative quantity. Any vertical displacement of an air parcel will produce a temperature change because, at any given height, the parcel will tend to equalize its pressure with that of the background. This pressure change produces adiabatic temperature changes; consequently, if we define a new quantity, that is the difference between the absolute temperature T and the changes in temperature with height caused by an adiabatic lapse rate, we shall obtain a conservative quantity. For this reason we shall analyze the fluctuations in the potential temperature $\Theta = T - \gamma_a z$, where $\gamma_a = 9.8°$ C/km, and relate the statistics of Θ to the velocity fluctuation statistics. Passive additives do not affect the turbulence dynamics; that is, when they are inserted into the turbulent medium they do not affect its statistics, so that the two-thirds law described in (2.6) still applies for the velocity fluctuations. It was first shown by *Corrsin* [2.13] that conservative passive additives also obey the two-thirds law;

thus, there is a well-defined universal structure function for the potential temperature variations, and it has the form

$$D_\Theta(r) = C_\Theta^2 r^{2/3} \tag{2.7}$$

in its inertial subrange. (The inertial subrange of the temperature fluctuations is defined by slightly different scale sizes (*Tatarskii* [2.11, 12]) than the velocity fluctuations. This is unimportant for our discussion.) All that remains in order to construct our statistical model of the refractivity variations, that is, to find its two-thirds law, is to relate the refractive index n to one or more conservative passive additives.

Equation (2.1) defines n in terms of the temperature T and pressure P. To calculate a structure function for n, we must know how to relate changes in refractive index to changes in pressure and temperature. We accomplish this by finding the differential of (2.1)

$$\delta n = \frac{79P}{T} \left(\frac{\delta P}{P} - \frac{\delta T}{T} \right) \times 10^{-6}. \tag{2.8}$$

[In calculating (2.8), we have assumed a value for λ in (2.1) corresponding to red light, $\lambda \approx 0.6 \times 10^{-6}$ m.] If the pressure fluctuations are measured at a point fixed with respect to the ground, they are relatively small and rapidly dispersed. Consequently, the observed refractive index variations will be produced almost exclusively by temperature fluctuations, and we can neglect the $\delta P/P$ term. From the definition of potential temperature Θ, we can see that refractivity variations δn at a fixed height will be a function of $\delta \Theta$ alone, so that

$$\delta n = -79P \frac{\delta \Theta}{T^2} \times 10^{-6}. \tag{2.9}$$

Finally, because $\delta \Theta$ is a conservative passive additive, δn is also, and we may write a two-thirds law for the refractivity fluctuations

$$D_n(r) = C_n^2 r^{2/3}. \tag{2.10}$$

To complete the calculation, from (2.9) we can write the refractive index structure constant C_n^2 directly in terms of C_Θ^2 as

$$C_n^2 = \left(\frac{79P}{T^2} \times 10^{-6} \right)^2 C_\Theta^2. \tag{2.11}$$

We have now shown that the refractive index fluctuations have the same statistical properties, that is, obey the same structure function law as velocity *turbulence*, and that variations in optical refractive index are caused by

variations in temperature. Equation (2.11) relates the refractivity and temperature structure parameters with a coefficient that depends only upon the mean
atmospheric pressure and temperature. Hereafter in the text, we shall follow the
usual convention and write C_T^2 in place of C_θ^2; however, it is important to
remember in the following sections that the significant physical quantity is the
structure constant of potential temperature.

Finally it should be noted that the structure constant is not really a
constant, but a function of both time and space. It is really a coefficient that
describes the strength of the refractive index turbulence. In practice, care must
be taken with problems concerned with appropriate averaging times when
determining this constant. Workers in optical propagation tend to use quite
different averaging times when determining C_T^2 or C_n^2.

2.1.4 Stationary Random Functions

The frequency content of a time-varying, physically realizable, deterministic
signal such as a pulse is easily determined by taking its Fourier transform. If the
pulse is $p(t)$, then its transform $W(\omega)$ is defined by

$$W(\omega) = \frac{1}{2\pi} \int_{-\infty}^{\infty} dt\, p(t)\, e^{-i\omega t}, \tag{2.12}$$

where $\omega = 2\pi f$ and f is the frequency in Hertz. If we wish to reconstruct $p(t)$
exactly, we merely invert the relation (2.12) in the conventional way,

$$p(t) = \int_{-\infty}^{\infty} d\omega\, W(\omega)\, e^{i\omega t}, \tag{2.13}$$

and obtain an exact point-for-point reconstruction of the time signal from
knowledge of $W(\omega)$. We are able to reconstruct $p(t)$ exactly because $W(\omega)$ is, in
general, a complex quantity that tells not only the relative amplitudes of the
different frequencies (power) but also how they must be shifted relative to each
other (phase) to make up the final time-domain function.

In a completely analogous way, we can write a stationary random function
in terms of a Fourier-Stieltjes transform (*Yaglom* [2.14])

$$p(t) = \int_{-\infty}^{\infty} dH(\omega)\, e^{i\omega t}. \tag{2.14}$$

Equation (2.14) expresses the relation between a given realization of a random
process and its spectrum in the same sense that (2.13) does for a deterministic
function; however, in this case, $p(t)$ and $dH(\omega)$ are both random variables and
therefore are not the same for each realization of the random process. The

function $dH(\omega)$ plays the same role as $W(\omega)d\omega$ in (2.13). In fact, if $H(\omega)$ were differentiable, we could write $dH(\omega) = H'(\omega)d\omega$ and have an exact analog to (2.13). The particular form $dH(\omega)$ allows us to use (2.14) even in the case in which the spectrum of our random process is not differentiable; for example, (2.14) is a useful generalization for random processes consisting of a finite number of random harmonic generators.

The notion of stationarity is the time-domain analogy of spatial homogeneity discussed in Section 2.1.2. Stationarity implies that statistical quantities, such as the mean and variance, do not vary with time. For example, if we find the mean value of a process at t_0, it should be identical with the value found at $t_0 + t$. The mean referred to is the ensemble mean or expected value, that is indicated by angle brackets, and we average over different realizations of the random process. Unfortunately, we do not have at our disposal an ensemble of identical atmospheres which we may observe. Usually, we have a single function of time, say, of atmospheric temperature fluctuations, and we must invoke the assumption of ergodicity to obtain estimates of ensemble averages. An ergodic random process is one whose infinite time average is equal to its ensemble average; and, if we can make this assumption, then we can substitute time averages wherever we have ensemble averages. This leads to another problem. We cannot average forever as required in the ergodic hypothesis. Our only hope is that the process is sufficiently band limited so that a reasonable averaging time will give a stable number. How long to average and what filtering effects the averaging time, record length, and sampling rate have on the observations are the subject of an entire area of research, time series analysis. For further details of this extremely interesting and rich field, consult [2.15–17], and, for application to atmospheric processes, [2.18]. These considerations have a bearing on how well you can measure the relevant statistical quantities discussed in the following chapters. It is interesting that the structure function, already introduced in Section 2.1.2, was created in response to these problems. The structure function behaves as a high pass filter because, as we shall see later, the subtraction process removes the slowly varying large-scale fluctuations that affect both points of the measurement, and provides a much more stable number than the related correlation function.

Equation (2.14) has associated with it a host of mathematical problems, some outlined in the above paragraph. A further related difficulty is that (2.14) and its inverse violate at least two conditions usually required for the existence of a Fourier transform: absolute integrability (finite area), and $dH(\omega)$ and $p(t)$ may have an infinite number of maxima and minima. Consequently, the interpretation of (2.14) for random functions is much more cloudy than the precise relatives (2.12, 13) for physically realizable deterministic functions. These mathematical difficulties are overcome when we consider ensemble-averaged properties of the random process. An example is the correlation function, defined by

$$B(t, t+\tau) = \langle p(t)p^*(t+\tau)\rangle, \tag{2.15}$$

where the asterisk indicates a complex conjugate. After substituting (2.14, 15) takes the form

$$B(t, t+\tau) = \int\limits_{-\infty}^{\infty} \int\limits_{-\infty}^{\infty} e^{i\omega_1 t - i\omega_2(t+\tau)} \langle dH(\omega_1) dH^*(\omega_2) \rangle, \tag{2.16}$$

where we have assumed the ability to interchange integration and averaging. From our discussion of stationarity, we know that if $p(t)$ is a stationary random function, then the correlation function cannot depend on the time t when the average is taken. The only mathematical way $B(t, t+\tau)$ can be independent of t for all possible t's is if $\langle dH(\omega_1) dH^*(\omega_2) \rangle$ takes the form,

$$\langle dH(\omega_1) dH^*(\omega_2) \rangle = \delta(\omega_1 - \omega_2) P(\omega_1) d\omega_1 d\omega_2, \tag{2.17}$$

where $\delta(\omega_1 - \omega_2)$ is the Dirac delta function. Then (2.16) reduces to

$$B(t, t+\tau) = B(\tau) = \int\limits_{-\infty}^{\infty} d\omega_1 \, e^{-i\omega_1 \tau} P(\omega_1). \tag{2.18}$$

The function $P(\omega_1)$ is called the power spectral density of the function $p(t)$. Note the important result in (2.18) that, by assuming that only equal frequencies $\omega_1 = \omega_2$ are correlated, all the phase information about the function $p(t)$ is lost. We cannot reconstruct the details of $p(t)$ from knowledge of $P(\omega)$. The information contained in $P(\omega)$ is the amount of power that exists in a given process per unit frequency interval. To see this, we note that because $[p(t)p^*(t)]$ is the instantaneous power, then from (2.15), $B(t, t)$ is its average value, that is,

$$B(t, t) = \langle p(t)p^*(t) \rangle, \tag{2.19}$$

and

$$\langle p(t)p^*(t) \rangle = \int\limits_{-\infty}^{\infty} P(\omega) d\omega. \tag{2.20}$$

Since the left side is the average power, $P(\omega)d\omega$ must be the power in a frequency band from ω to $\omega + d\omega$.

In summary, the only meaningful spectral measure of a stationary random process is its power spectral density. Unlike the case of realizable deterministic variables where complete reconstruction of the time function is possible from knowledge of its Fourier transform, knowledge of the power spectral density does not permit exact reconstruction of the random process because phase information has been lost.

2.1.5 The Power Spectral Density of the Refractive Index Fluctuations

In Section 2.1.3 we considered briefly the spatial structure of the random refractive index fluctuations. In Section 2.1.4 the temporal characteristics of a general random variable were considered, with a brief discussion of time series analysis problems. To connect these two points of view in our physical model of the atmosphere, we make the assumption of "frozen turbulence", which implies that temporal variations of meteorological variables at a point are produced by advection of these quantities by the mean flow and not by changes in the variables themselves ([2.11, 12]). This point of view forces a one-to-one correspondence between spatial and temporal fluctuations. For example, if we place a point temperature sensor in the atmosphere, record its fluctuating readings in time, and find its temporal power spectral density $P(\omega)$, then, because of the frozen turbulence assumption and with the knowledge of the mean windspeed v_0, we can directly convert to a one-dimensional spectral density in terms of the spatial wave number K. The spatial wave number for a given temporal frequency ω will be $K = \omega v_0$. In this chapter, we shall consider the spatial statistics as more fundamental and the temporal statistics as derivable from them using the mean windspeed. (A more detailed description of the assumption of "frozen turbulence" and its limitations is given by *Tatarskii* [2.12].)

We can write the refractive index as the sum of a mean plus a fluctuating part

$$n(r) = \langle n(r) \rangle + n_1(r), \tag{2.21}$$

where $\langle n_1(r) \rangle = 0$ by definition. (Note that this definition of n_1 is slightly different than the n_1 defined in Section 2.1.1.) The variable $n_1(r)$ can be decomposed in terms of a three-dimensional Fourier-Stieltjes integral

$$n_1(r) = \int dN(K) e^{iK \cdot r}, \tag{2.22}$$

where $K = (K_x, K_y, K_z)$ is the three-dimensional spatial wave number and dN is the random spectral amplitude [respectively, the analogs of ω and dH in (2.14)]. Following the analysis in Section 2.1.4, we consider the covariance function of the refractive index field at two positions separated by r,

$$B_n(r_1 + r, r_1) = \langle n_1(r + r_1) n_1(r_1) \rangle. \tag{2.23}$$

If we substitute (2.22) into (2.23), then we obtain the spatial analog of (2.16),

$$B_n(r_1 + r, r_1) = \int\int e^{iK \cdot (r_1 + r) - iK' \cdot r_1} \langle dN(K) dN^*(K') \rangle. \tag{2.24}$$

From our earlier discussion of homogeneity, we know that a statistically homogeneous random field cannot have its average properties depend upon the location in the field at which the average is computed, that is, the same result should obtain if we translate our sensors to any other position in the field and perform the same operation. The impact of this assumption on (2.23) is to make $B_n(r_1 + r, r_1) = B_n(r)$. For the double integral in (2.24) to depend only upon r as required by the homogeneity assumption, the quantity in the angle brackets must satisfy the relation

$$\langle dN(K)dN^*(K')\rangle = \delta(K - K')\Phi_n(K)d^3Kd^3K',$$ (2.25)

where now δ is the three-dimensional Dirac delta function and $\Phi_n(K)$ is the three-dimensional spectral density of the refractive index fluctuations. Substituting (2.25) into (2.24) and performing the K' integration, we obtain the well-known Fourier transform relation between the spectral density and the covariance function,

$$B_n(r) = \int d^3K \, e^{iK \cdot r} \Phi_n(K).$$ (2.26)

To invert (2.26), we Fourier transform both sides of the equation to obtain

$$\Phi_n(K) = \frac{1}{(2\pi)^3} \int d^3r B_n(r) e^{-iK \cdot r}.$$ (2.27)

We are at complete liberty to construct these same relations in one or two spatial dimensions ([2.11, 12]). The procedure is identical to that resulting in either (2.18) or (2.26). We shall make extensive use of the two-dimensional transform relations in later sections.

A further assumption we wish to impose upon the turbulent refractivity fluctuations is that they be isotropic, that is, that statistical quantities such as the covariance function depend merely on the distance between sensors and not on the orientation of the line joining them. Mathematically, this implies that $B_n(r) = B_n(r)$ and, from the inverse transform of (2.26), that $\Phi_n(K) = \Phi_n(K)$. If we substitute this latter result into (2.26), and change to spherical coordinates $K = (K, \theta, \phi)$ and $d^3K = K^2 \sin\theta d\theta d\phi dK$, the angular integrations may be performed to yield

$$B_n(r) = \frac{4\pi}{r} \int\limits_0^\infty dK K \Phi_n(K) \sin(Kr).$$ (2.28)

Similarly, we can integrate (2.27) over the angle coordinates to obtain

$$\Phi_n(K) = \frac{1}{2\pi^2 K} \int\limits_0^\infty dr r B_n(r) \sin(Kr).$$ (2.29)

Once we have assumed a homogeneous and isotropic turbulence model, we can generate expressions for $B_n(r)$ and $\Phi_n(K)$ from Kolmogorov's structure function law $D_n(r)$. First, we expand the definition of the isotropic form of the structure function

$$D_n(r) = \langle [n_1(r_1 + r) - n_1(r_1)]^2 \rangle$$ (2.30)

into three terms and note that the assumption of homogeneity requires $\langle n_1^2(r_1 + r)\rangle = \langle n_1^2(r)\rangle = \langle n_1^2 \rangle = B_n(0)$ and by definition $\langle n_1(r + r_1)n_1(r_1)\rangle = B_n(r)$; consequently,

$$B_n(0) - B_n(r) = \tfrac{1}{2}D_n(r). \tag{2.31}$$

Computing the relation between $\Phi_n(K)$ and $D_n(r)$ is more complicated. From (2.28), after taking the limit as $r \to 0$ and substituting into (2.31) for $B_n(0)$, we obtain the structure function in terms of the spectral density Φ_n in the form

$$D_n(r) = 8\pi \int\limits_0^\infty dK K^2 \Phi_n(K)\left[1 - \frac{\sin(Kr)}{(Kr)}\right]. \tag{2.32}$$

[It is clear from (2.32), as mentioned above, that $D_n(r)$ removes the influence of large-scale refractive index fluctuations. The $[1 - (Kr)^{-1}\sin(Kr)]$ term in (2.32) severely attenuates spatial frequencies in the range $K < r^{-1}$, that is, those spatial frequencies with periods larger than the spacing between measurement points. Note that there is no such high pass filtering in the definition of $B_n(r)$ in (2.28).] We now must invert (2.32) to obtain Φ_n in terms of D_n, which is not quite as straightforward as inverting a simple Fourier transform relation; however, it can be accomplished without too much difficulty in terms of derivatives of $D_n(r)$ (*Strohbehn* [2.19] and *Tatarskii* [2.12]),

$$\Phi_n(K) = \frac{1}{4\pi^2 K^2} \int\limits_0^\infty \frac{\sin(Kr)}{(Kr)} \frac{d}{dr}\left[r^2 \frac{d}{dr} D_n(r)\right] dr. \tag{2.33}$$

Now we can find the spectral density function Φ_n in terms of Kolmogorov's inertial subrange model,

$$D_n(r) = C_n^2 r^{2/3}; \quad l_0 \ll r \ll L_0. \tag{2.34}$$

By substituting (2.34) into (2.33) and performing the indicated operations, we get the integral

$$\Phi_n(K) = \frac{5}{18\pi} C_n^2 K^{-3} \int\limits_{l_0}^{L_0} dr \sin(Kr)r^{-1/3}. \tag{2.35}$$

If we let the limits go to 0 and ∞, this integral reduces to

$$\Phi_n(K) = 0.033 C_n^2 K^{-11/3}, \tag{2.36}$$

where we assume (2.36) is only valid for $2\pi L_0^{-1} \ll K \ll 2\pi l_0^{-1}$. Both (2.34) and (2.36) are equally valid and equivalent descriptions of the turbulent refractive index fields.

There are other common forms of Φ_n that were derived to effect a smooth transition at the limits of the inertial subrange instead of a sharp truncation as in (2.36). For eddy sizes less than the microscale of turbulence, l_0, the energy in the turbulence is dissipated due to viscosity effects. While the exact form of the spectrum in this region is still under investigation, it is clear that the spectrum must fall off more rapidly than $K^{-11/3}$. A different model that accomplishes this was described by *Tatarskii* [2.12], that is,

$$\Phi_n(K) = 0.033 \, C_n^2 K^{-11/3} \exp(-K^2/K_m^2),\tag{2.37}$$

where $K_m = 5.92/l_0$. Similarly, for eddy sizes greater than L_0, the energy in the eddies must be less than that predicted by the Kolmogorov model. This effect can be approximated to some degree by the Von Kármán spectrum (*Tatarskii* [2.11])

$$\Phi_n(K) = \frac{\Gamma(11/6)}{\Gamma(1/3)} \frac{\pi^{-9/2}}{8} \langle \delta_n^2 \rangle L_0^3 (1 + K^2/K_0^2)^{-11/6},\tag{2.38}$$

where $K_0 = 2\pi/L_0$, and $\langle \delta_n^2 \rangle$ is the variance of the refractivity fluctuations and is related to C_n^2 by $C_n^2 \approx 1.9 \langle \delta_n^2 \rangle K_0^{2/3}$. These spectra have the correct behavior only in the inertial subrange, that is, $\Phi_n(K) = K^{-11/3}$ for $2\pi L_0^{-1} \ll K \ll 2\pi l_0^{-1}$. Outside of this range, particularly for small values of K, there is no physical basis for their behavior; consequently, they should be used only as an indication of the effects of l_0 and L_0 on the propagation statistics that we shall calculate in later sections. Another way to state this difficulty is that if the propagation statistics strongly depend on the microscale l_0, or particularly the outer scale L_0, then the propagation problem probably requires more accurate information on the atmospheric effects than is presently available.

2.1.6 Behavior of C_n^2 in the Open Atmosphere

The central role of the refractive index structure constant C_n^2 will become evident in the next subsection when we calculate the statistics of the fluctuating parameters of an optical wave propagating through atmospheric turbulence. In this subsection, we describe some of the most recent information about the behavior of C_n^2 as a function of altitude and time of day.

Figure 2.1 illustrates a typical example of the behavior of C_n over a 24-h period at a height of 2 m above the ground. The data were taken on a flat mesa in the lee of the Rocky Mountains near Boulder, Colorado, where the vegetation and irregularities in the ground surface were less than 0.3 m. The refractive index structure constant was derived from C_T^2 using (2.11). We computed C_T^2 from measurements of the mean-square temperature difference of two fine-wire thermometers (*Lawrence* et al. [2.20]) by dividing this quantity by the thermometer separation to the two-thirds power in accordance with (2.7).

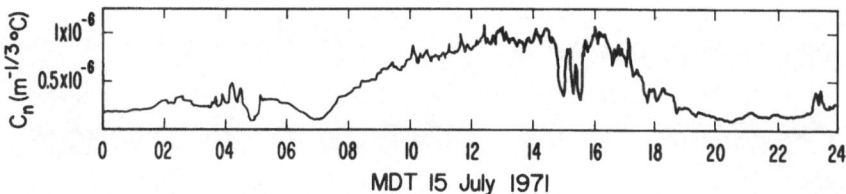

Fig. 2.1. Plot of the behavior of the refractive index structure parameter for a 24 h period on a clear sunny day

The diurnal cycle of C_n, evident in Fig. 2.1, is typical of its behavior in the surface layer of the earth (approximately the lowest 30–100 m). The intensity of the temperature fluctuations (hence C_n) is quite small at night and in the early morning hours until solar heating of the ground becomes significant enough to initiate convective instability (*Hess* [2.21]). This instability generates temperature fluctuations at a given height in the following manner. The sun-heated surface materials warm a thin layer of the air above. If air parcels from the layer become displaced upward, they find themselves warmer, and hence less dense and more buoyant, than the ambient air; therefore, they continue to accelerate upward. It is the mixing of these hot rising air parcels with cool descending air parcels that produces the observed temperature irregularities. This process continues, causing C_n to increase, until late afternoon when solar heating subsides, producing a consequent drop in C_n. A typical range of measurable C_n^2 values for a sunny day at 2 m above the ground is $10^{-17}\,m^{-2/3} < C_n^2 < 10^{-12}\,m^{-2/3}$. In Fig. 2.1, the sharp dips in C_n values, occurring between 1400 and 1600, are due to intermittent cloud cover. The sun heats only the top layer of the ground; when clouds temporarily obscure the sun, the ground rapidly cools, the temperature structure of the air above rapidly becomes more uniform, and C_n decreases precipitously.

There have been many measurements that have attempted to define the spatial variation of C_n^2. Investigators have obtained *in situ* data from temperature sensors that were tower mounted [2.22], aircraft mounted [2.20, 23, 24], and balloon borne [2.25]. Remote sensing of C_n^2 probably began with the measurement of stellar scintillations to infer high altitude C_n^2 values (see *Hufnagel* and *Stanley* [2.26]) and has progressed to the present-day use of acoustic sounders as remote sensors of C_T^2 and hence optical C_n^2 [2.22, 27]. Aircraft mounted and balloon-borne sensors gave the first indication of the complex behavior of C_n^2 with altitude. In fact, these measurements have been the primary force in changing the exponential C_n^2 profile [2.26] to a more realistic statistical model [2.28].

Acoustic sounder records provide the most complete picture of the spatial structure of C_n^2 in the planetary boundary layer (roughly the lower 1 km of the atmosphere). Although most of the early echo sounder results have been qualitative in nature, recently there has been progress in quantifying the sounder returns by comparing the sounder-derived C_T^2 values with those

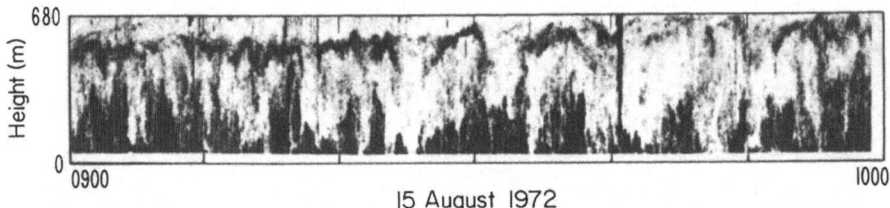

Fig. 2.2. Height-time plot of the strength of acoustic echo sounder returns

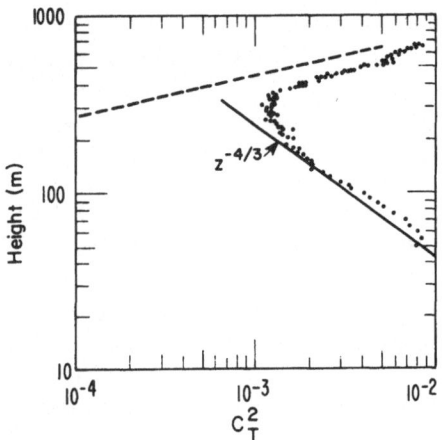

Fig. 2.3. Acoustic echo sounder derived values of C_T^2 versus height in the planetary boundary layer

measured by temperature sensors mounted on a nearby tower (*Neff* [2.22]). Figure 2.2 is a height-time plot of the strength of acoustic echo returns. The atmosphere was interrogated by a 20 ms acoustic pulse emitted from a vertically pointing antenna that also detected the echo return with an approximate vertical spatial resolution of 3–4 m. Height-time intervals of strong echo return (dark regions) indicate the presence of intense temperature fluctuations, whereas height-time intervals of weak echo return (light regions) indicate relatively weak temperature fluctuations. (The first 34 m are blank because of the time required to wait for the sounder to stop ringing from transmission and prepare to receive.) It is readily apparent from this record that, even in the space of the 60-min time pictured, the spatial structure of C_T^2 (hence C_n^2) is extremely varied. The dark vertical structures originating at the bottom of the record are associated with thermal plumes which are regions of warm rising air containing large thermal contrast. The light areas between plumes are thermally homogeneous regions of cool descending air.

Figure 2.3 contains a more quantitative estimate of the C_T^2 values. Each dot was obtained by inverting the formula for the acoustic scattering cross section and then obtaining C_T^2 from the measured cross-section value. (From the theory [2.11, 12], the backscattering cross-section is proportional to C_T^2. For a further discussion of the measurement technique and the assumptions involved in

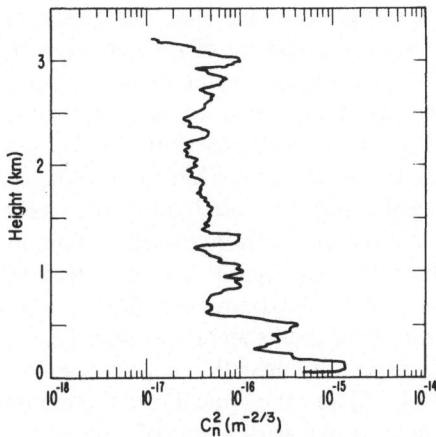

Fig. 2.4. Airplane-borne temperature sensor determinations of C_n^2 versus height

deriving Fig. 2.3, see *Neff* [2.22].) These data are averages over the period from 0900 to 0950. The facsimile record for the same period is shown in Fig. 2.2. From the data in Fig. 2.3, we see that $10^{-3}\,\mathrm{m}^{-2/3}\,°\mathrm{C}^2 < C_T^2 < 10^{-2}\,\mathrm{m}^{-2/3}\,°\mathrm{C}^2$, corresponding to approximate refractive index structure constant values of $10^{-15}\,\mathrm{m}^{-2/3} < C_n^2 < 10^{-14}\,\mathrm{m}^{-2/3}$. (The acoustic sounder data were also compared with tower-measured C_T^2 at a height of 92 m, and the measurements agreed to within 50 %.) The solid line represents a decrease of C_T^2 with height as $z^{-4/3}$, which is the theoretical prediction for the behavior of C_T^2 in the daytime surface layer (*Tsvang* [2.23] and *Wyngaard* et al. [2.29]). The extension of the "minus four-thirds" region above the first 100 m was noted earlier by *Tsvang* and confirmed by these data. There is, as of yet, no theoretical justification for the extension of the surface layer C_T^2 behavior to higher altitudes, although *Frisch* and *Ochs* [2.30] have shown that the $-4/3$ law should describe the C_T^2 behavior up to 20 % of the convective layer depth. The convective layer depth increases during the morning and reaches a value of 1–2 km in the afternoon when fully developed. For the data of Fig. 2.3, it was about 700 m thick. In the nighttime surface layer, the only behavior predicted theoretically (*Wyngaard* et al. [2.29]) is that C_T^2 will decrease with height more slowly than $z^{-2/3}$.

Above the first few hundred meters there is no theoretical prediction of C_n^2 behavior. Figure 2.4 illustrates the complicated behavior of the C_n^2 profile up to a height of 3.2 km. This curve is the average of data from three consecutive days, each day's data consisting of one profile taken during daylight hours (*Ochs* and *Lawrence* [2.24]). The three measured profiles were averaged together and then a 100 m running average applied to the resultant curve. (Each data point represents a 1 minute average.) Even after averaging, the curve illustrates an extensively varying C_n^2 structure which, except for the overall tendency to decrease with height, appears to be random in nature. Other data using balloon-borne sensors (*Bufton* et al. [2.25]) show the same random behavior of C_n^2 with a decreasing mean up to heights of 15 km. In the unsmoothed data of both *Ochs* and *Lawrence* and *Bufton* et al., a typical variation in C_n^2 in a 100–200 m altitude range is more than an order of magnitude. This indicates an extensively layered structure of turbulence with typical thicknesses of 100–200 m, instead of a smoothly varying profile as early models hypothesized. *Bufton* et al. found what appeared to be orographic effects (turbulence associated with mountain lee waves) present in the 3–7 km region, a decrease to a region of minimum values in the 7–9 km range, and, in the 9–15 km range, a region of sharp peaks superimposed on a relatively constant C_n^2 background value. They again attributed this increased spiking in the profile to the highly layered structure in this region, and cite this as confirmation of the theories of increased C_n^2 near the tropopause. *Bufton* et al. found general agreement with the earlier model of *Hufnagel* and *Stanley* [2.26] with the exception of the intense layering, especially near the tropopause.

Recently, in response to the data of [2.24, 25] and the optical measurements of [2.31–33], *Hufnagel* [2.28] derived a new model with several features not found in his earlier work. Based on these experimental observations, he suggests the following C_n^2 model,

$$C_n^2 = \{[(2.2 \times 10^{-53})h^{10}(W/27)^2] e^{-h/1000} + 10^{-16} e^{-h/1500}\} \exp[r(h,t)],$$

(2.39)

where h is the height in meters above sea level. The model is valid for $3 \, \text{km} < h < 24 \, \text{km}$. The variable r is a zero-mean, homogeneous, Gaussian random variable with a covariance function given by

$$\langle r(h+h_1, t+\tau)r(h,t)\rangle = A(h_1/100) e^{-\tau/5} + A(h_1/2000) e^{-\tau/80},$$

(2.40)

where

$$A(h/L) = \begin{pmatrix} 1 - |h/L|, & |h| < L \\ 0 & , \text{ otherwise} \end{pmatrix}.$$

(2.41)

(The interval τ is measured in minutes.) From (2.40), it follows that $\langle r^2 \rangle = 2$ and $\langle \exp(r) \rangle = e \approx 2.7$. These numbers may be substituted into (2.39), after finding the expected value $\langle C_n^2 \rangle$, to determine the behavior of the mean profile. Finally, the function W in (2.39) is a correlating factor defined by

$$W = [(1/15 \, \text{km}) \int_{5 \, \text{km}}^{20 \, \text{km}} v^2(h) dh]^{1/2},$$

(2.42)

where v is the wind speed at height h. The parameter W, the rms wind speed in the height range from 5–20 km, turns out to account for 60 % of the variance of the stellar scintillation data of *Mikesell*. [Data for $v(h)$ are readily obtained from the daily radiosonde profiles determined by the National Weather Service.] Note that W is a random function of time, and *Hufnagel* has found from Maryland data that W is normally distributed with a mean of 27 m/s and a standard deviation of $9\,\mathrm{m/s}$. To extend this model down to local ground level, we should add the surface layer C_T^2 dependence, for example, $z^{-4/3}$ for daytime, and account for orographic effects and any sharp inversion layers that may influence local values of C_T^2. (For further discussion, see *Hufnagel* [2.28].)

It should be noted that this new model is an empirical one based on a limited amount of data, but is probably the best model available at this time. In a certain sense, the model is also inconsistent, and hence it is important to understand the assumptions made in its development. First, when C_n^2 was first derived in (2.10), it was assumed that it represented an average quantity, but in (2.39) C_n^2 is a random variable. The distinction here is that in optical propagation experiments we are usually interested in reasonably short time scales, for example, seconds or minutes. Therefore optical workers usually consider C_n^2 values averaged over these time scales. In (2.39), r and W are parameters that represent slowly varying quantities, with time scales on the order of hours or days. With this interpretation, the model in (2.39) is consistent and highly useful in practical situations. An ultimate model for C_n^2 might also include parameters that account for seasonal, diurnal, and geographical effects.

Finally, we consider the statistics of the C_n^2 fluctuations as measured by three different techniques. From data taken on a 12 m tower, *Kallistratova* and *Timanovskiy* [2.34] determined C_n^2 from the mean profiles of wind speed and temperature, using a relation derived by *Obukov* [2.35]. During the daytime (1115–1515 LT), they observed that 84% of the values of C_n^2 were in the range $5.4 \times 10^{-14}\,\mathrm{m}^{-2/3} < C_n^2 < 5.4 \times 10^{-13}\,\mathrm{m}^{-2/3}$; whereas at night, 59% of the values were in the range $5.4 \times 10^{-15}\,\mathrm{m}^{-2/3} < C_n^2 < 5.4 \times 10^{-14}\,\mathrm{m}^{-2/3}$. Their observations also indicated a much larger spread of C_n^2 data in the nighttime than in the daylight hours. *Neff* [2.22] observed both tower-measured and sounder-derived values of C_n^2 and found a tendency for both sets of data to be log normally distributed in the early morning (0510–1000 LT) with a mean $\langle C_n^2 \rangle$ $\sim 7 \times 10^{-16}\,\mathrm{m}^{-2/3}$ and a standard deviation of $1.5 \times 10^{-15}\,\mathrm{m}^{-2/3}$. In the afternoon, the data deviated from log normality and had a mean $\langle C_n^2 \rangle$ $\sim 3.0 \times 10^{-15}\,\mathrm{m}^{-2/3}$ and a standard deviation of $2.0 \times 10^{-14}\,\mathrm{m}^{-2/3}$. *Neff* did not analyze any nighttime data so that comparison with the data of *Kallistratova* and *Timanovskiy* during that time is not possible; however, if we use the 4/3 rule to scale Neff's daytime data at 92 m down to the 12 m measurement height of *Kallistratova* and *Timanovskiy*, these data fall into the range of most probable C_n^2 values found by the Soviet workers. This agreement is quite remarkable, considering the completely different measurement techniques used in the two experiments. Both studies give C_n^2 data for the summer months, and there appears to be no equivalent study available for the other seasons of the year.

In this subsection we have sketched some of the most recent observations and attempts to model the behavior of the refractive index structure constant in the open atmosphere. Knowledge of this behavior is exceedingly important, as will become obvious in the next section where we define the central role of the refractive index fluctuations in atmospheric optical propagation.

2.2 The "Classical" Theory of Optical Propagation

There is an extensive literature on the subject of the theory of line-of-sight propagation through the atmosphere and other random media. The early work by *Chernov* [2.36] and by *Tatarskii* [2.11] solved the wave equation directly using perturbation techniques. More recently, *Lee* and *Harp* [2.37], using a highly physical approach employing a weak phase screen model, rederived the classical results, usually with much more ease than with the *Chernov* and *Tatarskii* approach. *Fante* [2.38] described the power of transport theory in connection with the propagation of beams through turbulent plasmas. This technique can also be applied to the atmospheric optical propagation problem and appears to have some advantages for beam wave propagation.

Because this chapter is somewhat historical, we shall develop the theoretical results directly from Maxwell's equation, using a perturbation technique that is perhaps best described in the early works of *Tatarskii* [2.11, 12]. To maintain simplicity in the derivation and consistency with Tatarskii's original work, we shall assume that a plane wave is incident upon the random medium and consider the resultant fluctuations of amplitude and phase only. Other important quantities, such as phase difference and angle-of-arrival fluctuations, were considered in *Lee* and *Harp* [2.37]. *Lawrence* and *Strohbehn* [2.39] and *Fante* [2.40] reviewed most of the significant work in this area and provided a good source of references. Chapter 5 gives a full mathematical treatment of the more general beam wave case.

2.2.1 The Wave Equation for Optical Propagation

We assume that the atmosphere has zero conductivity and unit magnetic permeability and that the electromagnetic field has a sinusoidal time dependence, given by $e^{-i\omega t}$. [The following results are directly applicable to the pulse propagation problem for pulses longer than about 100 ps (*Fante* [2.40]).] We can derive the amplitude and phase statistics for short pulses by Fourier decomposition of the waveform into a spectrum of plane waves. (Some relevant references are [2.41–47].) Under these circumstances, Maxwell's equations take the following form:

$$\boldsymbol{V} \cdot \boldsymbol{H} = 0 \tag{2.43}$$

$$\boldsymbol{V} \times \boldsymbol{E} = ik\boldsymbol{H} \tag{2.44}$$

$$\boldsymbol{V} \times \boldsymbol{H} = -ikn^2\boldsymbol{E} \tag{2.45}$$

$$\boldsymbol{V} \cdot (n^2\boldsymbol{E}) = 0, \tag{2.46}$$

where $i = \sqrt{-1}$, $k = \omega/c$ is the wave number of the radiation, ω is its radian frequency, and c is the speed of light in free space. The $\boldsymbol{V}$ operator is the well-known vector derivative $(\partial/\partial x, \partial/\partial y, \partial/\partial z)$. The quantities E and H are the vector amplitudes of the electric and magnetic fields and are a function of position alone; the assumed sinusoidal time dependence is contained in the wave number k. The atmospheric refractive index $n(r)$ is a random function of position with the statistical properties described in Section 2.1.

To generate the more familiar wave equation for the electric field, we take the curl of (2.44) and, after substituting (2.45), obtain

$$-\nabla^2 E + \nabla(\nabla \cdot E) = k^2 n^2 E. \tag{2.47}$$

Equation (2.46) is expanded and solved for $\nabla \cdot E$, and the result inserted into (2.47) and we obtain the final form of the vector wave equation

$$\nabla^2 E + k^2 n^2 E + 2\nabla(E \cdot \nabla \log n) = 0. \tag{2.48}$$

In (2.48) we have substituted the gradient of the natural logarithm for $\nabla n/n$.

The last term on the left-hand side is related to the change in polarization of the wave as it propagates. Various authors [2.19, 48–52] considered the effects of this term in great detail and deemed it negligible for $\lambda \ll l_0$. *Clifford* [2.52] also showed that the depolarization effects of the atmosphere are negligible even when the wavelength λ is greater than the inner scale l_0. These results permit us to drop the last term, and (2.48) then simplifies to

$$\nabla^2 E + k^2 n^2 E = 0. \tag{2.49}$$

Two final points about (2.49) will be considered before launching into its solution. First, from (2.44), we see that H is a function of E and that any results derived from (2.49) directly apply to H. Second, because (2.49) is easily decomposed into three scalar equations, one for each component of the electric field, we may solve one scalar equation and ignore the vector character of the wave until the final solution; therefore we need only analyze

$$\nabla^2 E + k^2 n^2 E = 0, \tag{2.50}$$

where E is either the magnitude of the electric field, if it is linearly polarized, or any of the vector components of an arbitrarily polarized field.

2.2.2 Solution by the Method of Small Perturbations

The method of small perturbations involves the expansion of E into a series of ever-decreasing terms that may or may not converge, that is,

$$E = E_0 + E_1 + \dots. \tag{2.51}$$

Each term E_m is assumed to be of the order of smallness $(n_1)^m$. Now, if the magnitude of successive terms is sufficiently small compared to the next lower order term $|E_{m+1}| \ll |E_m|$, the series converges and a finite number of terms will accurately describe the propagation problem. Unfortunately, this does not occur for all situations in our problem and the series may be divergent. The physical significance of this fact is related to the interpretation of higher order terms as multiple scatter. The zero-order term represents the unscattered wave, the first order represents single scattering, the second order double scattering, and so forth. It turns out that multiple scattering (higher order terms) becomes increasingly important as the strength of the refractive index fluctuations increases, either by n_1 becoming larger or by the thickness of the medium increasing so that the total integrated refractive index change becomes large. We shall ignore these problems and solve for the single scatter solution in this chapter, leaving the full discussion of the higher order terms for Chapter 3. However, we should remember that the single scatter results are valid in a wide range of practical problems.

The procedure that follows is to substitute $E = E_0 + E_1$ into (2.50) and to equate to zero each group of terms of the same order in n_1. This results in two equations,

$$\nabla^2 E_0 + k^2 E_0 = 0 \tag{2.52}$$

$$\nabla^2 E_1 + k^2 E_1 + 2k^2 n_1 E_0 = 0. \tag{2.53}$$

In (2.53), we have ignored terms like $n_1 E_1$ because, as stated above, E_1 is of order n_1 and therefore the product $n_1 E_1$ is of order n_1^2. Following *Tatarskii* [2.11, 12], we assume that the incident or unperturbed field is a unit amplitude plane wave propagating in the z direction, $E_0 = \exp[ikz]$, and, after noting that it satisfies (2.52), we substitute it into (2.53) where it becomes part of the source term. The resulting equation is

$$\nabla^2 E_1 + k^2 E_1 = -2k^2 n_1 \, e^{ikz}. \tag{2.54}$$

The major advantage of the perturbation procedure is that we have transformed (2.50), which is a homogeneous partial differential equation with random, space-dependent coefficients, to the nonhomogeneous partial differential equation with constant coefficients given in (2.54). Although we have sacrificed considerable generality in this process, (2.50) may not be solved without specifying the particular form of n_1 (not merely its statistics), whereas (2.54) may be solved as it stands. Its solution is simply the convolution of the source term on the right-hand side with the Green's function of the equation, i.e.,

$$E_1(r) = (4\pi)^{-1} \int_V d^3 r' \frac{e^{ik|r-r'|}}{|r-r'|} [2k^2 n_1(r') e^{ikz'}]. \tag{2.55}$$

Equation (2.55) states that the scattered field observed at r is that due to a spherical wave emitted at r', that is $|r-r'|^{-1}\exp(ik|r-r'|)$, whose amplitude is proportional to the product of the local refractive index fluctuation $n_1(r')$ and the strength of the incident radiation, and whose phase is determined by the total number of wavelengths along the path from source to scatterer to receiver. We then integrate these contributions from all points r' over the entire scattering volume V.

There are several simplifications of (2.55) allowed by the particular characteristics of laser propagation. Monochromatic light, scattered by weak, large-scale refractivity fluctuations, is contained in a narrow cone about the forward scatter $(+z)$ direction. From the laws of diffraction, the angle of scattering will be a maximum for the smallest eddies of interest, whose scale size is the microscale of turbulence l_0. The scattering angle in this case is of the order λ/l_0, and for typical values ($\lambda = 0.6 \times 10^{-6}$ m and $l_0 = 2 \times 10^{-3}$ m) is 3×10^{-4} radians. Hence, the maximum extent of the atmosphere perpendicular to the z direction, from which scattered radiation is incident on a receiver, is typically much less than the longitudinal distance from scatterer to receiver. Mathematically, if $z-z'$ is the longitudinal distance from scatterer to receiver and if $|\varrho-\varrho'|$ is the transverse displacement from the z axis, then we would expect that $|z-z'| \gg |\varrho-\varrho'|$ would be a good approximation and we may directly replace $|r-r'|$ by $z-z'$ in the denominator of (2.55).

For the argument of the exponential we must be more careful because what is important there is not just the relative size of the transverse and longitudinal displacements of the scatterers, but how many wavelengths are contained in these distances. We must retain terms that are large enough to make the argument go through more than 2π radians. Expanding $k|r-r'|$ in a binomial series, we have

$$k|r-r'| = k(z-z')\left[1 + \frac{(\varrho-\varrho')^2}{2(z-z')^2} - \frac{(\varrho-\varrho')^4}{8(z-z')^4} + \cdots\right]. \tag{2.56}$$

By assuming that the maximum transverse extent is approximately $\lambda L/l_0$ and $z-z' \sim L$, where L is the total path length, the second term in (2.56) is of the order $\pi\lambda L/l_0^2$. Therefore, if $l_0 \ll \sqrt{\lambda L}$, the second term is large and we must retain it. For $\lambda = 0.6 \times 10^{-6}$ m and for a typical path of 1 km, the Fresnel length $\sqrt{\lambda L}$ is ~ 2.5 cm; with $l_0 = 0.2$ cm, the condition for the second term being large is satisfied. (The third and successive terms are negligible by the same argument; for example, the third term is of the order $\lambda^3 L/l_0^4$ which is extremely small for the above parameters.) In summary, we need retain only $z-z'$ in the denominator of (2.55) and the first two terms of (2.56) in the exponential, and this results in the expression

$$E_1(r) = \frac{k^2 e^{ikz}}{2\pi} \int_V d^3r' \exp\left\{\frac{ik[(\varrho-\varrho')^2]}{2(z-z')}\right\} \frac{n_1(r')}{(z-z')}. \tag{2.57}$$

The Fresnel approximation (2.56) transforms the Green's function scattering integral (2.55) to the so-called Fresnel diffraction formula (2.57). This formula (2.57) is the exact solution to the diffusion equation with a source term $-2k^2 n_1$ (*Tatarskii* [2.11]). One further interesting point about (2.57) is that the exponential term is the equation for the location of the interference fringes produced in the receiving plane at z by the interference of the incident plane wave with the scattered spherical wave (Huygen's wavelet) emitted from the point (ϱ', z'). It is these fringes that produce the amplitude and phase fluctuations that we observe in atmospheric optical propagation.

Questions of interest in this problem are: What are the amplitude or intensity fluctuations relative to the free space values, that is, how much will the received amplitude fluctuate due to its interaction with the atmosphere? and what are the phase fluctuations relative to the phase of the unperturbed wave? Following *Tatarskii* [2.11, 12], we use a technique that takes advantage of the simplifications implicit in the weak scattering assumption to obtain expressions for the amplitude and phase fluctuations in terms of E_1. The field $E = E_0 + E_1$ by assumption; therefore, in terms of amplitude and phase, we have

$$\frac{E}{E_0} = 1 + \frac{E_1}{E_0} = \frac{A}{A_0} \exp[i(s - s_0)], \tag{2.58}$$

where we have written $E = A \exp(is)$ and $E_0 = A_0 \exp(is_0)$. Taking the natural logarithm of (2.58), we obtain

$$\log\left(1 + \frac{E_1}{E_0}\right) = \log\left(1 + \frac{A_1}{A_0}\right) + i(s - s_0), \tag{2.59}$$

where A_1 is the amplitude of E_1. Because $|E_1/E_0| \ll 1$ and $A_1/A_0 \ll 1$, we may expand the logarithm in a power series and retain only the first term so that (2.59) becomes

$$\frac{E_1}{E_0} \cong \frac{A_1}{A_0} + i(s - s_0). \tag{2.60}$$

This implies that we may obtain the amplitude ratio from the real part of (2.57) and the relative phase fluctuations from the imaginary part, provided that we normalize this equation by the initial wave E_0. Consequently, we may write

$$A_1/A_0 = \frac{k^2}{2\pi} \int_V d^3r' \cos\left[\frac{k(\varrho - \varrho')^2}{2(z - z')}\right] \frac{n_1(r')}{(z - z')} \tag{2.61}$$

and

$$(s - s_0) = \frac{k^2}{2\pi} \int_V d^3r' \sin\left[\frac{k(\varrho - \varrho')^2}{2(z - z')}\right] \frac{n_1(r')}{(z - z')}. \tag{2.62}$$

[In (2.61) and (2.62), we have used the fact that $n_1(r')$ is real, that is, there is no attenuation of the wave as it propagates.] By convention, we define $\chi = \ln(A/A_0) \cong A_1/A_0$ and $s - s_0 = s_1$. Hereafter, we shall be considering the log amplitude χ because that is often what is measured experimentally, but recall that, for the weak scatter point of view, it is identical to the normalized amplitude fluctuation A_1/A_0.

We still have not taken advantage of the path geometry to simplify our equations. In particular, because we are considering a plane wave, we expect statistical homogeneity of both the amplitude and phase fluctuations in the plane $z = $ constant. Of course, this is not true along the z axis because we expect the fluctuations of χ and s to increase with increasing z as the wave propagates through progressively more refractive turbulence. Consequently, it is useful to expand the refractive index field $n_1(r')$ as a two-dimensional Fourier-Stieltjes transform of the form

$$n_1(r') = \int dv(K, z') e^{iK \cdot \varrho'}, \tag{2.63}$$

where we have transformed the variation of the refractive index field only in the plane perpendicular to the propagation direction and allowed the random amplitude dv to be a function of z'. Equation (1.63) is the same, in principle, as those discussed in Section 2.1.5; specifically, it is analogous to the three-dimensional transform in (2.22). If we substitute this expression for n_1 into (2.61) and (2.62) and interchange the integration over dv with that over d^3r', then

$$\begin{bmatrix} \chi(r) \\ s_1(r) \end{bmatrix} = \frac{k^2}{2\pi} \int_0^z dz' \frac{dv(K, z')}{(z - z')} \int d^2\varrho' \, e^{iK \cdot \varrho'} \begin{Bmatrix} \cos \\ \sin \end{Bmatrix} \left[\frac{k(\varrho - \varrho')^2}{2(z - z')} \right], \tag{2.64}$$

where we have used the fact that $d^3r' = d^2\varrho' dz'$.

The integral over ϱ' is readily performed after substituting $\varrho'' = \varrho' - \varrho$. Because the cosine and sine terms are functions of the magnitude of ϱ'' alone, we can use polar coordinates $\varrho'' = (\varrho'', \theta)$ and integrate directly over θ. If we denote the inner integrals of (2.64) as $I_{1,2}$, we have

$$I_{1,2} = e^{iK \cdot \varrho} \int_0^\infty d\varrho'' \varrho'' \begin{Bmatrix} \cos \\ \sin \end{Bmatrix} \left[\frac{k\varrho''^2}{2(z - z')} \right] \int_0^{2\pi} d\theta \, e^{iK\varrho'' \cos(\theta - \psi)}, \tag{2.65}$$

where we let $K = (K, \psi)$ in the last term. The result of the inner integration is the Bessel function $2\pi J_0(K\varrho'')$ (Gradshteyn and Ryzhik [2.53]) which, when substituted into (2.65), leaves the integral

$$I_{1,2} = 2\pi \, e^{iK \cdot \varrho} \int_0^\infty d\varrho'' \varrho'' J_0(K\varrho'') \begin{Bmatrix} \cos \\ \sin \end{Bmatrix} \left[\frac{k\varrho''^2}{2(z - z')} \right]. \tag{2.66}$$

The integral in (2.66) is the Hankel transform of the cosine or sine term (*Erdélyi* [2.54] and *Gradshteyn* and *Ryzhik* [2.53]). After performing this integration, (2.66) becomes

$$I_{1,2} = \frac{2\pi(z-z')}{k} e^{i\mathbf{K}\cdot\varrho} \begin{Bmatrix} \sin \\ \cos \end{Bmatrix} \left[\frac{K^2(z-z')}{2k}\right],$$ (2.67)

and (2.67) substituted into (2.64) gives the final result

$$\begin{bmatrix} \chi(r) \\ s_1(r) \end{bmatrix} = \int e^{i\mathbf{K}\cdot\varrho} \left(k \int\limits_0^z dz' dv(\mathbf{K}, z') \begin{Bmatrix} \sin \\ \cos \end{Bmatrix} \left[\frac{K^2(z-z')}{2k}\right] \right).$$ (2.68)

This expression can be interpreted in terms of a two-dimensional Fourier expansion similar to (2.63), that is, the term in brackets in (2.68) plays the same role as $dv(\mathbf{K}, z')$ in (2.63). Therefore, we can interpret the quantity in brackets in (2.68) as the random spectral amplitudes of the amplitude and phase fluctuations.

As in any description of a random process, we must settle for averaged properties of the outcome of experiments. Consequently, we consider the two-dimensional covariance functions $B_\chi(\varrho)$ and $B_s(\varrho)$. (The first moments of χ and s_1, that is, $\langle\chi\rangle$ and $\langle s_1\rangle$, are both zero because, by definition, $\langle n_1\rangle = 0$. This result is due to retaining only the first-order perturbation term E_1 in the initial expansion of the field. If we retained second-order or higher terms, $\langle\chi\rangle$ and $\langle s_1\rangle$ would be nonzero, but small.) We use the two-dimensional covariance functions for the reasons alluded to earlier, namely, the statistical homogeneity of the amplitude and phase fluctuations in the plane $z=$constant. Normally, $B_\chi(\varrho)$ and $B_s(\varrho)$ are computed in the receiving plane, which we shall later assume is located at the position $z=L$, that is,

$$B_\chi(\varrho, z) = \langle\chi(\varrho_1 + \varrho, z)\chi^*(\varrho_1, z)\rangle,$$ (2.69)

with a similar definition for $B_s(\varrho, z)$. If we insert (2.68) into (2.69), we obtain the fourfold integral

$$\begin{bmatrix} B_\chi(\varrho) \\ B_s(\varrho) \end{bmatrix} = k^2 \iint e^{i\mathbf{K}\cdot(\varrho_1+\varrho) - i\mathbf{K}'\cdot\varrho_1} \int\limits_0^z dz' \int\limits_0^z dz'' \begin{Bmatrix} \sin \\ \cos \end{Bmatrix} \left[\frac{K^2(z-z')}{2k}\right] \begin{Bmatrix} \sin \\ \cos \end{Bmatrix} \left[\frac{K'^2(z-z'')}{2k}\right]$$

$$\cdot \langle dv(\mathbf{K}, z')dv^*(\mathbf{K}', z'')\rangle.$$ (2.70)

We already know how to handle the quantity in the angle brackets from the discussion in Section 2.1.5. For the particular case of two-dimensional random amplitudes, such as dv, the equation

$$\langle dv(\mathbf{K}, z')dv^*(\mathbf{K}', z'')\rangle = \delta(\mathbf{K} - \mathbf{K}')F_n(\mathbf{K}', z' - z'')d^2K d^2K'$$ (2.71)

pertains. The function $F_n(K', z' - z'')$ is the two-dimensional spectral density of the refractive index fluctuations defined by

$$F_n(K', z' - z'') = \int_{-\infty}^{\infty} dK_z \Phi_n(K', K_z) \cos[K_z(z' - z'')], \tag{2.72}$$

where Φ_n is the three-dimensional refractivity spectrum. After inserting (2.71) into (2.70) and performing the K' integration, (2.70) reduces to

$$\begin{bmatrix} B_\chi(\varrho) \\ B_s(\varrho) \end{bmatrix} = \int d^2K \exp^{iK \cdot \varrho}$$

$$\cdot \left(k^2 \int_0^z dz' \int_0^z dz'' \left\{ \begin{matrix} \sin \\ \cos \end{matrix} \left[\frac{K^2(z - z')}{2k} \right] \begin{matrix} \sin \\ \cos \end{matrix} \left[\frac{K^2(z - z'')}{2k} \right] \right\} F_n(K, z' - z'') \right). \tag{2.73}$$

Because the Fourier transform of the quantities in the brackets equals their respective covariance functions, these quantities must be the two-dimensional spectral densities of χ and s. To be consistent with the notation in (2.72), we shall denote these spectra as $F_\chi(K, 0)$ and $F_s(K, 0)$.

Equation (2.73) can be simplified greatly by considering some of the properties of F_n as well as the geometric characteristics of the optical propagation problem. (The remaining part of the analysis closely parallels [2.11, 12] where it is done in considerably more detail.) First, it is apparent from (2.72) that $F_n(K, z' - z'')$ is an even function of $z' - z''$, that is,

$$F_n(K, z' - z'') = F_n(K, z'' - z'). \tag{2.74}$$

Second, F_n expresses the correlation of n_1 on adjacent planes along the z axis, and clearly only those refractive index homogeneities whose size $2\pi K^{-1}$ is as large as or greater than the separation of the two planes $z' - z''$ will contribute significantly to that correlation. Smaller eddies will not intersect the two planes and will not contribute much to the correlation. From this argument and the assumption of isotropic turbulence $F_n(K, z' - z'') = F_n(K, z' - z'')$, we expect that F_n will decay rapidly for $K|z' - z''| > 1$. To take optimum advantage of these facts, we make the transformation of variables $\xi = z' - z''$, $2\eta = z' + z''$. After noting that the Jacobian of this transformation is unity and with F_n now even in ξ, we can write the amplitude and phase spectra in the following form:

$$\begin{bmatrix} F_\chi(K, 0) \\ F_s(K, 0) \end{bmatrix} = k^2 \int_0^L d\xi F_n(K, \xi) \int_{\frac{\xi}{2}}^{L - \frac{\xi}{2}} d\eta \left[\cos\left(\frac{K^2 \xi}{2k} \right) \mp \cos\left(\frac{K^2(L - \eta)}{k} \right) \right]. \tag{2.75}$$

In (2.75), we have dropped the vector notation because of the assumption of an isotropic spectrum for F_n. We also have replaced the position variable z by L, the *total* path length through the medium. To simplify (2.75) even further, we note, as above, that $F_n \to 0$ for $\xi > K^{-1}$, or equivalently, the important region of integration is for $\xi \lesssim K^{-1}$. This implies that the argument of the first cosine satisfies $K^2 \xi/(2k) \lesssim K/2k$. Now $K_{max} \sim 2\pi l_0^{-1}$ where l_0 is the microscale of turbulence; therefore, $K/2k$ is of the order λ/l_0 which, from our earlier assumption, is much less than unity. Consequently, $\cos[K^2 \xi/(2k)]$ is approximately unity for the entire range of integration over ξ. Further, if we make the assumption $\xi \ll L$, we can ignore its presence in the η integration limits. This is obviously true because in the important region of integration $\xi \lesssim K^{-1}$ and for the Kolmogorov spectrum, K^{-1} has a maximum value of $L_0/(2\pi)$; consequently, $\xi \lesssim L_0$. Therefore, if $L_0 \ll L$ then we can drop the ξ dependence inside the η integral altogether. Physically, this requires that our path length be much longer than the outer scale length. Because $L_0 \sim 1$ m near the ground, this is easily satisfied for reasonable optical paths. The final result of all these assumptions is that the two integrals in (2.75) become independent. The ξ integration is easily performed in the case $L_0 \ll L$ because, for $\xi \lesssim L_0$, the upper limit may be replaced by infinity with negligible error. The final integration follows from the definition (2.72), that is,

$$\int_0^\infty d\xi F_n(K, \xi) = \pi \Phi_n(0, K),$$ (2.76)

and for isotropic turbulence $\Phi_n(0, K) = \Phi_n(K)$. The η integration is readily performed, and the final result for the amplitude and phase spectra becomes

$$\begin{bmatrix} F_\chi(K, 0) \\ F_s(K, 0) \end{bmatrix} = \pi k^2 L \left[1 \mp \left(\frac{k}{K^2 L} \right) \sin \left(\frac{K^2 L}{k} \right) \right] \Phi_n(K).$$ (2.77)

2.2.3 Covariance and Structure Functions for Kolmogorov Turbulence

In (2.77), we found the power spectral density of the amplitude and phase fluctuations of a plane optical wave propagating through a random refractive index field whose power spectrum is $\Phi_n(K)$. The first factor in (2.77) is the optical transfer function for the atmosphere. When the minus sign applies (log-amplitude spectrum), this function approaches zero as $K \to 0$, or equivalently, it suppresses the influence of the large-scale refractive index irregularities on the log-amplitude fluctuations. On the other hand, when the plus sign applies (phase spectrum), as $K \to 0$, this function approaches a constant and does not suppress the large-scale inhomogeneities.

If we insert the Kolmogorov spectrum (2.36) in (2.77), the log-amplitude spectrum would be band limited with a peak at a spatial frequency $K \sim 2\pi(\lambda L)^{-1/2}$, whereas, the phase spectrum would depend quite strongly on the largest scales and, consequently, would have a maximum at $K \sim 2\pi L_0^{-1}$. This implies that the log-amplitude fluctuations are produced largely by

Fresnel zone size eddies with scale size of the order $\sqrt{\lambda L}$ and the phase fluctuations are produced by the largest scale eddies in the medium. This latter fact makes the measurement of the phase structure function preferable to the phase covariance function because, as discussed before, the structure function acts as a spatial filter to remove the influence of large-scale eddies. The difficulty in measuring the phase covariance function directly is that large spatial scales imply long measurement times to get adequate, stable statistics, and, if you wait too long, the nonstationarity of the atmosphere will contaminate the data.

From (2.77), we can compute more easily measured quantities such as the covariance of the log-amplitude and the structure function of the phase fluctuations. The relation between $F_\chi(K, 0)$ and the covariance function $B_\chi(\varrho)$ is given by the two-dimensional Fourier transform (2.73). Because we have assumed that F_χ is isotropic in K, this relation can be simplified by changing variables in (2.73) from $K = (K_x, K_y)$ to circular coordinates K and θ. Then we may perform the integral over the angular coordinate as we did in (2.28) and (2.29) to yield the isotropic transform

$$B_{\chi,s}(\varrho) = 2\pi \int\limits_0^\infty dK K J_0(K\varrho) F_{\chi,s}(K, 0), \tag{2.78}$$

where $J_0(K\varrho)$ is the zero-order Bessel function of the first kind, defined by the relation

$$J_0(K\varrho) = \frac{1}{2\pi} \int\limits_0^{2\pi} e^{iK\varrho\cos\theta} d\theta. \tag{2.79}$$

When (2.77) is substituted into (2.78), we obtain

$$B_{\chi,s}(\varrho) = 2\pi^2 k^2 L \int\limits_0^\infty dK K J_0(K\varrho)\left[1 \mp \left(\frac{k}{K^2 L}\right)\sin\left(\frac{K^2 L}{k}\right)\right]\Phi_n(K). \tag{2.80}$$

Since it is nearly impossible to measure $B_s(\varrho)$ because of its sensitivity to large spatial and temporal scales, the physically useful quantity is the phase structure function, defined in analogy to (2.31) as

$$D_s(\varrho) = 2[B_s(0) - B_s(\varrho)]. \tag{2.81}$$

Substituting (2.80) into (2.81) gives

$$D_s(\varrho) = 4\pi^2 k^2 L \int\limits_0^\infty dK K \left[1 + \left(\frac{k}{K^2 L}\right)\sin\left(\frac{K^2 L}{k}\right)\right][1 - J_0(K\varrho)]\Phi_n(K). \tag{2.82}$$

We are finally ready to calculate the exact form of $B_\chi(\varrho)$ and $D_s(\varrho)$ by inserting the Kolmogorov spectrum for Φ_n and integrating. Using (2.80), we

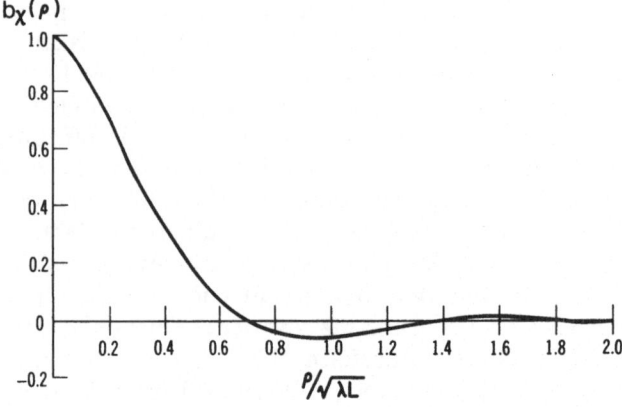

Fig. 2.5. Plot of the normalized covariance function of the log-amplitude fluctuations as a function of spacing in Fresnel zones

first calculate the variance of the log-amplitude σ_χ^2 from $B_\chi(0)$. After performing the integration (*Tatarskii* [2.11, 12]) and using the assumption $l_0 \ll \sqrt{\lambda L}$, we obtain

$$\sigma_\chi^2 = 0.31 k^{7/6} L^{11/6} C_n^2 . \tag{2.83}$$

As we shall see in later chapters, it is the breakdown of the validity of this formula for long paths and strong turbulence that caused a reexamination of the validity of the perturbation approach.

Figure 2.5 illustrates the behavior of the normalized covariance function $b_\chi(\varrho) = B_\chi(\varrho)/\sigma_\chi^2$ in the receiving plane $z = L$. Note again the preeminent role of the Fresnel length $\sqrt{\lambda L}$ and the negative tail indicative of the band limited nature of the spatial spectrum. The negative tail implies that for any two points in the receiving plane separated by approximately one Fresnel length, on the average, one point will be in a region that is brighter than the average irradiance and the other will be in a region that is darker. (The rather complicated expressions for B_χ/σ_χ^2, involving hypergeometric functions, are available in *Lawrence* and *Strohbehn* [2.39].)

Finally, inserting the Kolmogorov spectrum into (2.82), we can calculate directly the power law of the phase structure function. For the case where the separation ϱ satisfies the condition $l_0 \ll \varrho \ll \sqrt{\lambda L}$, we have ([2.11, 12])

$$D_s(\varrho) = 1.46 C_n^2 k^2 L \varrho^{5/3} \tag{2.84}$$

and, for $\sqrt{\lambda L} \ll \varrho \ll L_0$,

$$D_s(\varrho) = 2.92 C_n^2 k^2 L \varrho^{5/3} . \tag{2.85}$$

The behavior of $D_s(\varrho)$ for $\varrho \gg L_0$ is unknown because we have only specified the spectrum Φ_n in the inertial subrange. The detailed behavior of D_s for large ϱ

depends on the particular form of Φ_n in the low spatial frequency region, which is probably not a universal function.

The above formulas for the amplitude and phase statistics were derived using the Born approximation (2.55). The range of validity of this approximation is determined by the smallness of the scattered field with respect to the unperturbed field, i.e., $|E_1| \ll |E_0|$. In his earlier work accomplished in the 1950s, *Tatarskii* [2.11], using a technique first described by *Rytov* [2.55], derived these same results under apparently less restrictive conditions requiring only the smallness of the gradient of the complex phase of the scattered wave compared with the unperturbed phase gradient. Initially, many workers believed that this gave the amplitude and phase results a much greater range of validity in terms of C_n^2 and path length than the Born approximation derivation. Russian workers such as *Pisareva* [2.56], *Feinberg* [2.57], *Tatarskii* [2.58], and later American workers *Hufnagel* and *Stanley* [2.26] and *Brown* [2.59] challenged the claim for the extended range of validity of the Rytov method. It is now well established that the optimistic estimates of the range of validity of the Rytov approximations were unfounded and the results obtained by this method are valid only when $\sigma_\chi^2 \leq 0.3$. This failure is due to the fact that the Rytov technique fails to account for multiple scattering effects in a consistent manner.

It appears from some experimental work (*Gurvich* et al. [2.60]), where measurements have been made in strong integrated turbulence, that the phase results (2.84) and (2.85) and other related quantities such as angle-of-arrival fluctuations (*Strohbehn* and *Clifford* [2.48]) have a greater range of validity than the log-amplitude statistics (*Fante* [2.40]). It may then be true that measurements of the phase fluctuation statistics in the late afternoon on long paths where we expect $\sigma_\chi^2 > 0.3$ will agree with the first-order theory results even though the amplitude statistics will not.

2.2.4 A Qualitative Interpretation of the First-Order Scattering Theory

Figure 2.6 illustrates schematically the interaction of a plane optical wave with the turbulent atmosphere. The wave, initially having a uniform phase front at A, encounters refractive index irregularities between A and B. At location B, the wave has experienced phase changes which are due only to the speeding up and slowing down of different segments of the wave front. The bumpy nature of the wave front at B produces a deviation of the local wave normals so that rays, emanating from different portions of the wavefront, will eventually interfere and produce irradiance fluctuations (scintillations) at C. It is this process that produces the observed log-amplitude variance (2.83) and the spatial structure described statistically by the covariance function in Fig. 2.5.

There are several important characteristics of the scintillation-producing process that can be gleaned from a simple geometric analysis of the propagation problem. In Fig. 2.7, a refractive index irregularity of radius l, located at z, is illuminated by a plane optical wave. A detector at path position L

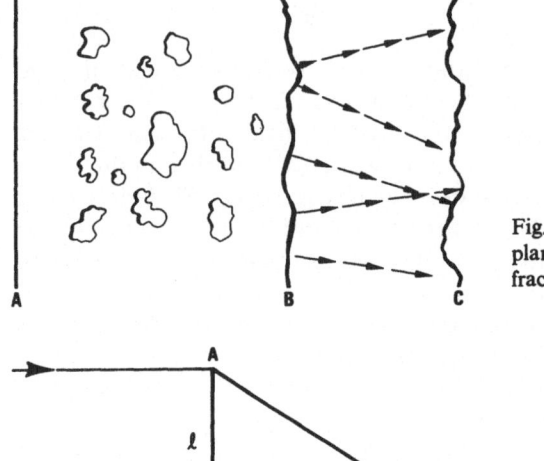

Fig. 2.6. Schematic of the interaction of a plane optical wave with atmospheric refractive index fluctuations

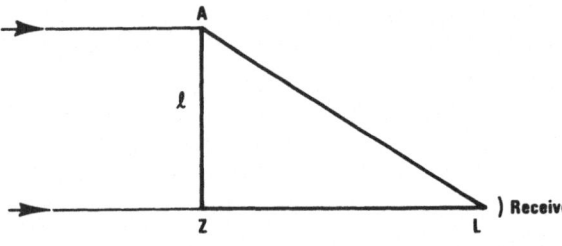

Fig. 2.7. Diagram of the geometry for determining the size l of the most effective eddy located at path position z on a total optical path of length L

observes the scintillation pattern that the irregularity produces. The criterion that determines the effectiveness of this eddy in producing irradiance fluctuations at L is the amount of excess phase path between AL and ZL, that is, for this eddy to be completely effective in producing scintillation, the ray paths AL and ZL should differ by at least one-half wavelength of the incoming plane wave; this produces strong interference of the signal at L. From the geometry of Fig. 2.7, this criterion requires that $l = \sqrt{\lambda(L-z)}$. This minimum size irregularity also turns out to be the most effective size for producing scintillation; smaller eddies at z contribute less because of the weaker refractivity fluctuations associated with them in the Kolmogorov spectrum, and hence, because the amplitude of the field scattered by the eddy is proportional to its refractivity fluctuation, proportionally less scattered energy. Larger eddies at z will not produce strong scintillation because they do not diffract light through a large enough angle to reach the observer at $z = L$. Since an eddy of size l diffracts light through an angle λ/l, in order to be observed at $z = L$ the eddy must diffract energy through an angle $l/(L-z)$. However, by assumption, the eddy is larger than a Fresnel zone and, as a consequence, the inequality $\lambda/l < l/(L-z)$ pertains and no energy reaches the observer. Therefore, as predicted by the theoretical results (2.77) and confirmed by our geometric argument, the most effective eddy for producing scintillations at the receiver is the eddy whose size is equal to that of a Fresnel zone for the rest of the optical path.

Because the refractive index irregularities are illuminated by a plane wave, the disturbance they produce at the receiver will have the same scale size, and hence, the predominant size of the spatial structure in the scintillation pattern also will be the size of a Fresnel zone for the whole path. This explains why the

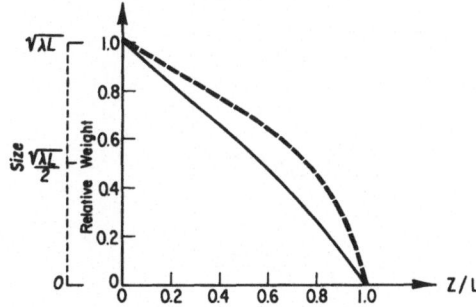

Fig. 2.8. The dashed curve is the size (in Fresnel zones $\sqrt{\lambda L}$) of the most effective eddy for producing scintillation on an optical path of length L in the z direction; the solid curve is the relative importance of refractive turbulence in producing scintillation as a function of position

covariance function (Fig. 2.5) depends only on the Fresnel length and not on the entire refractive index spectrum.

Finally, the dashed curve in Fig. 2.8 shows the behavior of the radius of the most effective irregularity as a function of path position. Note that the largest (smallest) scale structure in the scintillation pattern originates from the transmitter (receiver) end of the path. One further piece of information readily available from the diagram is the relative effectiveness of different portions of the path for producing scintillation. Because we are performing a one-dimensional calculation, we assume the Kolmogorov spectrum in its one-dimensional form, that is, $K^{-5/3}$, and we insert the most effective eddy size $l = \sqrt{\lambda(L-z)}$ into this form of the spectrum. Then we obtain the optical scintillation-weighting function for the path $W \sim [(L-z)]^{5/6}$. This function describes the relative effectiveness of each portion of the propagation path for producing optical scintillation. The solid curve in Fig. 2.8 shows this function; note that the beginning of the path is most effective in contributing to the scintillation and that the end near the receiver has a very small influence. The dominance of the Fresnel zone sized eddies occurs because of the geometry of the propagation problem and not because of any predominant size of refractive index eddies in the atmosphere.

2.3 The Early Experimental Work

Perhaps the first experiments in line-of-sight optical propagation along near-ground paths were carried out by *Tatarskii* et al. [2.61] and *Gurvich* et al. [2.62]. There have also been many experiments with stellar scintillations that predated this work, for example [2.63–67], but their relevance in direct comparison with the theory is limited. The advantage of the near-ground, horizontal-path measurements is the ability to control the experimental path length, and more importantly to measure C_n^2 directly with temperature probes.

Tatarskii et al. [2.61] and *Gurvich* et al. [2.62] found quite good agreement between the theoretical results for the variance (2.83) and covariance (Fig. 2.5) of the fluctuations in the log-amplitude. In their experiments, they also found

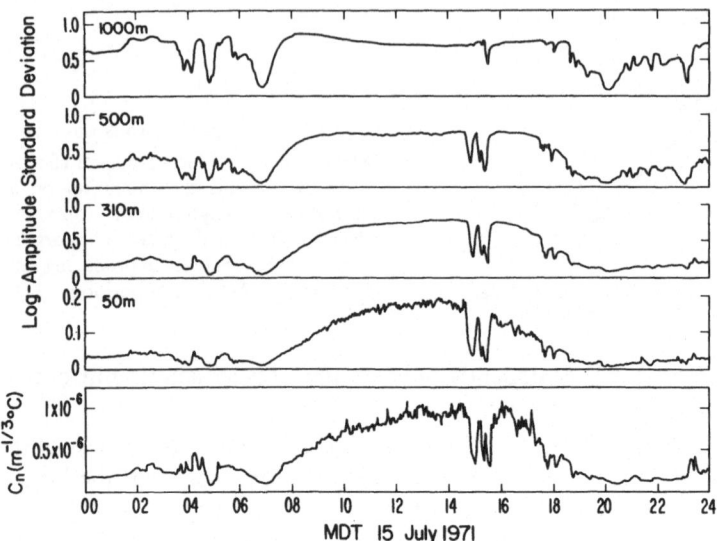

Fig. 2.9. Variance of the log-amplitude standard deviation σ_χ at each of four path lengths compared to the refractive index structure parameter C_n

that the irradiance fluctuations were log normally distributed, that is, the logarithm of the irradiance was normally or Gaussian distributed. Their actual technique was to measure the log irradiance fluctuations $\sigma^2 = (\log I/I_0)^2$ and use the relation $\sigma^2 = 4\sigma_\chi^2$ to check the theory. It is perhaps expected that they would obtain good agreement with the first-order theory because the maximum observed scintillation of their incandescent source was $\sigma_\chi^2 \sim 0.1$, which is sufficiently weak to prevent the effects of multiple scattering from dominating. Most recent experiments show the multiple-scattering regime sets in at about $\sigma_\chi^2 \gtrsim 0.3$.

Following the initial success of the theory, Soviet scientists *Gracheva* and *Gurvich* [2.68] were the first to observe the so-called saturation phenomenon. Initially, the log-amplitude variance would increase as predicted by (2.83) with increasing path length or C_n^2, only to stop increasing and saturate when $k^{7/6}L^{11/6}C_n^2 > 1$. Further observations of this phenomenon were made in the Soviet Union [2.69, 70] and later in the United States [2.71, 72]. Some authors [2.73, 74] even noted a decrease in σ_χ^2 with further increases in integrated turbulence when $k^{7/6}L^{11/6}C_n^2 \gg 1$.

Figure 2.9 illustrates some recent data on the saturation phenomenon (*Clifford* et al. [2.75]). Here we have simultaneous measurements of σ_χ for 24 h over four different optical paths—50, 310, 500, and 1000 m. Also shown for the same period is a direct point measurement of C_n derived from fine-wire thermometer measurements of C_T. The theory (2.83) indicates that σ_χ is proportional to C_n. This is clearly evident on the 50 m path where there appears to be detailed correlation between C_n and σ_χ. On the longer paths, the increase in refractive turbulence does not produce a corresponding increase in σ_χ. In fact, on the 1000 m path as C_n increases between 800 and 1400 MDT, σ_χ actually decreases. The path length dependence in (2.83) also appears to be inadequate to explain the data, that is, $\sigma_\chi \sim L^{11/12}$. Again, the log-amplitude standard deviation, instead of being roughly proportional to path, tends at times to level off and even decrease.

Interestingly, almost all experimenters noted that the probability distribution remained log normal even when the weak scattering theory failed. This remains especially surprising today because it is well demonstrated experimentally that the structure of the two-dimensional scintillation pattern, as indicated by the covariance function, is drastically altered in strong turbulence from the prediction (Fig. 2.5). This was the state of the theory in the late 1960s. Severe doubt had been cast on the validity of the weak scattering theory, especially with regard to the Rytov assumption (*Strohbehn* [2.19]). This stimulated a prodigious amount of theoretical attempts, both purely mathematical, using techniques from quantum electrodynamics to solve the scattering equation (2.50) and new physical models of propagation to bring the theoretical predictions closer in line with experimental reality. These results will be discussed in more detail in Chapters 3 and 4.

References

2.1   J.C.Owens: Appl. Opt. **6**, 51 (1967)
2.2   M.L.Wesely, E.C.Alcaraz: J. Geophys. Res. **78**, 6224 (1973)
2.3   C.A.Friehe, J.C.LaRue, F.H.Champagne, C.H.Gibson, C.F.Dreyer: J. Opt. Soc. Am. **65**, 1502 (1975)
2.4   G.K.Batchelor: *The Theory of Homogeneous Turbulence* (Cambridge University Press, London 1953)
2.5   J.L.Lumley, H.A.Panofsky: *The Structure of Atmospheric Turbulence* (Interscience, New York 1964)
2.6   A.S.Monin, A.M.Yaglom: *Statistical Fluid Mechanics: Mechanics of Turbulence*, Vol. 1 (MIT Press, Cambridge, Mass. 1971)
2.7   H.Tennekes, J.L.Lumley: *A First Course in Turbulence* (MIT Press, Cambridge, Mass. 1972)
2.8   A.Kolmogorov: In *Turbulence, Classic Papers on Statistical Theory*, ed. by S.K.Friedlander and L.Topper (Interscience, New York 1961) p. 151
2.9   L.O.Myrup: Tellus **21**, 341 (1969)
2.10 J.C.Kaimal, J.C.Wyngaard, Y.Izumi, O.R.Coté: Quart. J. Roy. Meteorol. Soc. **98**, 563 (1972)
2.11 V.I.Tatarskii: *Wave Propagation in a Turbulent Medium*, translated by R.A.Silverman (McGraw-Hill, New York 1961)
2.12 V.I.Tatarskii: *The Effects of the Turbulent Atmosphere on Wave Propagation* (Translated by Israel Program for Scientific Translations; originally published in 1967) U.S. Dept. of Commerce, National Technical Information Service, Springfield, Va. 1971)
2.13 S.Corrsin: J. Appl. Phys. **22**, 469 (1951)
2.14 A.M.Yaglom: *An Introduction to the Theory of Stationary Random Functions*, translated by R.A.Silverman (Prentice Hall, Englewood Cliffs, New York 1962)
2.15 N.Wiener: *Extrapolation, Interpolation, and Smoothing of Stationary Time Series, with Engineering Applications* (Wiley and Sons, New York 1949)

2.16 R.B.Blackman, J.W.Tukey: *The Measurement of Power Spectra, from the Point of View of Communications Engineering* (Dover Publ., New York 1958)

2.17 G.M.Jenkins, D.G.Watts: *Spectral Analysis and Its Applications* (Holden-Day, San Francisco 1968)

2.18 F.Pasquill: *Atmospheric Diffusion; The Dispersion of Windborne Material from Industrial and Other Sources* (Van Nostrand, New York 1962)

2.19 J.W.Strohbehn: Proc. IEEE **56**, 1301 (1968)

2.20 R.S.Lawrence, G.R.Ochs, S.F.Clifford: J. Opt. Soc. Am. **60**, 826 (1970)

2.21 S.L.Hess: *Introduction to Theoretical Meteorology* (Holt, New York 1959)

2.22 W.D.Neff: NOAA Tech. Rept. ERL 322-WPL **38** (1975)

2.23 L.R.Tsvang: Radio Sci. **4**, 1175 (1969)

2.24 G.R.Ochs, R.S.Lawrence: NOAA Tech. Rept. ERL **251**-WPL **22** (1972)

2.25 J.L.Bufton, P.O.Minott, M.W.Fitzmaurice, P.J.Titterton: J. Opt. Soc. Am. **62**, 1068 (1972)

2.26 R.E.Hufnagel, N.R.Stanley: J. Opt. Soc. Am. **54**, 52 (1964)

2.27 D.W.Beran, W.H.Hooke, S.F.Clifford: Boundary-Layer Meteorol. **4**, 133 (1973)

2.28 R.E.Hufnagel: "Proc. Topical Meeting on Optical Propagation Through Turbulence" (Boulder, Colo. 1974)

2.29 J.C.Wyngaard, Y.Izumi, S.A.Collins,Jr.: J. Opt. Soc. Am. **61**, 1646 (1971)

2.30 A.S.Frisch, G.R.Ochs: J. Appl. Meteorol. **14**, 415 (1974)

2.31 A.H.Mikesell: Publ. U.S. Naval Obs., 2nd Series **17** (4), 137 (1955)

2.32 A.N.Demidova: Izv. Glav. Astronom. Obs. **21** (6), 2 (1960)

2.33 J.L.Bufton: NASA Tech. Rept., TR-R-**369** (1971)

2.34 M.A.Kallistratova, D.F.Timanovskiy: Izv. Akad. Nauk. SSSR, Atmos. Ocean. Phys. **7**, 46; Russ.: **1**, 73 (1971)

2.35 A.M.Obukov: Izv. Akad. Nauk. SSSR **13**, 96 (1949)

2.36 L.A.Chernov: *Wave Propagation in a Random Medium*, translated by R.A.Silverman (McGraw-Hill, New York 1960) p. 168

2.37 R.W.Lee, J.C.Harp: Proc. IEEE **57**, 375 (1969)

2.38 R.L.Fante: IEEE Trans. AP-**21**, 750 (1973)

2.39 R.S.Lawrence, J.W.Strohbehn: Proc. IEEE **58**, 1523 (1970)

2.40 R.L.Fante: Proc. IEEE **63**, 1669 (1975)

2.41 H.Su, M.Plonus: J. Opt. Soc. Amer. **61**, 256 (1971)

2.42 A.Ishimaru: IEEE Trans. AP-**20**, 10 (1972)

2.43 L.W.Pickering, R.E.McIntosh: IEEE Trans. AP-**20**, 528 (1972)

2.44 M.Plonus, H.Su, C.Gardner: IEEE Trans. AP-**20**, 801 (1972)

2.45 C.Gardner, M. Plonus: J. Opt. Soc. Am. **64**, 68 (1974)

2.46 I.Sreenivasiah, A.Ishimaru: EE Dep. Univ. Wash: AFCRL-TR-74-02045 (1974)

2.47 C.Liu, A.Wernik, K.Yeh: IEEE Trans. AP-**22**, 624 (1974)

2.48 J.W.Strohbehn, S.F.Clifford: IEEE Trans. AP-**15**, 416 (1967)

2.49 A.A.M.Saleh: IEEE J. QE-**3**, 540 (1967)

2.50 V.I.Tatarskii: Izv. VUZ, Radiofiz. **10**, 1762 (1967); [English transl.: Radiophys. Quant. Electron **10**, 987 (1967)]

2.51 Y.A.Kravtsov: Izv. VUZ, Radiofiz. **13**, 281 (1970); [English transl.: Radiophys. Quant. Electron **13**, 217 (1970)]

2.52 S.F.Clifford: PhD Thesis, Dartmouth Coll., Hanover, N.H. (1969)

2.53 I.S.Gradshteyn, I.M.Ryzhik: *Table of Integrals, Series, and Products*, translated by Scripta Technica, Inc. (Academic Press, New York 1965)

2.54 A.Erdélyi: *Tables of Integral Transforms*, Vol. 2, ed. by A.Erdélyi (McGraw-Hill, New York 1954)

2.55 S.M.Rytov: Izv. Akad. Nauk SSSR (Ser. Fiz.) **2**, 223 (1937)

2.56 V.V.Pisareva: Soviet Phys.-Acoust. **6**, 87 (1960)

2.57 Y.L.Feinberg: Moscow: Academy of Sciences USSR (1961)

2.58 V.I.Tatarskii: Izv. VUZ, Radiofiz. **5**, 490 (1962)

2.59 W.P.Brown,Jr.: J. Opt. Soc. Am. **56**, 1045 (1966)

2.60 A.S.Gurvich, M.Kallistratova, N.Time: Radiophys. Quant. Electron. **11**, 771 (1968)
2.61 V.I.Tatarskii, A.S.Gurvich, M.A.Kallistratova, L.V.Terenteva: Astronom. Zh. **35**, 623 (1958)
2.62 A.S.Gurvich, V.I.Tatarskii, L.R.Tsvang: Dokl. Akad. Nauk. SSSR **123**, 655 (1958)
2.63 E.Gaviola: Astron. J. **54**, 155 (1949)
2.64 A.H.Mikesell, A.A.Hoag, J.S.Hall: J. Opt. Soc. Am. **41**, 689 (1951)
2.65 M.A.Ellison, H.Seddon: Monthly Notices Roy. Astron. Soc. **112**, 73 (1952)
2.66 I.G.Kolchinski: Astronom. Zh. **34**, 638 (1957)
2.67 V.A.Krasilnikov: Dokl. Akad. Nauk. SSSR **65**, 291 (1949)
2.68 M.E.Gracheva, A.S.Gurvich: Izv. VUZ, Radiofiz. **8**, 717 (1965); [English transl.: Radiophys. Quant. Electron **8**, 511 (1965)]
2.69 M.E.Gracheva: Izv. VUZ, Radiofiz. **10**, 775 (1967); [English transl.: Radiophys. Quant. Electron **10**, 424 (1967)]
2.70 A.S.Gurvich, M.A.Kallistratova, N.S.Time: Izv. VUZ, Radiofiz. **11**, 1360 (1968); [English transl.: Radiophys. Quant. Electron **11**, 771 (1968)]
2.71 P.H.Deitz, N.J.Wright: Ballistic Res. Labs., Aberdeen Proving Ground, Md. BAL-Memo Rept.-**1941**-Rev. (1968)
2.72 G.E.Mevers, D.L.Fried, M.P.Keister,Jr.: J. Opt. Soc. Am. **55**, 1575 (1965), Abstract
2.73 G.R.Ochs, R.S.Lawrence: J. Opt. Soc. Am. **59**, 226 (1969)
2.74 M.E.Gracheva, A.S.Gurvich, M.A.Kallistratova: Izv. VUZ, Radiofiz. **13**, 56 (1970); [English transl.: Radiophys. Quant. Electron **13**, 40 (1970)]
2.75 S.F.Clifford, G.R.Ochs, R.S.Lawrence: J. Opt. Soc. Am. **64**, 148 (1974)

3. Modern Theories in the Propagation of Optical Waves in a Turbulent Medium

J. W. Strohbehn

With 18 Figures

In Chapter 2, Clifford discussed in some detail the classical approach for describing the propagation of electromagnetic waves in turbulent media, including different models for describing atmospheric turbulence. In addition, he showed the results of a number of experiments that clearly demonstrate that the classical theories, which are based on some type of perturbation approach, are inadequate to explain the experimental measurements. In this chapter we shall concentrate on modern theories, all of which have been motivated by the desire to explain the discrepancy between the classical theory and experiment, in particular the so-called saturation effect demonstrated in Fig. 2.9.

In this section, we shall try to give an overview of the different theoretical approaches, their similarities and differences, and the results that they are able to predict. At least as of this writing, the author feels it is fair to state that there does not exist any rigorously derived theory that adequately explains all the experimental data. There appear to be two primary directions that presently exist in the field. A large number of workers have attempted to develop a rigorous mathematical approach. Many of these workers, using different routes, have arrived at the same set of equations for describing the moments of an optical wave. However, to date only the simplest of these equations have been solved analytically, while the equations of the most immediate interest have only been solved approximately. The correctness of these solutions is still being investigated, both theoretically and experimentally. The second approach, which has recently been pursued by a number of workers in the U.S., is an attempt to understand in a physical sense the important observed phenomena and to develop a somewhat heuristic theory to explain these phenomena. This latter approach seems to be reasonably successful, and presently needs a more rigorous mathematical basis to justify certain ad hoc steps.

3.1 Overview

There have been a number of excellent review articles over the past few years, especially those by *Barabanenkov* et al. [3.1], *Prokhorov* et al. [3.2], and *Fante* [3.3]. In this chapter we shall try to present an overview of the various theoretical approaches, a comparison of the results of the different theories, and how they relate to experimental data.

In essence, the problem that we are trying to solve involves a stochastic wave equation. Naturally, as in any problem involving random variables, we concentrate on determining certain average or mean quantities, such as the mean irradiance, the phase structure function, the variance of angle of arrival, or the average beam width. The problem is to start with the stochastic wave equation and solve for the mean value of the quantity of interest. The important characteristic of our equation from a theoretical viewpoint is that it is a parametric stochastic equation. As pointed out by Clifford (see Sect. 2.2.1) [2.19, 48–51], the depolarization terms are negligible for propagation of optical waves in the earth's atmosphere. Therefore, if we assume a monochromatic wave, and that the time variations in the random dielectric constant $\varepsilon(r, t)$ are slow enough so that a quasi-steady state approach can be used[1], then the wave equation becomes

$$\nabla^2 E(r, t) + k_0^2 \varepsilon(r, t) E(r, t) = 0, \tag{3.1}$$

where the actual electric field is given by $E(r, t)\exp(i\omega t)$. The difficulty in solving this equation is a consequence of the random nature of $\varepsilon(r, t)$ which is a coefficient of $E(r, t)$, the quantity for which we wish to solve. Note that the problem would be much simpler mathematically if the random term entered as an independent source term on the right-hand side. In both of the classical perturbation methods, i.e., the Born approximation or the method of smooth perturbations (MSP), the effect of the approximations is to produce a new equation of the form

$$\nabla^2 u(r, t) + k^2 u(r, t) = f(r, t), \tag{3.2}$$

where the random dielectric constant appears as a factor in $f(r, t)$. Techniques for solving this equation are well-known and are discussed in Chapter 2. Since it is not known how to solve (3.1) exactly, approximate methods must be applied to simplify the problem. Approximate methods normally are based on expansions in which one parameter of interest is small compared to another. In actual fact, we have several choices open to us and the differences in approaches rely on these choices.

First of all, we know that in the earth's atmosphere the fluctuating part of the dielectric constant is much smaller than the mean value, i.e., if we write

$$\varepsilon(r, t) = \varepsilon_0(r) + \varepsilon_1(r, t) \tag{3.3}$$

then $\langle \varepsilon_1^2(r, t) \rangle \ll \varepsilon_0^2(r)$. For simplicity, we shall assume that the medium is homogeneous and that $\varepsilon_0^2(r) = 1$. A second important consideration is that the wavelength is very small compared to the size of a typical dielectric constant inhomogeneity. This means that most of the radiation will be scattered in a

---

[1] Several authors have investigated this assumption [3.4–6].

small cone about the original propagation direction. Another relation that may be of importance is that a typical path length is large compared to almost all other linear parameters, e.g., the wavelength, the size of an inhomogeneity, or the diameter of a Fresnel zone.

When attempting to classify the different approaches to solving (3.1), we find there are different ways that this can be done, and any particular method may be a combination of a number of different techniques. For example, we may start with the scalar wave equation (3.1), or we may immediately let

$$E = \exp(\Phi). \tag{3.4}$$

This substitution transforms the wave equation to the Ricatti equation

$$\nabla^2\Phi + \nabla\Phi \cdot \nabla\Phi + k^2\varepsilon = 0, \tag{3.5}$$

whose primary advantage over the original wave equation is that it is in a nonparametric form. However, the price is a nonlinear equation, and we still do not have any method to obtain an exact solution. No matter which equation is used as the starting point, both equations can be simplified somewhat by assuming that the direction of propagation is a preferred coordinate. If we assume the initial direction of propagation is along the z axis, let

$$E(r, t) = u(r, t)\exp(ikz)$$

and assume $|\partial^2 u/\partial z^2| \ll 2k|\partial u/\partial z|$, then (3.1) transforms approximately to

$$2ik\frac{\partial u}{\partial z} + \nabla_T^2 u + k^2\varepsilon_1 u = 0, \tag{3.6}$$

where

$$\nabla_T^2 = \frac{\partial^2}{\partial x^2} + \frac{\partial^2}{\partial y^2}.$$

Equation (3.6) is known as the parabolic equation or the quasi-optical approximation. Similarly, if we let $\Phi = \psi + ikz$, and again assume $|\partial^2\psi/\partial z^2|$ and $|\partial\psi/\partial z|^2 \ll 2k|\partial\psi/\partial z|$, then (3.5) becomes

$$2ik\frac{\partial\psi}{\partial z} + \nabla_T^2\psi + (\nabla_T\psi)^2 = -k^2\varepsilon_1. \tag{3.7}$$

Equation (3.7) was originally derived by *Rytov* [3.7] when considering a problem of the diffraction of light by ultrasound[2].

---

[2] For a more historical account of these and other developments in this field, see [3.1].

Equations (3.6) and (3.7) both depend on the same physical assumption. Mathematically they are based on the assumption that $\lambda \ll l$, where λ is the wavelength and l a typical scale size for an inhomogeneity. Physically, this assumption implies that there exists only small-angle scattering, i.e., the direction of the scattered wave is very close to the original direction of propagation, which also implies no backscattering. Therefore, in integral representations of (3.6) or (3.7) we need to integrate only from $z = 0$ to $z = L$, where L is the location of the received signal, e.g., see (2.55), (2.57), or (2.68).

Given that $|\varepsilon_1| \ll \varepsilon_0$ (or in statistical terms, $\sigma_\varepsilon^2 \ll \varepsilon_0^2$), we now have four equations to which we could apply a perturbation method. If we expand E in (3.1) and keep only the first term, then we have a single-scattering approximation or the Born approximation from quantum mechanics. A similar expansion could be applied to Φ in (3.5), but to the author's knowledge is rarely used. Since (3.6) and (3.7) are simpler, but hopefully equally valid, versions of (3.1) and (3.5), it is mathematically more convenient to perform a perturbation expansion on them. If a perturbation expansion is applied to (3.7), we have the method of smooth perturbations, which is discussed in detail in *Tatarskii's* monographs [3.8, 9].

Knowledge of the region of validity of these various perturbation techniques is of course critical when applying the results. In general, determining the region of validity of a particular approximation is difficult. First of all, the only sure method of verifying the region of validity of an approximate solution is to compare it with the exact solution. However, if the exact solution is known, the approximate case is no longer of interest. Furthermore, when defining a region of validity, it is necessary to specifically state what quantity is being discussed. For example, it is well known that for optical propagation through atmospheric turbulence, the geometrical optics solution for amplitude fluctuations is limited by the inequality $\sqrt{(\lambda L)} \ll l_0^2$, where L is the path length and l_0 the size of the smallest inhomogeneity of interest. This condition may imply path lengths between 1 and 100 m, depending on l_0. However, the geometrical optics solution for phase fluctuations is within a factor of two of the wave-optics solution for much longer path lengths, and may be a good approximation even in the saturation region. There has been a great deal of discussion of the region of validity of the MSP and a comparison of its region of validity with that of the Born approximation [3.8–30]. To summarize the discussion in the above references succinctly, there appear to be theoretical differences between the two methods, but for the problem of concern here the differences are not very significant. More precisely, expressions derived for σ_I^2 using the Born approximation do not appear to have a much different range of validity than the expression for $\sigma_{\ln I}^2$ found from the MSP. However, there is one definite experimental result that certainly favors the MSP. In the region of weak fluctuations, the probability distribution of the irradiance, according to almost all experimental evidence, is very close to a log normal distribution. This result agrees with conclusions based on the MSP, but contradicts the implications of a

Born approximation, which would predict a Rayleigh or Rice-Nakagami probability distribution.

A second decision that needs to be made when attempting to derive results for certain average values is when in the derivation the averaging should be done. In the perturbation methods already mentioned, the approach is to attempt to find a good approximate solution to the partial differential equations for the instantaneous field. From this expression the average values are found. An alternative strategy is to attempt to find differential equations for the average values of interest. Almost all of the efforts devoted to finding the field describing the optical wave in the region of multiple scatter have concentrated on finding either partial differential equations or integral equations for such quantities as the mean field, the mutual coherence function, and high order moments. Unfortunately, it is not possible to get closed equations for these averaged quantities directly from (3.1, 5, 6) or (3.7), since when averaging these equations for a given moment, higher order moments always appear.

Historically, the first approach was to attempt to find solutions of (3.1) using a Green's function method. If we rewrite (3.1) in the form

$$V^2 E + k^2 E = -k^2 \varepsilon_1 E, \tag{3.8}$$

and assume $g(r, r')$ is the Green's function for a homogeneous medium, then (3.8) may be written as an integral equation

$$E(r) = E_0(r) - k^2 \int g(r, r') \varepsilon_1(r') E(r') dr', \tag{3.9}$$

where $E_0(r)$ is the field if $\varepsilon_1(r) = 0$, and $g(r, r') = -\exp(ik|r - r'|)/(4\pi|r - r'|)$. An iterative solution of (3.9) gives

$$E(r) = E_0(r) - k^2 \int g(r, r') \varepsilon_1(r') E_0(r') dr' + k^4 \iint g(r, r') \varepsilon_1(r') g(r', r'')$$

$$\cdot \varepsilon_1(r'') E_0(r'') dr' dr'' - \dots. \tag{3.10}$$

Note that (3.10) is an expression for the instantaneous field and results from iteratively applying the perturbation method. However, (3.10) is of little practical value unless it is simplified (if we keep the first two terms we have the Born approximation). The approach, which was originally applied by *Bourret* [3.31, 32] to the scattering of waves in a medium with slowly varying dielectric constant, was to use the graphical technique of *Feynman* [3.33, 34]. In general, the results lead to the *Dyson* equation [3.35] to describe the mean field and the *Bethe-Salpeter* equation [3.36] for the mutual coherence function. These are integral equations, basically in the same form as (3.9), but for averaged quantities instead of the instantaneous field.

Since *Bourret*'s original work this technique has been applied by numerous workers, and we shall discuss some of the results in more detail below. As we shall see, while the equations look simpler, they still cannot be solved except by approximate methods.

A second approach to obtaining equations for the mean values of the moments is to average the original equations directly. This technique is most often applied to the parabolic equation, (3.6). If (3.6) is averaged in an attempt to find an equation for the mean field, we have

$$-2ik\frac{\partial\langle u\rangle}{\partial z}+V_T^2\langle u\rangle+k^2\langle\varepsilon_1(r)u(r)\rangle=0. \tag{3.11}$$

The mathematical difficulty with this approach is immediately apparent. Equation (3.11) contains not only the mean field, which we are trying to find, but a second unknown $\langle\varepsilon_1(r)u(r)\rangle$. Therefore, (3.11) is one equation in two unknowns, and the problem is how to determine $\langle\varepsilon_1(r)u(r)\rangle$. If (3.6) is multiplied by $\varepsilon_1(r)$ and averaged, we then get an equation in $\langle\varepsilon_1(r)u(r)\rangle$ and in $\langle\varepsilon_1^2(r)u(r)\rangle$; hence again we introduce a new unknown. The problem is how to close the set of equations so that they may be solved, or at least are complete in a mathematical sense. *Tatarskii* and *Klyatskin* [3.9, 37–40] found an elegant method to accomplish exactly this by assuming that the dielectric constant is uncorrelated in the direction of propagation. With these two assumptions, i.e., approximating the wave equation by the parabolic equation and assuming the correlation function of the dielectric constant is a delta function in the z direction, the optical propagation problem may be treated as a Markov random process, and closed equations for the moments of u may be derived. This approach will be discussed in Section 3.3.

A second approach for closing the set of equations has been called "the local method of small perturbations" (LMSP) [3.2, 41–47]. In this method, the random medium is divided into thin slabs oriented perpendicular to the direction of propagation and the various terms in the averaged equations are investigated as the wave propagates through a single slab. For example, in (3.11) the term $\langle\varepsilon_1(r)u(r)\rangle$ is studied as the wave propagates through one slab. Under suitable conditions, this term may be approximated by a term involving the variance of the dielectric constant fluctuations and the mean field, and hence (3.11) becomes a closed equation. Similar approximations can be made for higher order moments. The results are identical to those found assuming a Markov random process.

Another approach to derive the average moments is to first derive the mean value of the functional of u, and then find the various moments from the functional. This approach has been used by *Furutsu* [3.48] and *Klyatskin* [3.49].

Before discussing the different approaches for describing multiple-scatter effects in detail, it is worthwhile to examine the four approximations that are used in all the methods, though sometimes included in a subtle manner. The four critical approximations, all of which appear reasonable for laser propagation in a turbulent medium are:

1) depolarization effects are negligible,

2) backscattering may be neglected,

3) the full wave equation, (3.1), may be approximated by the parabolic equation, (3.6),

4) the correlation function of the dielectric constant fluctuations may be assumed to be delta correlated in the direction of propagation.

The first two approximations follow from the same assumption, i.e., the size of a typical dielectric inhomogeneity is much larger than a wavelength. These assumptions are discussed in some detail by *Tatarskii* [3.8, 9] when deriving the equations for the smooth-perturbation method; however, the same arguments are equally valid in the multiple-scattering region. Physically, in order to get significant backscattering it is necessary to have a significant change in the dielectric constant in a distance short compared to (or at least on the order of) a wavelength. This situation does not occur in the earth's turbulent clear atmosphere. The mathematical simplification provided by this assumption is that it is not necessary to integrate over the entire space, but only over the region between the transmitter and receiver. In addition, the Green's function only needs to include the forward-scattered term.

The third assumption, which states that (3.1) may be replaced by (3.6) is also based on the fact that $\lambda \ll l$ for any scale size of importance. Again this leads to small angle scattering, and hence in the region of interest the Green's function, given in (3.9), may be replaced by the Green's function

$$G(r, r') = \frac{1}{4\pi(z_1 - z_2)} \exp\left[-ik\frac{(x-x')^2 + (y-y')^2}{2(z-z')}\right], \tag{3.12}$$

where z is the propagation direction. This assumption was considered in detail by *Tatarskii* [3.9], both for weak and strong scattering.

While the first three assumptions usually are stated clearly, the last assumption, which is usually referred to as the Markov approximation, is often not as clearly brought out. However, *Tatarskii* and *Klyatskin* stated this assumption quite clearly in their recent work [3.37–40]. If we assume the dielectric constant fluctuations have a covariance function given by

$$\langle \varepsilon_1(r_1)\varepsilon_1(r_2) \rangle = B_\varepsilon(|r_1 - r_2|) = \int\!\!\int\!\!\int_{-\infty}^{\infty} \Phi_\varepsilon(\kappa) \exp[i\kappa \cdot (r_1 - r_2)]\, d^3\kappa, \tag{3.13}$$

then we wish to approximate $B_\varepsilon(|r_1 - r_2|)$ by

$$B_\varepsilon(|r_1 - r_2|) \cong \delta(z_1 - z_2) A_\varepsilon(|\varrho_1 - \varrho_2|), \tag{3.14}$$

where $\varrho = (x, y)$. Note that

$$A_\varepsilon(|\varrho_1 - \varrho_2|) = \int_{-\infty}^{\infty} B_\varepsilon[|(\varrho_1, z_1) - (\varrho_2, z_2)|] dz_1,$$

$$= 2\pi \int\int_{-\infty}^{\infty} \Phi_\varepsilon(\boldsymbol{\kappa}, 0) \exp(i\boldsymbol{\kappa} \cdot \varrho) d^2\kappa, \tag{3.15}$$

and

$$\frac{A_\varepsilon(0)}{B_\varepsilon(0)} = \frac{1}{B_\varepsilon(0)} \int_{-\infty}^{\infty} B_\varepsilon[|(0, z_1) - (0, z_2)|] dz_1$$

$$= \frac{1}{B_\varepsilon(0)} \int_{\infty}^{\infty} B_\varepsilon(\varrho) d\varrho \equiv l_1, \tag{3.16}$$

where l_1 is defined as the integral scale[3]. The justification for this approximation is based on the assumption that l_1 is much smaller than any scale of interest in the longitudinal direction (direction of propagation). In particular, the correlation length of the field $u(r)$ in the longitudinal direction is much larger than l_1. This approximation is critical in closing the various expressions for the moments, but is applied in a different manner by different authors. In actual fact, it is used in the method of smooth perturbations when obtaining expressions that contain $\Phi(\kappa_x, \kappa_y, 0)$.

3.2 The Diagram Method

As mentioned in Section 3.1, one of the first techniques used in an attempt to correctly include multiple-scattering effects was the graph or diagram technique of *Feynman* [3.33, 34], which *Bourret* [3.31, 32] first applied to the propagation of waves in a continuous random medium. Since then the approach has been used by numerous workers in a variety of ways. *Tatarskii* [3.9] presents this method in detail; therefore in this section we shall apply the method only to derive the equation for the mean field and let the interested reader consult the references for more complete discussions.

Generally, all diagram or perturbation theory methods follow the same major steps, but differ in the initial equations from which they start or in the approximations they use to get closed form solutions. We can write the initial equation in the form

$$w(r) = w_0(r) + \int G(r, r_1) m(r_1) w(r_1) dr_1, \tag{3.17}$$

---

[3] Note that κ may stand for (κ_x, κ_y) as in (3.15) or for $(\kappa_x, \kappa_y, \kappa_z)$ as in (3.13). It should be clear from the equations which is implied.

where $\int dr_1 = \int_0^\infty dz_1 \int\int_{-\infty}^\infty dx_1 dy_1$ unless otherwise noted. Equation (3.17) corresponds to (3.9) where $w(r) = E(r)$; $G(r, r_1) = g(r, r_1)$, the Green's function for the free space scalar wave equation; and $m(r) = -k^2 \varepsilon_1(r)$. This starting point was used by various authors [3.31, 32, 50–54]. If we use the parabolic equation, (3.6) is the initial equation, $w(r)$ corresponds to $u(r)$, the Green's function is

$$G(r, r_1) = -\frac{1}{4\pi(z - z_1)} \exp\left[ik\frac{(\varrho - \varrho_1)^2}{2(z - z_1)}\right], \tag{3.18}$$

and the z integration is from 0 to z. This is equivalent to the approach used by *Shishov* [3.55]. *Brown*'s [3.52, 53] approach was somewhat different in that he worked in the spatial frequency domain, and hence the form of the equations is somewhat different, though clearly equivalent. *Molyneux* [3.54] started with (3.1); hence our $w(r)$ represent his $E(r)$, but used the same transformation as in (3.6) to obtain a new equation where $w(r) = u(r)$, and

$$G(r, r_1) = \exp[-ik(z - z_1)]g(r, r_1). \tag{3.19}$$

Solving (3.17) iteratively, we get an infinite series

$$w(r) = \sum_{n=0}^\infty w_n(r) \tag{3.20}$$

where

$$w_{n+1}(r) = \int G(r, r_1)m(r_1)w_n(r_1)dr_1 \tag{3.21}$$

or

$$w(r) = w_0(r) + \int G(r, r_1)m(r_1)w_0(r_1)dr_1$$
$$+ \int\int G(r, r_1)m(r_1)G(r_1, r_2)m(r_2)w_0(r_2)dr_1 dr_2 + \int\int\int \dots . \tag{3.22}$$

From these equations, we wish to obtain the statistical moments of $w(r)$. The equation for the mean field $\langle w(r) \rangle$ can be obtained quite easily, the second moment $\langle w(r)w^*(r) \rangle$ requires somewhat more labor, and higher order moments involve considerable effort.

As we proceed with this development a number of approximations will be made. The first is that $m(r)$ is statistically homogeneous and isotropic. Since in most formulations, $m(r)$ is nonzero only in the half-space $z > 0$, this assumption is not strictly true. However, since the path length is much longer than the correlation scale of $m(r)$, effects near $z = 0$ are not of great importance in the final result.

In addition we assume that $m(r)$ is a gaussian random function. This assumption also cannot be strictly true because $m(r) = k^2\varepsilon_1(r)$ must be greater than $-k^2\varepsilon_0(r)$, but a true gaussian random variable is not bounded. However, $\sigma_m^2 \ll \langle m \rangle^2$, and hence modeling $m(r)$ by a gaussian random variable is not unreasonable. Other authors [3.9, 56] examined this question in more detail.

The gaussian assumption for $m(r)$ permits a great simplification in the mathematics because any moment for $m(r)$ can be expressed in terms of the second moment. Note that no matter what formulation was used for (3.17), they all assumed $\langle m(r) \rangle = 0$. We shall make use of the following properties of gaussian random variables:

$$\langle m(r_1)m(r_2)\dots m(r_k) \rangle = 0 \quad k \quad \text{odd} \tag{3.23}$$

$$\begin{aligned}\langle m(r_1)m(r_2)m(r_3)m(r_4) \rangle &= \langle m(r_1)m(r_2) \rangle \langle m(r_3)m(r_4) \rangle \\ &+ \langle m(r_1)m(r_3) \rangle \langle m(r_2)m(r_4) \rangle \\ &+ \langle m(r_1)m(r_4) \rangle \langle m(r_2)m(r_3) \rangle. \end{aligned} \tag{3.24}$$

The extension for higher order even moments should be obvious; it is the sum of all possible combinations of products of second-order moments. In general, the $2n$th moment contains $(2n-1)!! = (2n-1)(2n-3)\dots 3 \cdot 1$ terms, e.g., the 6th moment contains 15 terms. We will let

$$\langle m(r_1)m(r_2) \rangle = \sigma_m^2 R(|r_1 - r_2|), \tag{3.25}$$

where the isotropy assumption implies that $\langle m(r_1)m(r_2) \rangle$ depends only on $|r_1 - r_2|$.

If we average (3.22) we get the mean value of $w(r)$ directly, i.e.,

$$\begin{aligned}\langle w(r) \rangle &= w_0(r) + \sigma_m^2 \iint G(r, r_1)G(r_1, r_2)R_{12}w_0(r_2)dr_1 dr_2 \\ &+ \sigma_m^4 \iiiint G(r, r_1)G(r_1, r_2)G(r_2, r_3)G(r_3, r_4)w_0(r_4) \\ &\quad \cdot (R_{12}R_{34} + R_{13}R_{24} + R_{14}R_{23})dr_1 dr_2 dr_3 dr_4 + \dots, \end{aligned} \tag{3.26}$$

where $R_{ij} = R(|r_i - r_j|)$. Writing higher order terms becomes extremely tedious, and hence a graphical shorthand is adopted. In Fig. 3.1 the various symbols that will be used are defined. The rules for the construction of the integral equation from a diagram or vice versa are reasonably obvious. We shall write down the diagram for the last integral given in (3.26), which in actual fact consists of three terms. For each term, we put down points or vertices representing the different spatial coordinates, in this case, r, r_1, r_2, r_3, and r_4. Between the different vertices, fill in the appropriate diagrams from Fig. 3.1. Integrations are performed over all internal vertices of a diagram; hence in Fig. 3.2 the integrations are over r_1, r_2, r_3, and r_4. Therefore, we represent (3.26) as shown in Fig. 3.3. In the various terms in Fig. 3.3, we can differentiate between two basic types of diagrams. If a term can be broken into two parts by

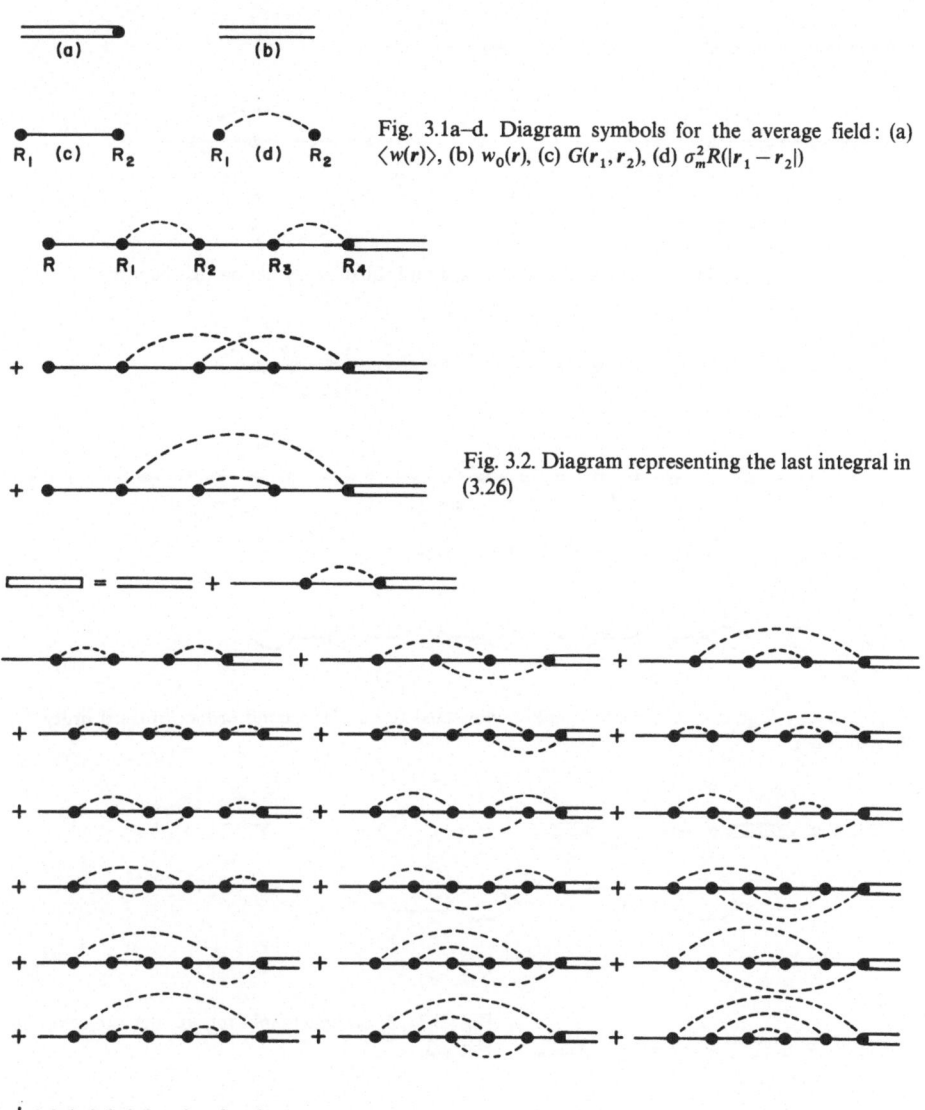

Fig. 3.1a–d. Diagram symbols for the average field: (a) $\langle w(r)\rangle$, (b) $w_0(r)$, (c) $G(r_1, r_2)$, (d) $\sigma_m^2 R(|r_1 - r_2|)$

Fig. 3.2. Diagram representing the last integral in (3.26)

Fig. 3.3. The diagram equation for $\langle w(r)\rangle$

breaking only one solid line it is called weakly connected, otherwise it is strongly connected. In Fig. 3.4 are examples of weakly and strongly connected diagrams. Let the sum of all strongly connected diagrams be given by the first symbol shown in Fig. 3.5, which we shall call $Q(r, r_2)$. By writing out the terms, it should be obvious that all weakly connected diagrams consisting of two strongly connected diagrams are given by the combination of two Q factors separated by a G factor, as shown in Fig. 3.6a. All weakly connected terms

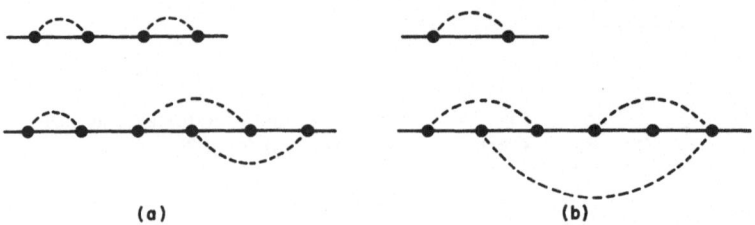

Fig. 3.4a and b. Examples of (a) weakly connected and (b) strongly connected diagrams

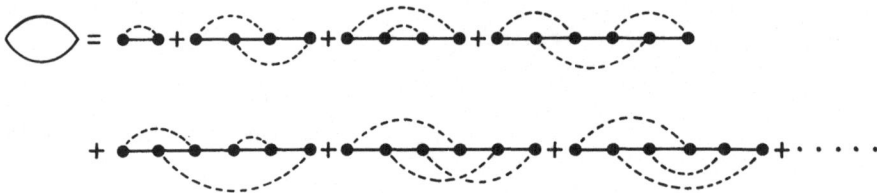

Fig. 3.5. The sum of all strongly connected diagrams

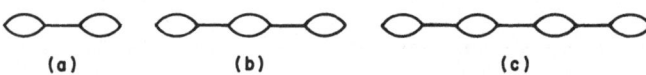

Fig. 3.6a–c. Diagrams for different weakly connected terms: (a) second order, (b) third order, (c) fourth order

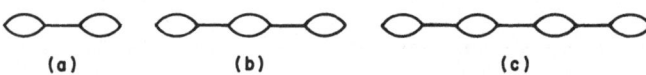

(a)

Fig. 3.7. A compact solution of the equation in Fig. 3.3

(b)

consisting of combinations of three strongly connected terms are given in Fig. 3.6b. Therefore, the mean field as given in Fig. 3.3 can be represented as in Fig. 3.7a, or more compactly as in Fig. 3.7b. Analytically, Fig. 3.7c can be written as

$$\langle w(r) \rangle = w_0(r) + \iint G(r, r_1) Q(r_1, r_2) \langle w(r_2) \rangle \, dr_1 \, dr_2. \tag{3.27}$$

Equation (3.27), which is known as *Dyson*'s equation [3.35], looks like a linear integral equation and could be solved if $Q(r_1, r_2)$ were known. However, Q is an

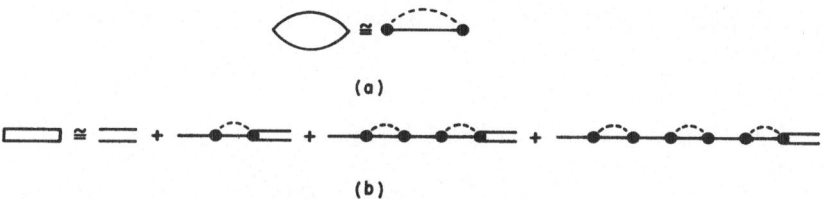

Fig. 3.8. An approximate solution of the equation in Fig. 3.7

infinite series of strongly coupled terms, and the summation cannot in actual fact be carried out. As *Barabanenkov* et al. [3.1] pointed out, even if we could sum the series for Q, we would obtain Q as a functional of $\langle w(r) \rangle$, and hence in reality we have a nonlinear equation. The value of this equation is that it is exact, and hence we may be able to discuss more carefully the convergence properties or the effects of particular approximations. Under certain approximations, linear integral equations will be obtained. A more analytical approach to deriving (3.27) is given in [3.9].

The simplest nontrivial approximation we can make to (3.27) or Fig. 3.7 is to approximate $Q(r_1, r_2)$ by the first term in Fig. 3.4 (see Fig. 3.8a); therefore (3.27) becomes (Fig. 3.8b)

$$\langle w(r) \rangle \cong w_0(r) + \sigma_m^2 \iint G(r, r_1) G(r_1, r_2) R(|r_1 - r_2|) \langle w(r_2) \rangle \, dr_1 dr_2 . \qquad (3.28)$$

Note that while Q is approximated by only one term, multiple scattering effects are taken into consideration to some degree, as is evident from Fig. 3.8b. The solution of the equation has been discussed in detail by numerous workers [3.1,9]. Generally it is assumed that $w_0(r)$ is a point source, therefore $\langle w(r) \rangle$ becomes the average Green's function, $\langle G(r, r_0) \rangle$. The solution for the averaged Green's function is of the form

$$\langle G(r, r_0) \rangle = - \frac{\exp(-ik_e R)}{4\pi R}, \qquad (3.29)$$

where $R = |r - r_0|$, and k_e is the effective wave number. However, most derivations get this result by assuming $kl \ll 1$, where the magnitude of l is of the order of the outer scale of turbulence. This condition is clearly violated for optical waves in the atmosphere. A second weaker condition is $\sigma_\varepsilon^2 k^2 l^2 \ll 1$; however, even this condition is often not met in the atmosphere. While there is a great deal of theoretical discussion of the mean field in the literature, primarily because it is the simplest quantity to calculate, this author knows of no experimental attempts to measure this quantity and hence of no comparison with the theoretical predictions.

Molyneux [3.54] notes that (3.28) can be reduced to a partial differential equation provided we accept the four assumptions already discussed earlier.

Starting with (3.28), the last three approximations (see p. 51) introduce the following changes:

$$\int d\mathbf{r} \equiv \int_0^\infty dz \int\int_{-\infty}^\infty dxdy \cong \int_0^z dz \int\int_{-\infty}^\infty dxdy \equiv \int_0^z dz \int d\boldsymbol{\varrho}, \tag{3.30}$$

$$G(\mathbf{r}_1, \mathbf{r}_2) = -\frac{\exp(ik|\mathbf{r}_1 - \mathbf{r}_2|)}{4\pi|\mathbf{r}_1 - \mathbf{r}_2|} \cong -\frac{\exp\left[ik\dfrac{|\boldsymbol{\varrho}_1 - \boldsymbol{\varrho}_2|^2}{2(z_1 - z_2)}\right]}{4\pi(z_1 - z_2)}, \tag{3.31}$$

$$\sigma_m^2 R(|\mathbf{r}_1 - \mathbf{r}_2|) = k^4 B_\varepsilon(|\mathbf{r}_1 - \mathbf{r}_2|) \cong k^4 \delta(z_1 - z_2) A_\varepsilon(|\boldsymbol{\varrho}_1 - \boldsymbol{\varrho}_2|). \tag{3.32}$$

Also note that

$$\lim_{z_1 \to z_2} -\frac{\exp\left[ik\dfrac{|\boldsymbol{\varrho}_1 - \boldsymbol{\varrho}_2|^2}{2(z_1 - z_2)}\right]}{4\pi(z_1 - z_2)} = \frac{i}{2k}\delta(\boldsymbol{\varrho}_1 - \boldsymbol{\varrho}_2). \tag{3.33}$$

Introducing (3.30–32) into (3.28) and carrying out the integral over z_2 gives

$$\langle w(\mathbf{r}) \rangle = w_0(\mathbf{r}) + \frac{1}{2}k^4 \int_0^z dz_1 \int d\boldsymbol{\varrho}_1 \int d\boldsymbol{\varrho}_2 \frac{\exp\left[ik\dfrac{|\boldsymbol{\varrho} - \boldsymbol{\varrho}_1|^2}{2(z - z_1)}\right]}{4\pi(z - z_1)}$$
$$\frac{i}{2k}\delta(\boldsymbol{\varrho}_1 - \boldsymbol{\varrho}_2) A_\varepsilon(\boldsymbol{\varrho}_1 - \boldsymbol{\varrho}_2) \langle w(z_1, \boldsymbol{\varrho}_2) \rangle.$$

Carrying out the integral over $\boldsymbol{\varrho}_2$ gives

$$\langle w(\mathbf{r}) \rangle = w_0(\mathbf{r}) + \frac{i}{4}k^3 A_\varepsilon(0) \int_0^z dz_1 \int d\boldsymbol{\varrho}_1 \frac{\exp\left[ik\dfrac{|\boldsymbol{\varrho} - \boldsymbol{\varrho}_1|^2}{2(z - z_1)}\right]}{4\pi(z - z_1)} \langle w(\mathbf{r}_1) \rangle. \tag{3.34}$$

We also note that the first two approximations give

$$w_0(\mathbf{r}) \cong -ik \int d\boldsymbol{\varrho}_1 \frac{\exp\left[ik\dfrac{|\boldsymbol{\varrho} - \boldsymbol{\varrho}_1|^2}{2(z - z_1)}\right]}{2\pi(z - z_1)} P(\boldsymbol{\varrho}_1), \tag{3.35}$$

where $P(\boldsymbol{\varrho}_1)$ is the incident field at $z = 0$. As shown by *Molyneux* [3.54], (3.34) and (3.35) satisfy the partial differential equation

$$2ik\frac{\partial \langle w(\mathbf{r}) \rangle}{\partial z} + V_T^2 \langle w(\mathbf{r}) \rangle + \frac{i}{4}k^3 A_\varepsilon(0) \langle w(\mathbf{r}) \rangle = 0, \tag{3.36}$$

where

$$\langle w(0, \varrho) \rangle = P(\varrho).$$

Equation (3.36) is the desired equation for the mean field. Note that mathematically it is well posed. Its solution is straightforward and is discussed in Section 3.6. Using this method, similar equations can be derived for the mutual coherence function [3.9, 31, 32, 50–53] or for higher order moments [2.54, 55]. Instead of deriving these results here, we shall discuss other approaches which lead to the same results in a somewhat more efficient manner.

In conclusion, we find that the elegant mathematical techniques of quantum mechanics, in particular diagram and renormalization methods, can lead to exact integral equations for the various moments. However, the kernels of these equations are sums of infinite series, which must be approximated to obtain explicit solutions. While this approach leads to reasonable solutions for the mean field and the mutual coherence function, these results had been obtained earlier via other methods. The equations for the more interesting higher order moments become extremely cumbersome, and are only beginning to be studied in any detail. The major value of these equations is that the integral equations are exact, and hence, formally, it is possible to study the effect of various approximations. However, from a practical point of view such an investigation is extremely difficult, except for the lowest order moments. A second advantage of the method is that even for the simplest approximations, multiple scattering of some type is being included.

3.3 The Markov Approximation

Tatarskii and *Klyatskin* [3.9, 37–40, 57] in a series of publications beginning in 1969 introduced a new approach for deriving the equations for the moments of the field. From a conceptual point of view, their method is directly opposite to the diagram method. In the diagram method, exact equations were first derived, and then approximate equations or solutions were obtained. In the work of *Klyatskin* and *Tatarskii* the approximations are explicitly stated at the outset and used to obtain a closed set of partial differential equations for the moments, but the equations obtained are the same as those derived by *Brown, Shishov,* and *Molyneux* using the diagram method. This is not unexpected, since in both approaches the four approximations discussed in Section 3.1 are eventually used. A detailed discussion of this method, which we shall call the Markov approximation, is given in the book by *Tatarskii* [3.9]. Since the method is one of the more important new developments in this field, we shall give a summary of the approach and some of the major results, referring the interested reader to the more complete discussion cited above. In this section there are two mathematical concepts that must be understood. The first is the concept of a variational derivative, and the second is a relationship between a random function $f(r)$ and a functional of f, i.e., $\Phi[f(r)]$. Both of these concepts are discussed in Ref. [3.9], Appendix 1; [3.57].

In their development *Tatarskii* and *Klyatskin* start with the parabolic equation, (3.6). As already discussed, this implies at the outset the acceptance of the first three assumptions discussed in Section 3.1, i.e., that depolarization, backscattering, and wide-angle scattering are insignificant. Second, they assume that the propagating wave may be considered to be a Markov random process. The assumption necessary for this approximation is that the dielectric constant fluctuations are delta correlated in the z direction. With these assumptions they were able to obtain explicitly closed equations for the moments of the field. In addition, in most of their derivations they assumed that the dielectric constant is a gaussian random process, though they also showed results for the nongaussian case [3.9, 38]. We shall discuss the derivations for the mean field, the mutual coherence function, and the fourth-order moment.

3.3.1 The Derivation for the Mean Field

If we average (3.6) directly, we obtain an equation of the form

$$2ik\frac{\partial\langle u(\varrho,z)\rangle}{\partial z} + V_T^2\langle u(\varrho,z)\rangle + k^2\langle\varepsilon_1(\varrho,z)u(\varrho,z)\rangle = 0. \tag{3.37}$$

As discussed before, the difficulty with this equation is the unknown term $\langle\varepsilon_1 u\rangle$. The problem then is to obtain an expression for $\langle\varepsilon_1 u\rangle$ without introducing another unknown term. This problem was solved by *Furutsu* [3.58] and *Novikov* [3.59]. They showed that for a gaussian random process $f(r)$ with mean zero,

$$\langle f(r)\Phi[f(\Sigma)]\rangle = \int\langle f(r)f(r')\rangle\left\langle\frac{\delta\Phi[f(\Sigma)]}{\delta f(r')}\right\rangle dr', \tag{3.38}$$

where $\delta\Phi/\delta f$ is a variational derivative[4]. Since $u(r)$ is given by (3.22), where $w(r)$ is replaced by $u(r)$ and $m(r)$ by $k^2\varepsilon_1(r)$, $u(r)$ is clearly a functional of $\varepsilon_1(r)$; $\varepsilon_1(r)$ is assumed to be a gaussian random variable with zero mean. Furthermore, since backscattering is neglected, and the medium is nonrandom for $z<0$, $u(\varrho,z)$ depends on $\varepsilon_1(\varrho',z')$ only for $0\leqq z'\leqq z$; therefore

$$\frac{\delta u(\varrho,z)}{\delta\varepsilon_1(\varrho',z')} = 0 \quad \text{if } z<0, \text{ or } z'>z. \tag{3.39}$$

---

[4] The notation $\Phi[f(\Sigma)]$ means that Φ may involve the variable $f(r)$ over the whole space.

Therefore, from (3.38), we have

$$\langle \varepsilon_1(\varrho, z) u(\varrho, z) \rangle = \int_0^z dz' \int\!\!\int_{-\infty}^{\infty} d\varrho' \langle \varepsilon_1(\varrho, z) \varepsilon_1(\varrho', z') \rangle \left\langle \frac{\delta u(\varrho, z)}{\delta \varepsilon_1(\varrho', z')} \right\rangle. \tag{3.40}$$

But from (3.13) and (3.14), the Markov approximation gives

$$\langle \varepsilon_1(\varrho, z) \varepsilon_1(\varrho', z') \rangle = \delta(z_1 - z_2) A_\varepsilon(|\varrho - \varrho'|),$$

which when inserted in (3.40) and integrated over z' leads to

$$\langle \varepsilon_1(\varrho, z) u(\varrho, z) \rangle = \tfrac{1}{2} \int\!\!\int_{-\infty}^{\infty} d\varrho' A_\varepsilon(|\varrho - \varrho'|) \left\langle \frac{\delta u(\varrho, z)}{\delta \varepsilon_1(\varrho', z)} \right\rangle. \tag{3.41}$$

The factor of $\tfrac{1}{2}$ is due to the integral of $\delta(z - z')$, which is only from 0 to z, and to the evenness of the covariance function of ε_1, which justifies the assumption of the evenness of the delta function.

We now need an expression for $\delta u/\delta \varepsilon_1$, which can be obtained by returning to the parabolic equation for u. If we integrate (3.6) over z from 0 to z, we have

$$2ik[u(\varrho, z) - u(\varrho, 0)] + \nabla_T^2 \int_0^z u(\varrho, z'') dz''$$

$$+ k^2 \int_0^z \varepsilon_1(\varrho, z'') u(\varrho, z'') dz'' = 0. \tag{3.42}$$

We take the variational derivative of (3.42) with respect to $\varepsilon_1(\varrho', z')$ and assume $0 < z' < z$. We use the relations [3.9, 57]

$$\frac{\delta \varepsilon_1(\varrho, z)}{\delta \varepsilon_1(\varrho', z')} = \delta(\varrho - \varrho') \delta(z - z')$$

and

$$\frac{\delta(\varepsilon_1 u)}{\delta \varepsilon_1} = u \frac{\delta \varepsilon_1}{\delta \varepsilon_1} + \varepsilon_1 \frac{\delta u}{\delta \varepsilon_1},$$

to find

$$2ik \frac{\delta u(\varrho, z)}{\delta \varepsilon_1(\varrho', z')} + \nabla_T^2 \int_0^z \frac{\delta u(\varrho, z'')}{\delta \varepsilon_1(\varrho', z')} dz'' + k^2 \delta(\varrho - \varrho') u(\varrho, z')$$

$$+ k^2 \int_0^z \varepsilon_1(\varrho, z'') \frac{\delta u(\varrho, z'')}{\delta \varepsilon_1(\varrho', z')} dz'' = 0. \tag{3.43}$$

Using (3.39), the lower limit of integration can be replaced by z'. In addition, we are interested in $\delta u(\varrho, z)/\delta \varepsilon_1(\varrho', z')$ for $z' = z$; therefore the $\int\limits_0^z \to \int\limits_{z'}^z \to \int\limits_z^z = 0$, and (3.43) becomes

$$\frac{\delta u(\varrho, z)}{\delta \varepsilon_1(\varrho', z)} = \frac{ik}{2} \delta(\varrho - \varrho') u(\varrho, z). \tag{3.44}$$

If we average (3.44) we have the desired relation for $\langle \delta u/\delta \varepsilon_1 \rangle$. Substituting the average of (3.44) into (3.41), and integrating over ϱ', we have

$$\langle \varepsilon_1(\varrho, z) u(\varrho, z) \rangle = \frac{ik}{4} A_\varepsilon(0) \langle u(\varrho, z) \rangle. \tag{3.45}$$

Substituting (3.45) into (3.37), we finally obtain a closed equation for $\langle u \rangle$, i.e.,

$$2ik \frac{\partial \langle u(\varrho, z) \rangle}{\partial z} + V_T^2 \langle u \rangle + \frac{ik^3}{4} A_\varepsilon(0) \langle u(\varrho, z) \rangle = 0, \tag{3.46}$$

$$\langle u(\varrho, 0) \rangle = u(\varrho, 0).$$

This is identical to (3.36) derived by *Molyneux* using the diagram method, but employing the same assumptions. The solution of (3.46) is well known and will be given in Section 3.6. Clearly, the key to closing (3.37) was the derivation of (3.45) for $\langle \varepsilon_1 u \rangle$. There are two critical steps in deriving (3.45). First, the assumption of a gaussian random process for $\varepsilon_1(r)$ permits using *Furutsu*'s and *Novikov*'s expression, which gives $\langle \varepsilon_1 u \rangle$ as a function of the covariance function of ε_1 and a variational derivative of u with respect to ε_1. Second, it is necessary to use the original parabolic equation to find an expression for the variational derivative. If we did not assume a Markov process, then (3.40) would not be integrable, and (3.46) would become an integral equation.

3.3.2 The Derivation for the Mutual Coherence Function

The derivation for the mutual coherence function is similar to that for the mean field. Defining $\Gamma(\varrho_1, \varrho_2; z) \equiv \langle u(\varrho_1, z) u^*(\varrho_2, z) \rangle$, we start by setting $\varrho = \varrho_1$ in (3.6) and multiplying by $u^*(\varrho_2, z)$ to get

$$2ik \frac{\partial u(\varrho_1, z)}{\partial z} u^*(\varrho_2, z) + V_{T1}^2 u(\varrho_1, z) u^*(\varrho_2, z)$$

$$+ k^2 \varepsilon_1(\varrho_1, z) u(\varrho_1, z) u^*(\varrho_2, z) = 0. \tag{3.47}$$

If we take the complex conjugate of (3.6), set $\varrho = \varrho_2$, and multiply by $u(\varrho_1, z)$, we have

$$-2iku(\varrho_1, z)\frac{\partial u^*(\varrho_2, z)}{\partial z} + \nabla_{T2}^2 u(\varrho_1, z)u^*(\varrho_2, z)$$

$$+ k^2\varepsilon_1(\varrho_2, z)u(\varrho_1, z)u^*(\varrho_2, z) = 0. \tag{3.48}$$

Subtracting (3.48) from (3.47), combining the two terms involving $\partial/\partial z$, and averaging, we find

$$2ik\frac{\partial \Gamma(\varrho_1, \varrho_2; z)}{\partial z} + (\nabla_{T1}^2 - \nabla_{T2}^2)\Gamma + k^2\langle[\varepsilon_1(\varrho_1, z)$$

$$- \varepsilon_1(\varrho_2, z)]u(\varrho_1, z)u^*(\varrho_2, z)\rangle = 0. \tag{3.49}$$

Again we must obtain a new expression for the last term in order to close the equation. In (3.38) we let $f(r) = \varepsilon_1(r)$ and $\Phi = uu^*$, and again use (3.13) and (3.14) to get

$$\langle\varepsilon_1(\varrho_1, z)u(\varrho_1, z)u^*(\varrho_2, z)\rangle = \frac{1}{2}\int\!\!\int_{-\infty}^{\infty} d\varrho' A_\varepsilon(\varrho_1 - \varrho')\left\langle\frac{\delta[u(\varrho_1, z)u^*(\varrho_2, z)]}{\delta\varepsilon_1(\varrho', z)}\right\rangle. \tag{3.50}$$

But we know

$$\left\langle\frac{\delta[u(\varrho_1, z)u^*(\varrho_2, z)]}{\delta\varepsilon_1(\varrho', z)}\right\rangle = \left\langle\frac{\delta u(\varrho_1, z)}{\delta\varepsilon_1(\varrho', z)}u^*(\varrho_2, z)\right\rangle + \left\langle u(\varrho_1, z)\frac{\delta u^*(\varrho_2, z)}{\delta\varepsilon_1(\varrho', z)}\right\rangle$$

$$= \frac{ik}{2}[\delta(\varrho_1 - \varrho') - \delta(\varrho_2 - \varrho')]\Gamma(\varrho_1, \varrho_2; z), \tag{3.51}$$

where the last step follows from (3.44). Substitution of (3.51) into (3.50) gives

$$\langle\varepsilon_1(\varrho_1, z)u(\varrho_1, z)u^*(\varrho_2, z)\rangle = \frac{ik}{4}[A_\varepsilon(0) - A_\varepsilon(\varrho_1 - \varrho_2)]\Gamma(\varrho_1, \varrho_2; z). \tag{3.52}$$

The expression for $\langle\varepsilon_1(\varrho_2, z)u(\varrho_1, z)u^*(\varrho_2, z)\rangle$ is the same except that i is replaced by $-i$. Substitution of these results in (3.49) gives

$$2ik\frac{\partial \Gamma(\varrho_1, \varrho_2; z)}{\partial z} + (\nabla_{T1}^2 - \nabla_{T2}^2)\Gamma + \frac{ik^3}{2}[A_\varepsilon(0)$$

$$- A_\varepsilon(\varrho_1 - \varrho_2)]\Gamma = 0 \tag{3.53}$$

with the initial condition $\Gamma(\varrho_1, \varrho_2; 0) = u(\varrho_1, 0)u^*(\varrho_2, 0)$.

This equation is identical to the equation obtained by *Molyneux* using the diagram method. As pointed out by *Dolin* [3.60] and *Tatarskii* [3.9], by an appropriate Fourier transformation, (3.53) can be made analogous to the radiation transfer equation. Solutions of this equation will be discussed in more detail in Section 3.6.2. However, note that if $\varrho_1 = \varrho_2$, $\Gamma(\varrho_1, \varrho_2; z) = \langle I(\varrho_1, z) \rangle$, i.e., the mutual coherence function is equal to the mean value of the intensity.

3.3.3 The Derivation of Higher Order Moments

In order to discuss higher order moments, it is useful to slightly alter our notation. We will define the m, nth moment as

$$\Gamma_{m,n}(z) = \langle u_1 \ldots u_m u_1^* \ldots u_n^* \rangle = \left\langle \prod_{i=1}^{m} u_i \prod_{j=1}^{n} u_j^* \right\rangle, \tag{3.54}$$

where $u_i = u(\varrho_i, z)$, $u_j^* = u^*(\varrho_j', z)$.

Tatarskii [3.9] derived expressions for the higher order moments by defining a characteristic functional

$$\psi_z[v, v^*] = \langle \exp(iR_z) \rangle = \left\langle \exp\left\{ i \iint\limits_{-\infty}^{\infty} [u(\varrho', z)v(\varrho') + u^*(\varrho', z)v^*(\varrho')] d\varrho' \right\} \right\rangle. \tag{3.55}$$

Following a method proposed by *Novikov* [3.59], he derived an equation analogous to the Einstein-Fokker-Planck equation for $\psi_z[v, v^*]$, from which partial differential equations for $\Gamma_{m,n}$ are derived. This method was described in detail by *Tatarskii*, and two more recent reviews have been given [3.49, 61]. The method of characteristic functionals was also used by *Furutsu* [3.48], and his results will be discussed below. While the characteristic functional method is very elegant, for our purposes, an extension of the procedure used for the first two moments leads to the same final equations for the moments, and in a more direct manner.

To construct an equation for $\Gamma_{m,n}$, we again start with the parabolic equation, (3.6). We form m equations of the form

$$2ik \frac{\partial u_k}{\partial z} \prod_{\substack{i=1 \\ i \neq k}}^{m} u_i \prod_{j=1}^{n} u_j^* + \nabla_{Tk}^2 u_k \prod_{\substack{i=1 \\ i \neq k}}^{m} u_i \prod_{j=1}^{n} u_j^* + k^2 \varepsilon_1(\varrho_k, z) \prod_{i=1}^{m} u_i \prod_{j=1}^{n} u_j^* = 0. \tag{3.56}$$

Similarly we get n equations of the form

$$-2ik \prod_{i=1}^{m} u_i \prod_{\substack{j=1 \\ j \neq l}}^{n} u_j^* \frac{\partial u_l}{\partial z} + \prod_{i=1}^{m} u_i \prod_{\substack{j=1 \\ j \neq l}}^{n} u_j^* \cdot \nabla_{Tl}^2 u_l^*$$

$$+ k^2 \varepsilon_1(\varrho_l', z) \prod_{i=1}^{m} u_i \prod_{j=1}^{n} u_j^* = 0. \tag{3.57}$$

For $k=1$ to m we add the m equations of the form of (3.56) and subtract (for $l=1$ to n) the n equations of the form (3.57). We get

$$2ik\frac{\partial \prod\limits_{i=1}^{m} u_i \prod\limits_{j=1}^{n} u_j^*}{\partial z} + \sum_{k=1}^{m} V_{Tk}^2 \left(\prod_{i=1}^{m} u_i \prod_{j=1}^{n} u_j^* \right) - \sum_{l=1}^{n} V_{Tl}^2 \left(\prod_{i=1}^{m} u_i \prod_{j=1}^{n} u_j^* \right)$$

$$+ k^2 \left[\sum_{k=1}^{m} \varepsilon_1(\varrho_k, z) - \sum_{l=1}^{n} \varepsilon_1(\varrho_l', z) \right] \prod_{i=1}^{m} u_i \prod_{j=1}^{n} u_j^* = 0. \qquad (3.58)$$

Averaging (3.58), we get

$$2ik\frac{\partial \Gamma_{mn}}{\partial z} + \left(\sum_{k=1}^{m} V_{Tk}^2 - \sum_{l=1}^{n} V_{Tl}^2 \right) \Gamma_{mn}$$

$$+ k^2 \left\langle \left[\sum_{k=1}^{m} \varepsilon_1(\varrho_k, z) - \sum_{l=1}^{n} \varepsilon_1(\varrho_l', z) \right] \prod_{i=1}^{m} u_i \prod_{j=1}^{n} u_j^* \right\rangle = 0. \qquad (3.59)$$

To find a closed expression for Γ_{mn}, we proceed as before with the last term, i.e., consider

$$\left\langle \varepsilon_1(\varrho_k, z) \prod_{i=1}^{m} u_i \prod_{j=1}^{n} u_j^* \right\rangle = \frac{1}{2} \iint_{-\infty}^{\infty} d\varrho A_\varepsilon(\varrho_k - \varrho) \left\langle \frac{\delta \prod\limits_{i=1}^{m} u_i \prod\limits_{j=1}^{n} u_j^*}{\delta \varepsilon_1(\varrho, z)} \right\rangle. \qquad (3.60)$$

The variational derivative of the product can be expanded as

$$\frac{\delta \prod\limits_{i=1}^{m} u_i \prod\limits_{j=1}^{n} u_j^*}{\delta \varepsilon_1(\varrho, z)} = \sum_{k=1}^{m} \frac{\delta u_k}{\delta \varepsilon_1(\varrho, z)} \prod_{\substack{i=1 \\ i\neq k}}^{m} u_i \prod_{j=1}^{n} u_j^* + \sum_{l=1}^{n} \prod_{i=1}^{m} u_i \prod_{\substack{j=1 \\ j\neq l}}^{n} u_j^* \frac{\delta u_l^*}{\delta \varepsilon_1(\varrho, z)}. \qquad (3.61)$$

Again substituting (3.44) into (3.61) and averaging, we get

$$\left\langle \frac{\delta \prod\limits_{i=1}^{m} u_i \prod\limits_{j=1}^{n} u_j^*}{\delta \varepsilon_1(\varrho, z)} \right\rangle = \frac{ik}{2} \left[\sum_{k=1}^{m} \delta(\varrho_k - \varrho) - \sum_{l=1}^{n} \delta(\varrho_l' - \varrho) \right] \Gamma_{mn}. \qquad (3.62)$$

Substituting (3.62) into (3.60) gives

$$\left\langle \varepsilon_1(\varrho_k, z) \prod_{i=1}^{m} u_i \prod_{j=1}^{n} u_j^* \right\rangle$$

$$= \frac{ik}{4} \left[\sum_{k'=1}^{m} A_\varepsilon(\varrho_k - \varrho_{k'}) - \sum_{l'=1}^{n} A_\varepsilon(\varrho_k - \varrho_{l'}') \right] \Gamma_{mn}. \qquad (3.63)$$

Deriving a similar expression for the last term in (3.59), we have

$$2ik\frac{\partial \Gamma_{mn}}{\partial z} + \left(\sum_{k=1}^{m} V_{Tk}^2 - \sum_{l=1}^{n} V_{Tl}^2\right)\Gamma_{mn} + \frac{ik^3}{4}Q_{mn}\Gamma_{mn} = 0,\tag{3.64}$$

where

$$Q_{mn} = \sum_{k=1}^{m}\sum_{l=1}^{n}[A_\varepsilon(\varrho_k-\varrho_l)-A_\varepsilon(\varrho_k-\varrho_l')-A_\varepsilon(\varrho_l'-\varrho_k)+A_\varepsilon(\varrho_k'-\varrho_l')].\tag{3.65}$$

Equations (3.63) and (3.64) are identical to (66.16) and (66.17) of *Tatarskii* [3.9], derived using characteristic functionals. For $m=n=2$, we have the fourth-order coherence function

$$\Gamma_{22}(\varrho_1,\varrho_2;\varrho_1',\varrho_2')=\langle u(\varrho_1,z)u(\varrho_2,z)u^*(\varrho_1',z)u^*(\varrho_2',z)\rangle.\tag{3.66}$$

This function is of particular interest because for $\varrho_1=\varrho_1'$ and $\varrho_2=\varrho_2'$, it reduces to the irradiance correlation function

$$\Gamma_{22}(\varrho_1,\varrho_2;\varrho_1,\varrho_2)=\langle I(\varrho_1,z)I(\varrho_2,z)\rangle.\tag{3.67}$$

The equation for the fourth-order moment is

$$2ik\frac{\partial \Gamma_{22}(\varrho_1,\varrho_2;\varrho_1',\varrho_2';z)}{\partial z}+(V_{T1}^2+V_{T2}^2-V_{T1'}^2-V_{T2'}^2)\Gamma_{22}$$

$$+\frac{ik^3}{4}Q_{22}\Gamma_{22}=0,\tag{3.68}$$

$$Q_{22}=2[2A_\varepsilon(0)+A_\varepsilon(\varrho_2-\varrho_1)+A_\varepsilon(\varrho_2'-\varrho_1')-A_\varepsilon(\varrho_1-\varrho_1')-A_\varepsilon(\varrho_1-\varrho_2')$$

$$-A_\varepsilon(\varrho_2-\varrho_1')-A_\varepsilon(\varrho_2-\varrho_2')]\tag{3.69}$$

with the initial condition

$$\Gamma_{22}(\varrho_1,\varrho_2;\varrho_1',\varrho_2';0)=P(\varrho_1)P(\varrho_2)P^*(\varrho_1')P^*(\varrho_2').$$

Solution of this equation is extremely difficult, and different approximate solutions will be discussed in Section 3.6.3. *Tatarskii* [3.9] discussed the implications of several of the approximations used in deriving these equations for the mean values, including the assumption of the Markov process, the use of the parabolic equation, and the effect of a non-gaussian random medium. To summarize his conclusions, the Markov approximation is valid for the mean field if

a) $z>L_0$,
b) $\lambda \ll l_0$, and
c) $kL_0\sigma_\varepsilon^2 \ll 1$;

and for the mutual coherence function if

a) $|\varrho_1 - \varrho_2| \ll z$,
b) $\varrho \ll L_0$,
c) $k^2 \varrho^2 \gg 1$,
d) $C_\varepsilon^2 k^2 L_0^{5/3} \ll 1$,
e) $C_\varepsilon^2 l_0^{-1/3} z \ll 1$, and
f) $C_\varepsilon^4 k^2 L_0^{5/3} l_0^{-1/3} z^2 \ll 1$.

The above conditions are almost always met for optical waves in the turbulent atmosphere. A similar investigation of the parabolic approximation does not lead to any new restrictions.

3.4 The Local Method of Small Perturbations (LMSP)

A quite different conceptual approach to finding the equations for the moments of the field has recently been called "the local method of small perturbations" (LMSP). In their excellent review article *Prokhorov* et al. [3.2] summarized the history of this method, presented the derivation, and then solved the relevant equations for a number of problems of interest. Apparently the method was first suggested by *Chernov* in 1964 [3.41] and used by *Shishov* in 1967 [3.42] to obtain the equation for the fourth-order coherence function. Unfortunately their work appeared in untranslated publications and hence was not widely appreciated in the United States. Besides the workers in the Soviet Union [3.41–44], essentially the same method was used by *Beran* in the United States [3.45 – 47, 62, 63]. In this section we shall review the derivation and show that the results agree with those given in Sections 3.2 and 3.3.

The initial steps in this derivation are identical to those used in the Markov approximation method in the previous section; therefore we start with (3.59). This means that we have already assumed that depolarization and backscatter effects are negligible, and have started from the parabolic equation (3.6). The difference between the method based on a Markov random process and the LMSP is the procedure for obtaining a closed equation, i.e., for expressing the last term in (3.59) in terms of Γ_{mn}. Under the Markov approximation, the last term is evaluated by assuming the dielectric constant is delta correlated. Here this term is considered in more detail. In essence the assumption on the delta correlation is justified. To evaluate this term, consider what happens as the field propagates between two planes, one located at z_0 and the other at $z_0 + \Delta z$, where we assume Δz is large compared to the largest inhomogeneity of interest, say L_0, but small enough so that we can use a perturbation method to calculate the field as it propagates from z_0 to $z_0 + \Delta z$. Consider the term

$$\left\langle \varepsilon_1(\varrho_k, z) \prod_{i=1}^{m} u(\varrho_i, z) \prod_{j=1}^{n} u^*(\varrho_j', z) \right\rangle. \tag{3.70}$$

In the region between the planes z_0 and $z_0 + \Delta z$ assume

$$u_i = u_i^{(0)} + u_i^{(1)}, \tag{3.71}$$

where the superscript $^{(0)}$ denotes the fields that would exist if there were not a turbulent medium between z_0 and $z_0 + \Delta z$, and the superscript $^{(1)}$ refers to the first order perturbation to the field caused by the turbulent layer between z_0 and $z_0 + \Delta z$. That is, we are going to apply the small perturbation method locally near z. Equation (3.6) can be solved for the case when $\varepsilon_1 = 0$ to give

$$u^{(0)}(\varrho_i, z) = \frac{k}{2\pi i(z - z_0)} \iint d\varrho' \exp\left[\frac{ik(\varrho_i - \varrho')^2}{2(z - z_0)}\right] u(\varrho', z_0), \tag{3.72}$$

and the first-order perturbation term is

$$u^{(1)}(\varrho_i, z) = \frac{k^2}{4\pi} \int_{z_0}^{z} \frac{dz'}{z - z_0} \iint d\varrho' \exp\left[\frac{ik(\varrho_i - \varrho')^2}{2(z - z_0)}\right] \varepsilon_1(\varrho', z') u^{(0)}(\varrho', z'). \tag{3.73}$$

Inserting (3.71) into (3.70), we have

$$\left\langle \varepsilon_1(\varrho_k, z) \prod_{i=1}^{m} u_i \prod_{j=1}^{n} u_j^* \right\rangle = \left\langle \varepsilon_1(\varrho_k, z) \prod_{i=1}^{m} u_i^{(0)} \prod_{j=1}^{n} u_j^{(0)*} \right\rangle$$

$$+ \left\langle \varepsilon_1(\varrho_k, z) u_\alpha^{(1)} \prod_{\substack{i=1 \\ i \neq \alpha}}^{m} u_i^{(1)} \prod_{j=1}^{n} u_j^{(0)*} \right\rangle$$

$$+ \left\langle \varepsilon_1(\varrho_k, z) u_\beta^{(1)*} \prod_{i=1}^{m} u_i^{(0)} \prod_{\substack{j=1 \\ j \neq \beta}}^{n} u_j^{(0)*} \right\rangle, \tag{3.74}$$

where terms in $\varepsilon_1 u_j^{(1)} u_k^{(1)*} u_l^{(0)} u_m^{(0)*}$ have been dropped as being of higher order. First of all we wish to show that the term

$$\left\langle \varepsilon_1(\varrho_k, z) \prod_{i=1}^{m} u_i^{(0)} \prod_{j=1}^{n} u_j^{(0)*} \right\rangle$$

is equal to zero. Note that all the u terms arise from perturbations due to inhomogeneities in the region $z < z_0$, while $\varepsilon_1(\varrho_k, z)$ is evaluated in the plane z located a distance Δz from z_0. Therefore, if Δz is much larger than the scales associated with the inhomogeneities, the $u^{(0)}$ terms and the ε_1 terms should be independent (note we are assuming no significant backscatter), and hence

$$\left\langle \varepsilon_1(\varrho_k, z) \prod_{i=1}^{m} u_i^{(0)} \prod_{j=1}^{n} u_j^{(0)*} \right\rangle = \langle \varepsilon_1(\varrho_k, z) \rangle \left\langle \prod_{i=1}^{m} u_i^{(0)} \prod_{j=1}^{n} u_j^{(0)*} \right\rangle = 0.$$

The remaining terms have the same basic form; therefore we will consider just one of them, e.g.,

$$\left\langle \varepsilon_1(\varrho_k, z) u_\alpha^{(1)} \prod_{\substack{i=1 \\ i \neq \alpha}}^{m} u_i^{(0)} \prod_{j=1}^{n} u_j^{(0)*} \right\rangle$$

$$= \frac{k^2}{4\pi} \int_{z_0}^{z} \frac{dz'}{z-z'} \iint d\varrho' \exp\left[\frac{ik(\varrho_\alpha - \varrho')^2}{2(z-z')}\right]$$

$$\cdot \left\langle \varepsilon_1(\varrho_k, z) \varepsilon_1(\varrho', z') u^{(0)}(\varrho', z') \prod_{\substack{i=1 \\ i \neq \alpha}}^{m} u_i^{(0)} \prod_{j=1}^{n} u_j^{(0)*} \right\rangle. \tag{3.75}$$

First let us consider the integral over ϱ'. The term in the exponential oscillates very rapidly as a function of ϱ' compared to the term in angular brackets; therefore the method of stationary phase gives

$$\left\langle \varepsilon_1(\varrho_k, z) u_\alpha^{(1)} \prod_{\substack{i=1 \\ i \neq \alpha}}^{m} u_i^{(0)} \prod_{j=1}^{n} u_j^{(0)*} \right\rangle$$

$$= \frac{ik}{2} \int_{z_0}^{z} dz' \left\langle \varepsilon_1(\varrho_k, z) \varepsilon_1(\varrho_\alpha, z') u^{(0)}(\varrho_\alpha, z') \prod_{\substack{i=1 \\ i \neq \alpha}}^{m} u_i^{(0)} \prod_{j=1}^{n} u_j^{(0)*} \right\rangle. \tag{3.76}$$

Again $\varepsilon_1(\varrho_k, z)$ is statistically independent of the terms in $u^{(0)}$. Now divide the integral over z' into two regions $(z_0, z_0 + a)$ and $(z_0 + a, z)$, where a is on the order of the largest dielectric inhomogeneities of interest, e.g., L_0. In the first region $\varepsilon_1(\varrho_k, z)$ and $\varepsilon_1(\varrho_\alpha, z')$ are independent and hence the integrand is zero. For the second region both $\varepsilon_1(\varrho_k, z)$ and $\varepsilon_1(\varrho_\alpha, z')$ are independent of the $u^{(0)}$ terms; therefore the angular bracket becomes

$$\langle \varepsilon_1(\varrho_k, z) \varepsilon_1(\varrho_\alpha, z') \rangle \left\langle u_0^{(0)}(\varrho_\alpha, z') \prod_{\substack{i=1 \\ i \neq \alpha}}^{m} u_i^{(0)} \prod_{j=1}^{n} u_j^{(0)*} \right\rangle.$$

However, this term is also close to zero over the region $(z_0, z_0 + a)$, and hence may be used over the entire region in (3.75). Finally, we note that Δz is small compared to the longitudinal correlation length of the field $u^{(0)}$, and hence we replace $u^{(0)}(\varrho_\alpha, z')$ by $u^{(0)}(\varrho_\alpha, z)$. Therefore, using (3.13–14) we have

$$\left\langle \varepsilon_1(\varrho_k, z) u_\alpha^{(1)} \prod_{\substack{i=1 \\ i \neq \alpha}}^{m} u_i^{(0)} \prod_{j=1}^{n} u_j^{(0)*} \right\rangle = \frac{ik}{2} \Gamma_{mn}^{(0)} \int_{z_0}^{z} B_\varepsilon [|(\varrho_k, z) - (\varrho_\alpha, z')|] \, dz'$$

$$= \frac{ik}{4} \Gamma_{mn}^{(0)} A_\varepsilon(\varrho_k - \varrho_\alpha). \tag{3.77}$$

To carry out the last step, we note that Δz is much larger than the correlation length of B_ε, and that B_ε is an even function of $z - z'$; hence we let the integral go from $-\infty$ to ∞, and use (3.15). The last term in (3.74) will lead to a term identical to (3.77) except i should be replaced by $-$ i. Finally, we note that over the layer of thickness Δz, Γ_{mn} should not change much, so we replace $\Gamma_{mn}^{(0)}$ by Γ_{mn}.

If we substitute (3.77) into (3.74) and the result into (3.59), we obtain (3.64), exactly the same result derived by the method based on a Markov process and by renormalization.

3.5 Heuristic Theories for the Saturation Region

As pointed out in Section 3.1, the major goal in the development of new theories for optical propagation in the atmosphere has been the explanation of the so-called saturation region for the variance of log-amplitude fluctuations. In this region it is well-known that multiple scattering effects are important and that simple perturbation theories are not adequate. The objective of the mathematical theories presented in the preceding three sections was to derive expressions that will adequately explain the observed experimental results. However, except for the mean field and the mutual coherence function, the theory has led to expressions (either partial differential equations or integro-differential equations) that are extremely complex and to date have not been solved in a manner completely useful for comparison with experiment, or in predicting new results. A discussion of some of the results is given in Section 3.6. Furthermore, the approaches have not led to a deeper understanding of the basic physical mechanisms underlying the saturation process.

Recently several workers in the United States have taken a much more heuristic approach to this problem. While the developments are not as rigorous mathematically, they are based on reasonable physical assumptions, and do lead to predictions that seem to agree with experiment. The most important breakthroughs in this area appear to be due to *Clifford* et al. [3.64, 65] and *Yura* [3.66–69]. However, early developments along these lines were discussed by *Tatarskii* [3.9] and *Young* [3.70].

In all of the work referenced above, the objective was to explain the observed phenomena from a physical viewpoint, and hence the results are qualitative. Therefore, we shall ignore factors on the order of unity. We shall first review the early explanations of *Tatarskii*, which are limited to geometrical optics and weak-scattering theories, and then discuss how *Clifford*, *Young*, and *Yura* extended these ideas to multiple-scattering phenomena.

First, we need a heuristic model of the random medium, i.e., of the atmosphere. For our purposes, we shall consider the atmosphere to be composed of random eddies or blobs of different dielectric constant and of different sizes, varying between some minimum size l_0, called the microscale of turbulence, to some maximum size L_0, the outer scale of turbulence. We know that in some sense the bigger the eddy the larger the dielectric constant variation in the eddy. Also, statistical isotropy of the medium suggests these eddies are roughly spherical in shape. From a geometrical optics viewpoint, the refractive index is a more direct quantity of interest than the dielectric constant. Therefore, we shall model the medium by the relations for the refractive index structure function ($\Delta n = 1/2\Delta\varepsilon$):

$$\langle \Delta n^2(\varrho) \rangle \cong \langle [n(\varrho_1) - n(\varrho_1 + \varrho)]^2 \rangle$$

$$\sim C_n^2 l_0^{2/3}(\varrho/l_0)^2 \qquad \varrho \ll l_0 \tag{3.78}$$

$$\sim C_n^2 \varrho^{2/3} \qquad l_0 \ll \varrho \ll L_0. \tag{3.79}$$

These relations arise from the well-known Kolmogorov-Obukhov model for turbulence [3.9].

We should note that while in some sense amplitude and phase effects are intimately related, in another sense they are quite separate. The phase fluctuations arise directly from the fact that the refractive index is a random function of space and time, which from the simple relation $v = c/n$ produces a random velocity in the propagating wave. The random velocity makes the effective path length between transmitter and receiver a random variable, therefore giving rise to a random phase variation. Consequently, the phase difference between two parallel rays is a function of the refractive index difference over the two parallel paths. The inhomogeneities having the major effect on the phase difference are those whose dimensions are roughly equal to the separation of the two rays. Inhomogeneities larger than this separation will tend to have an equal effect on both paths, and hence contribute little to the phase difference. Inhomogeneities much smaller than the separation will tend to be uncorrelated on the two paths and hence average to zero. For this reason we divide the path into segments of length ϱ, where ϱ is the separation of the two paths. The phase difference over a single segment is $\Delta s_1 \sim k\varrho\Delta n(\varrho)$, where the subscript $_1$ refers to the phase difference over a single segment. Since $\langle \Delta n(\varrho) \rangle = 0$, $\langle \Delta s \rangle = 0$. The phase structure function $\langle \Delta s_1^2(\varrho) \rangle \sim k^2 \varrho^2 \langle \Delta n^2(\varrho) \rangle$. Using (3.79) we get $\langle \Delta s_1^2 \rangle \sim k^2 C_n^2 \varrho^{8/3}$ for $l_0 \ll \varrho \ll L_0$. If we average over the entire path, there are approximately N segments, where $N = L/\varrho$. If we assume the phase differences over different segments are uncorrelated, then the mean square difference should be proportional to N, i.e.,

$$\langle \Delta s^2(\varrho) \rangle \sim N \langle \Delta s_1^2(\varrho) \rangle \sim k^2 C_n^2 L\varrho^{5/3}, \qquad l_0 \ll \varrho \ll L_0. \tag{3.80}$$

Except for a constant the above equation is the phase structure function as calculated by either geometrical optics or a first-order perturbation theory.

For $\varrho \ll l_0$, the situation changes since the smallest inhomogeneities of any importance are of size l_0. Therefore the path should be divided into segments of length l_0, i.e., $N = L/l_0$. In this region we use (3.78) for $\langle \Delta n^2(\varrho) \rangle$; therefore

$$\langle \Delta s^2(\varrho) \rangle \sim k^2 C_n^2 L l_0^{-1/3}\varrho^2, \qquad \varrho \ll l_0. \tag{3.81}$$

While the above arguments are basically very simple, to the best of our knowledge they give adequate results even into the multiple-scattering region, i.e., phase variations due to multiple-scattering processes appear to be of minor importance compared to the effect of velocity changes along a straight line in the medium. For example, calculations of the additional change in phase due to the fact that the ray follows a random trajectory instead of a straight line show this term to be unimportant. (However, *Tatarskii* [3.71] showed that the inclusion of this effect in amplitude fluctuations leads to a saturation type effect.) This conclusion supports the small angle scattering assumption discussed in Section 3.1. We shall discuss the phase fluctuations in the saturation region more fully near the end of this section.

While we know that amplitude effects are a direct result of the phase variations, the process is much more complicated than for phase alone. First of all, amplitude variations are a propagated effect, i.e., a phase variation at some point z produces a significant amplitude effect only at some distance $z' > z$. If we return to our model of the random medium consisting of blobs or eddies of refractive index variations, then we may think of these eddies as random lenses. The lenses are randomly distributed in space, have random size and shape (though they are roughly spherical), and change randomly with time. Therefore, the medium acts as if it were a random telescope. By examining the behavior of a single lens, then accounting for the large number of random lenses, we can develop a physical understanding of many of the processes involved.

Assume that we have a lens of diameter l and refractive index $\Delta n = n - 1$. Then this lens will have a focal length

$$F \cong l/\Delta n. \tag{3.82}$$

If $\Delta n > 0$, the focal point is beyond the lens and the lens tends to focus the light; while if $\Delta n < 0$, there is a virtual focal point behind the lens and the light is defocused. Since in our model, $l \sim l_0 - L_0 \sim 10^{-3}\,\mathrm{m} - 1\,\mathrm{m}$, and $\Delta n \sim 10^{-6} - 10^{-8}$, we expect $F \sim 10^3\,\mathrm{m} - 10^8\,\mathrm{m}$, i.e., these are very weak lenses with very long focal lengths. Using geometrical optics, we can calculate the amplitude variations due to these lenses. Assume that we have a plane wave incident on a lens located a distance z from the transmitter, and we are interested in the amplitude fluctuations at a distance L. For cylindrical symmetry, and the geometry shown in Fig. 3.9a, conservation of energy gives

$$A_0^2(l/2)^2 = A^2 r^2, \tag{3.83}$$

where A_0 is the amplitude of the wave before the lens, A is the amplitude at L, and r is shown in the figure. Defining $\delta A = A - A_0$, we have

$$\delta A = \left(\frac{l}{2r} - 1\right) A_0 = \frac{L - z}{F - L + z} A_0, \tag{3.84}$$

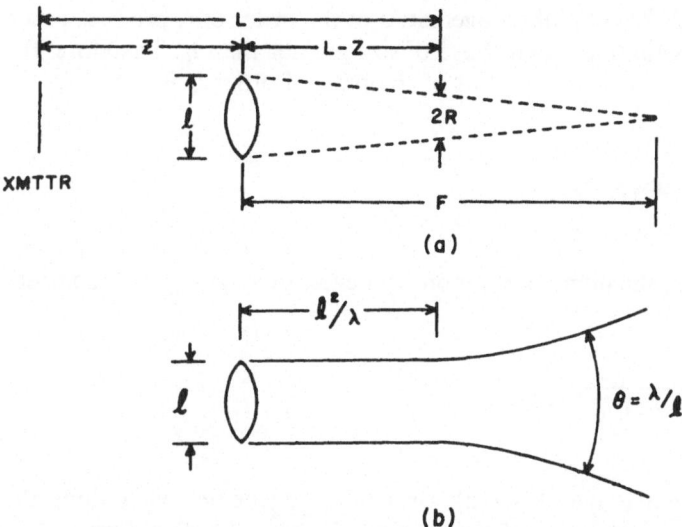

Fig. 3.9. (a) Amplitude fluctuations due to geometrical optics effect of a single lens: L = receiving plane, F = focal length of the lens, R = radius of the beam at z; (b) diffraction effects of a simple lens: l = diameter of the lens, $l_c = l^2/\lambda$ = distance when geometrical optics is valid, $\theta = \lambda/l$ = diffraction angle

where from the geometry of Fig. 3.9a, $l/2r = F/(F - L + z)$. Given the values of F estimated above, $L - z$ is usually much less than F; therefore,

$$\frac{\delta A}{A_0} \cong \frac{L - z}{F}. \tag{3.85}$$

The effect of a turbulent medium filled with such lenses but with random focal lengths and random positions will now be considered. We need to relate F to the observed experimental variables. From (3.82), $F = l/\Delta n$. We also know from (3.78) and (3.79) that $\langle \Delta n^2(l) \rangle \sim C_n^2 l_1^{2/3}$, where $l_1 = l$ for $l_0 \ll l \ll L_0$ and $l_1 = l^3/l_0^2$ for $l \ll l_0$. Assuming $\Delta n \sim C_n l_1^{1/3}$, we find $F = l/(C_n l_1^{1/3})$, and hence

$$\frac{\delta A}{A_0} \sim \frac{C_n l_1^{1/3}}{l}(L - z), \tag{3.86}$$

For the two cases we get

$$\frac{\delta A}{A_0} \sim C_n l_0^{-2/3}(L - z), \quad l \ll l_0; \tag{3.87}$$

$$\frac{\delta A}{A_0} \sim C_n l^{-2/3}(L - z), \quad l_0 \ll l \ll L_0. \tag{3.88}$$

Clearly, due to the $l^{-2/3}$ law, inhomogeneities of the order of l_0 produce much larger amplitude fluctuations than those of size greater than l_0. Therefore, the amplitude fluctuations from a segment of the path of order l_0 is

$$\left\langle \left(\frac{\delta A}{A_0}\right)^2 \right\rangle_1 \sim C_n^2 l_0^{-4/3}(L-z)^2 .$$

In a path of length L, the number of inhomogeneities of length l_0 is of the order $N = L/l_0$; therefore

$$\left\langle \left(\frac{\delta A}{A_0}\right)^2 \right\rangle \sim C_n^2 \overline{(L-z)^2} l_0^{-7/3} L ,$$

where $\overline{(L-z)^2}$ is the average of $(L-z)^2$ over inhomogeneities lying along the path. Within a numerical factor of order unity $\overline{(L-z)^2} \sim L^2$; therefore

$$\left\langle \left(\frac{\delta A}{A_0}\right)^2 \right\rangle \sim C_n^2 L^3 l_0^{-7/3} . \tag{3.89}$$

The above result is a geometrical optics approximation since it was calculated on the basis of the change in amplitude due to focusing or defocusing of a ray bundle by a random lens, and no attempt was made to include diffraction effects. As we shall see, the influence of diffraction is to weaken or smear out the focusing effect of the random lenses.

In Fig. 3.9b are shown the key parameters when diffraction is considered. The distance over which the geometrical optics approximation is valid is

$$L_g(l) = l^2/\lambda , \tag{3.90}$$

while the diffraction angle for a collimated beam is

$$\theta = \lambda/l .$$

For our atmospheric model of random lenses, l corresponds to the sizes of the different inhomogeneities, which vary between l_0 and L_0. Since l_0 is on the order of 10^{-2}–10^{-3} m, $L_g(l_0) \sim 2$–200 m for visible light, while for $L_0 \sim 1$ m or greater, $L_g(L_0) \gtrsim 10^6$ m. Therefore, we conclude that for typical optical propagation path lengths between 1 and 10 km, the effects of the larger inhomogeneities may be modeled using geometrical optics, but for the smaller inhomogeneities diffraction effects must be included. Furthermore, since (3.89) resulted from using a geometrical optics approximation to describe the effects of the smallest inhomogeneities, it clearly is not valid for typical path lengths in the earth's atmosphere. For path lengths greater than l_0^2/λ, the inhomogeneities can be divided into two groups: those for which diffraction effects are important, and those for which geometrical optics approximations still hold. The selection criterion is given by $l_c \sim \sqrt{(\lambda L)}$. For eddies such that $l > l_c$, geometrical optics is valid, but for $l < l_c$, diffraction must be considered. The primary effect of diffraction is to spread the beam out, no matter whether the beam was originally focused or defocused by the inhomogeneity. This spreading in effect counteracts the geometrical focusing effect. However, the diffraction angle is bounded by $\theta \leq \lambda/l_0 = 0.5 \cdot 10^{-3}$ rad, and hence does not cause much of an amplitude change. Therefore, we shall assume the inhomogeneities that are in the diffraction regime cause essentially no amplitude effects, and consider only the effect of inhomogeneities for which $l > l_c$. For these inhomogeneities we can still use a geometrical optics approach; therefore (3.89) should be valid, except that the dominant inhomogeneity is no longer of dimension l_0, but one whose size is governed by $l_c = \sqrt{(\lambda L)}$. Hence we replace l_0 by $\sqrt{(\lambda L)}$ in (3.89). With this substitution, and replacing λ^{-1} with k, we find that in the diffraction region

$$\left\langle \left(\frac{\delta A}{A_0}\right)^2 \right\rangle \sim C_n^2 k^{7/6} L^{11/6}, \qquad l_0 < \sqrt{(\lambda L)}. \tag{3.91}$$

This equation agrees to within a constant with the more rigorous derivation discussed in Chapter 2 by *Clifford*. Clearly the above argument, while including diffraction, corresponds to a single scattering approximation, since we did not study the effect of the scattering of lenses before and after the particular lens we were considering.

Yura [3.66] followed through the above argument on a somewhat more quantitative basis. In his development, he assumed that the medium is composed of spherically symmetric turbulent eddies whose refractive index varies in a gaussian manner as a function of position. He contended that this assumption is not critical, and it permitted him to derive a mathematical expression for the amplitude fluctuations from a single eddy. For an eddy of radius l located in the plane z, the field at $z = L$ is

$$\delta A(\varrho) \approx i l A_0 \left(\frac{L-z}{F}\right) \exp(i\phi) \left\{ 1 - \frac{\varrho^2}{2\left[l^2 + \left(\dfrac{L-z}{kl}\right)^2\right]} \right\}$$

$$\cdot \frac{1}{\left(il + \dfrac{L-z}{kl}\right)} \exp\left\{ -\frac{\varrho^2[1 - ikl/(L-z)]}{2\left[l^2 + \left(\dfrac{L-z}{kl}\right)^2\right]} \right\}, \tag{3.92}$$

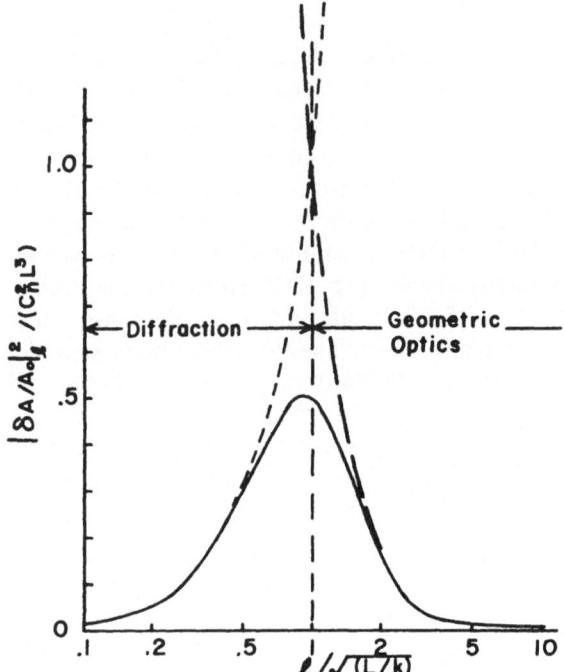

Fig. 3.10. The relative contribution of different size eddies to the amplitude fluctuations: the solid curve represents the exact solution, the long dash lines represents the geometrical optics approximation, and the short dashes the diffraction approximation

where $\phi = kL + \dfrac{k\varrho^2}{2(L-z)}$, $F \sim l/\Delta n$, and ϱ is the location of the eddy in the layer located at z. Since δA is maximum for $\varrho = 0$,

$$\left|\frac{\delta A}{A_0}\right|_l^2 \sim \frac{(L-z)^2}{F^2} \frac{l^2}{\left[l^2 + \left(\dfrac{L-z}{kl}\right)^2\right]}. \tag{3.93}$$

Comparing (3.93) with (3.85), we see that in the geometrical optics region, i.e., for $\sqrt{(\lambda L)} \ll l$, (3.93) and (3.85) are identical. In the diffraction region, for $\sqrt{(\lambda L)} \gg l$, (3.93) reduces to

$$\left|\frac{\delta A}{A_0}\right|_l^2 \sim \frac{k^2 l^4}{F^2}. \tag{3.94}$$

Recalling that $F = l/\Delta n$ and $\Delta n \sim C_n l^{1/3}$, (3.93) becomes

$$\left|\frac{\delta A}{A_0}\right|_l^2 \sim \frac{C_n^2 (L-z)^2 l^{2/3}}{\left[l^2 + \left(\dfrac{L-z}{kl}\right)^2\right]},$$

$$\sim C_n^2 (L-z)^2 l^{-4/3}, \qquad \sqrt{(\lambda L)} \ll l,$$

$$\sim C_n^2 k^2 l^{8/3}, \qquad \sqrt{(\lambda L)} \gg l. \tag{3.95}$$

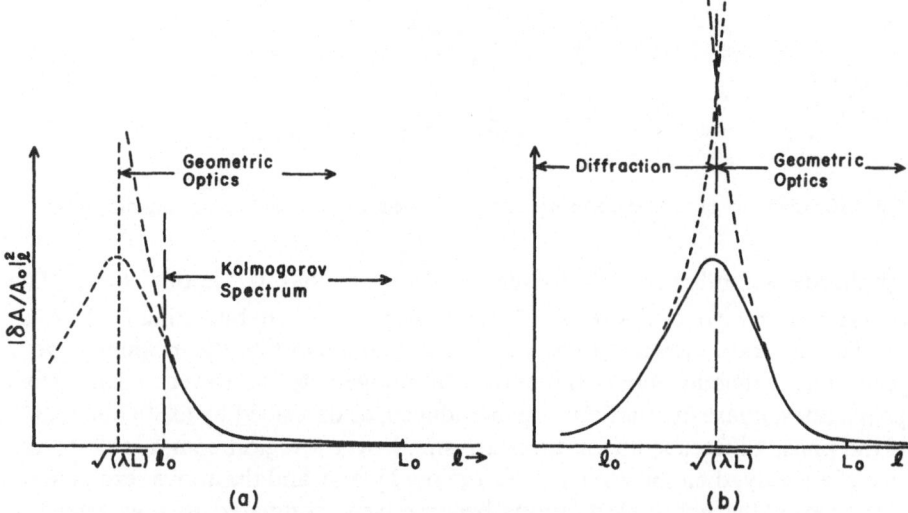

Fig. 3.11. (a) The situation when $\sqrt{(\lambda L)} < l_0$: geometrical optics holds, and the scintillations are dominated by the scale size l_0; (b) the situation when $\sqrt{(\lambda L)} > l_0$; diffraction effects are important and the scintillations are dominated by the scale size $\sqrt{(\lambda L)}$

The above expression holds for a single inhomogeneity located at z. To find the effect of all the inhomogeneities of a given size along the entire path, we multiply by the number of inhomogeneities, which is of the order of L/l, and let $(L-z)^2 \sim L^2$. Therefore,

$$\left|\frac{\delta A}{A_0}\right|_l^2 \sim \frac{C_n^2 L^3 l^{-1/3}}{\left(l^2 + \dfrac{L^2}{k^2 l^2}\right)}, \tag{3.96}$$

$$\sim C_n^2 L^3 l^{-7/3}, \qquad \sqrt{(L/k)} \ll l, \tag{3.96a}$$

$$\sim C_n^2 k^2 L l^{5/3}, \qquad \sqrt{(L/k)} \gg l. \tag{3.96b}$$

In Fig. 3.10, the value of $|\delta A/A_0|_l^2$ is shown as a function of l/l_c, $l_c = \sqrt{(L/k)}$. The solid curve represents (3.96), i.e., the complete solution for eddies with gaussian refractive index profiles, while the long dashes represent the geometrical optics approximation, (3.96a), and the short dashes the diffraction approximation (3.96b). If $\sqrt{(\lambda L)} \ll l_0$, then the situation is as shown in Fig. 3.11a. In this case, since the width of a Fresnel zone is much less than l_0, i.e., $\sqrt{(\lambda L)} \ll l_0$, geometrical optics holds and the scintillations are dominated by the microscale of turbulence, i.e., l_0. As the path length increases, $\sqrt{(\lambda L)}$ increases, and the scintillations are dominated by the scale size of a Fresnel zone, i.e., $\sqrt{(\lambda L)}$, as seen in Fig. 3.11b. The geometrical optics result for

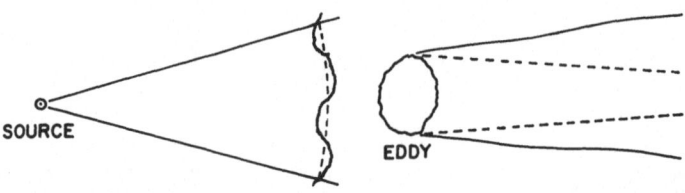

Fig. 3.12. The effect of turbulence before and after an eddy on the eddy's focusing properties

amplitude scintillations, (3.89), comes from (3.96a) by letting $l = l_0$. The diffraction or wave optics result, (3.91), comes from (3.96) by letting $l = \sqrt{(L/k)}$.

The situation shown in Figs. 3.10 and 3.11 assumes that the incident wave is plane and coherent across the entire inhomogeneity. However, as the wave propagates, phase fluctuations are introduced, as described by (3.80) and (3.81). If the phase difference fluctuations are small over a region comparable to the size of an eddy, then the wave can be assumed plane, and the above results hold. If, however, the phase fluctuations become large compared to a wavelength, then the wave incident on the eddy is no longer coherent, and the results given above must be reexamined. The physical situation is shown in Fig. 3.12. Clearly, the more turbulence between the original incident wave and the eddy under consideration, the greater the loss of coherence. To quantify this concept, if $\langle \Delta s^2(\varrho) \rangle$ is greater than π^2, we shall assume the wave has lost its coherence. In addition, we shall assume that when a wave loses its coherence, it no longer can be effectively focused or defocused by an inhomogeneity. Therefore, the criterion for an eddy to produce significant scintillations is that

$$\langle \Delta s^2(\varrho) \rangle < \pi^2 \,, \tag{3.97}$$

where $\varrho = l$, the size of the eddy. Let $l = \varrho_0$ denote the largest eddy for which the wave may be considered coherent, then from (3.80) and (3.97)

$$\varrho_0 \approx [\pi^2/(k^2 C_n^2 L)]^{3/5} \,. \tag{3.98}$$

A more rigorous calculation of the mutual coherence function, $\Gamma(\varrho, z)$, gives

$$\Gamma(\varrho, z) = \exp(-\varrho^2/\varrho_c^2), \quad l_0 \ll \varrho \ll L_0 \tag{3.99}$$

where

$$\varrho_c = (1.45 k^2 C_n^2 L)^{-3/5} \,. \tag{3.100}$$

Clearly (3.100) and (3.98) agree within an order of magnitude. Note the significance of ϱ_0. For eddies whose scale size l is less than ϱ_0, the results shown in Figs. 3.10 and 3.11 and given by (3.89), (3.91), and (3.96) are applicable. For

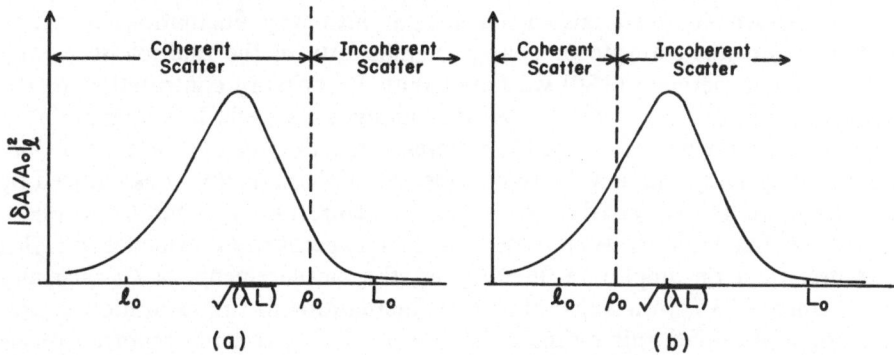

Fig. 3.13. (a) $\varrho_0 > \sqrt{(\lambda L)}$ is the coherent scattering case, and scintillations are dominated by $l \sim \sqrt{(\lambda L)}$; (b) $\varrho_0 < \sqrt{(\lambda L)}$ is the incoherent scattering case, and scintillations are dominated by $l \sim \varrho_0$

$l > \varrho_0$, an eddy essentially produces no scintillations. In Fig. 3.13 we illustrate these conditions assuming $\sqrt{(\lambda L)} > l_0$. If $\varrho_0 > \sqrt{(\lambda L)}$, then the coherence length is greater than a Fresnel zone, which is the size of the dominant in-homogeneities producing scintillations, and Fig. 3.11b and (3.91) are valid. If $\varrho_0 < \sqrt{(\lambda L)}$, then eddies whose size is of the order $\sqrt{(\lambda L)}$ are no longer causing scintillations, and the dominant scale size is ϱ_0. Noting that $\varrho_0 < \sqrt{(L/k)}$, we may let $l \sim \varrho_0$ and substitute (3.98) into (3.96b) to find

$$\left|\frac{\delta A}{A_0}\right|^2_{\varrho_0} \sim C_n^2 k^2 L \left[\left(\frac{\pi^2}{k^2 C_n^2 L}\right)^{3/5}\right]^{5/3} \sim 1. \tag{3.101}$$

In essence, this explains the saturation effect, i.e., when the coherence length is shorter than the width of a Fresnel zone, and the scintillations are dominated by eddies whose size is given by ϱ_0, the scintillations are independent of the turbulence intensity and the path length, in agreement with Figs. 2.9 and 4.5.

The above arguments also give the basis for the major scales important in the covariance function of the amplitude or irradiance scintillations. For plane wave propagation, the dominant scale in the covariance function should be of the same order as the dominant scale producing scintillations. Therefore, in the geometrical optics region, the transverse scale should be of the order of l_0; in the diffraction region, the transverse scale should be of the order of the width of a Fresnel zone, i.e., $\sqrt{(\lambda L)}$; and in the multiple scatter region the transverse scale should be of the order of ϱ_0. Note that this prediction has been well substantiated in the diffraction region, where the transverse scale increases with increasing path length. In the saturation region, ϱ_0 will decrease with increasing k, C_n, or L, as seen from (3.98). This prediction is in qualitative agreement with the experimental results of Gracheva et al. in Chapter 4 (see Fig. 4.8) and *Ochs* et al. [3.72] and *Clifford* et al. [3.64].

The above argument also lends insight into why fluctuations in phase difference do not saturate or change significantly in the multiple-scattering region. When deriving (3.80) we noted that the primary contribution to the phase difference is the change in velocity along a ray path. It is immaterial'in this calculation whether or not the coherence length is smaller or greater than an inhomogeneity; there will still be substantial phase fluctuations induced by the medium. Hence we do not expect a saturation-type effect for phase difference fluctuations. While there is less experimental evidence on this question than on amplitude fluctuations, the measurements of *Gurvich* and *Kallistratova* [3.73] on angle-of-arrival fluctuations in the saturation region tend to substantiate this estimate. *Born* et al. [3.74] recently reported phase measurements in the saturation region and, while they noted some other effects, such as phase front breakup, the basic relations seem to be valid.

Yura's derivation of (3.101) for saturation is more detailed than given here. He proved more rigorously one assumption made here. In the diffraction region, the extended Huygens-Fresnel principle [3.75, 76] was used to show that the angular spread of a beam of radius l is of the order

$$\theta^2 \approx \frac{1}{k^2 l^2} + \frac{1}{k^2 \varrho_0^2(z)}, \tag{3.102}$$

where ϱ_0^2 is given by (3.98). This equation is valid if the inhomogeneity of size l is located at some distance z into the random medium. Therefore the wave incident on the inhomogeneity is only partially coherent, and (3.102) describes the increased scattering due to the loss of coherence. Note that as $\varrho_0^2(z)$ decreases and becomes much smaller than l^2, the angular spread is completely dominated by the scattering process. To estimate the angular spread, we can assume an inhomogeneity of size l is replaced by an inhomogeneity with effective size l_e, i.e.,

$$\frac{1}{l_e^2} = \frac{1}{l^2} + \frac{1}{\varrho_0^2(z)}.$$

As the wave propagates from the lens or eddy to the receiving plane, reciprocity [3.77] indicates that the identical effect will take place. Therefore to model the entire propagation path, an additional term is added to get

$$\frac{1}{l_e^2} = \frac{1}{l^2} + \frac{1}{\varrho_0^2(z)} + \frac{1}{\varrho_0^2(L-z)}. \tag{3.103}$$

In (3.95) we replace $[(L-z)/kl]^2$ by $[(L-z)/kl_e]^2$, since this is the term that corresponds to the angular spread due to diffraction, but do not change the other term in l.

With this substitution we have

$$\left.\left|\frac{\delta A}{A_0}\right|^2\right|_l \sim \frac{C_n^2 L^2 (1-z/L)^2 l^{2/3}}{\left[l^2 + \frac{L^2(1-z/L)^2}{k^2 l_e^2}\right]},$$

(3.104)

where l_e is given by (3.103). This expression gives the amplitude fluctuations from eddies of a single size l, located in a slab oriented perpendicular to the propagation direction and at a distance z from the origin. To include the effects from all slabs, we multiply (3.104) by dz/l and integrate from 0 to L. The factor dz/l gives the weighting of a slab of thickness l to the total scintillations. Letting $\eta = 1 - z/L$, we find

$$\left.\left|\frac{\delta A}{A_0}\right|^2\right|_l = \int_0^1 \frac{C_n^2 L^3 \eta^2 l^{-1/3}}{[l^2 + L^2 \eta^2/(k^2 l_e^2)]}\, d\eta.$$

(3.105)

Finally, to find the scintillations due to eddies of all sizes, we integrate (3.105) over dl, from l_0 to L_0, i.e.,

$$\sigma_A^2(L) \sim \int_{l_0}^{L_0} \left.\left|\frac{\delta A}{A_0}\right|^2\right|_l \frac{dl}{l} = \int_{l_0}^{L_0} dl \int_0^1 d\eta \frac{C_n^2 L^3 \eta^2 l^{-4/3}}{[l^2 + L^2 \eta^2/(k^2 l_e^2)]}.$$

(3.106)

Inserting (3.103), we have

$$\sigma^2(L) \sim C_n^2 L^3 \int_0^1 d\eta \eta^2 \int_{l_0}^{L_0} \frac{dl}{l^4} \frac{l^{2/3}}{\left(1 + \eta^2 \frac{L^2}{k^2 l^2}\left\{\frac{1}{l^2} + \frac{1}{\varrho_0^2(L)}[\eta^{6/5} + (1-\eta)^{6/5}]\right\}\right)}.$$

(3.107)

Using the approximate relation $K \sim 1/l$, (3.107) may be converted to a spectral form, i.e.,

$$\sigma^2(L) \sim Q_0^{5/3} \int_{Q_1}^{Q_2} dQ Q^{4/3} \int_0^1 \frac{d\eta \eta^2}{1 + \eta^2 Q^2 [Q^2 + Q_0^2(\eta)]},$$

(3.108)

where $Q \equiv (L/k)^{1/2} K$ is a wave number normalized to a Fresnel zone wave number, and

$$\begin{aligned}
Q_0^2 &= L/[k\varrho_0^2(L)], \\
Q_0^2(\eta) &= Q_0^2[\eta^{6/5} + (1-\eta)^{6/5}], \\
Q_1 &= [L/(kL_0^2)]^{1/2}, \\
Q_2 &= [L/(kl_0^2)]^{1/2}.
\end{aligned}$$

(3.109)

Yura rewrote these results in order to facilitate comparison with the classical results using the smooth perturbation method. He found

$$\sigma^2(L) \sim Q_0^{5/3} \int\limits_{Q_1}^{Q_2} \Phi_n(Q) f_A(Q) Q dQ \tag{3.110}$$

where $\Phi_n(Q) = Q^{-11/3}$, and

$$f_A(Q) = Q^4 \int\limits_0^1 \frac{d\eta\eta^2}{1 + \eta^2 Q^2 [Q^2 + Q_0^2(\eta)]} \tag{3.111}$$

is the amplitude filter function. The filter function gives the importance of an inhomogeneity of wave number Q in affecting the amplitude variance. Also note that $Q_0^{5/3} = C_n^2 k^{7/6} L^{11/6}$, the smooth perturbation variance. We are interested in the case $Q_0 \geqq 1$, i.e., the saturation region. Assuming $Q_0^2(\eta) \sim Q_0^2$ in this region, (3.111) can be integrated and gives

$$f_A(Q) = \frac{Q^2}{Q^2 + Q_0^2} \left\{ 1 - \frac{\tan^{-1}[Q(Q_0^2 + Q^2)^{1/2}]}{Q(Q_0^2 + Q^2)^{1/2}} \right\}. \tag{3.112}$$

Inserting this result in (3.110) and using the condition $Q_0 \gg 1$, *Yura* estimated $\sigma^2(L) \sim \pi$ in the saturation region, in general agreement with (3.101).

Yura used (3.110) to make two generalizations. First, while (3.110) was derived assuming a specific form for $\Phi_n(K)$, he assumed that it is valid for any $\Phi_n(K)$. More importantly, he assumed that the form can be immediately generalized to find the amplitude covariance function, $B_A(\varrho) \equiv \langle A_1(\varrho_1) A_1(\varrho_1 + \varrho) \rangle$ where $A_1 = A(\varrho) - \langle A \rangle$. We know that for weak perturbations $B_A(\varrho)$ is given by (2.80)

$$B_A(\varrho) = \int\limits_0^\infty F_A(K) J_0(K\varrho) K dK$$

and hence $\sigma_A^2 = B_A(0) = \int\limits_0^\infty F_A(K) K dK$. Assuming the same is true for strong perturbations

$$B_A(\varrho) \sim Q_0^{5/3} \int\limits_{Q_1}^{Q_2} \Phi_n(Q) f_A(Q) J_0(Q\varrho_n) Q dQ, \tag{3.113}$$

where $\varrho_n = \varrho/(L/k)^{1/2}$. Equation (3.113) appears reasonable in terms of the physical model given in Fig. 3.13. Lenses with size $l > \varrho_0$ will not have a strong effect on the scattering and hence will not be important in the covariance function. Therefore, $B_A(\varrho)$ will be dominated by scale sizes less than ϱ_0, just as for the variance, and hence the same filter function $f_A(Q)$ comes into play. Figure 3.13b indicates what changes in the covariance function should be expected. The correlation radius is no longer dominated by $\sqrt{(\lambda L)}$ but by ϱ_0; therefore the covariance should fall off more rapidly with ϱ as the strength of the turbulence increases.

It should be noted that *Yura* calculates the amplitude fluctuations; however, measurements are normally made of $\sigma_{\ln I}^2$ or of $\beta^2 \equiv \langle (I - \bar{I})^2 \rangle / \bar{I}^2$. While conservation of energy dictates that the amplitude fluctuations must saturate, it does not put limits on $\sigma_{\ln I}^2$ or β^2. However, if the above heuristic arguments are interpreted as being equally valid for log-amplitude fluctuations, the approach leads to similar conclusions for $\sigma_{\ln I}^2$.

The approach of *Clifford* et al. [3.64] was quite similar but they worked in the spatial frequency domain. Furthermore, instead of considering the effect of coherence on the scattering characteristics of an eddy or lens, they considered a convolution of the weak perturbation covariance function with a modified form of the irradiance profile. They found

$$B_\chi^2(\varrho_n, \sigma_t^2) = 2.95\sigma_t^2 \int_0^1 du [u(1-u)]^{5/6} \int_0^\infty dy \frac{\sin^2 y}{y^{11/6}}$$

$$\cdot \exp\{-\sigma_t^2 [u(1-u)]^{5/6} f(y)\} J_0 \left[\left(\frac{4\pi y u}{1-u} \right)^{1/2} \varrho_n \right], \qquad (3.114)$$

where

$$f(y) = 7.02 y^{5/6} \int_{0.7y}^\infty d\xi \xi^{-8/3} [1 - J_0(\xi)],$$

$$\sigma_t^2 = 0.124 k^{7/6} L^{11/6} C_n^2,$$

$$y = k^2 u(1-u) L/(2k).$$

A comparison of the results predicted by (3.113) and (3.114) will be given in the next section.

3.6 Results

There presently exists excellent theoretical agreement on the form of the moment equations for all orders, as the same equations have been derived using quite different methods, but using essentially the same approximations. In this section we shall describe the results that are available for the solution of these equations, particularly for the mean field, the mutual coherence function, and the fourth-order coherence function. From the last quantity, the spectra and covariance of the irradiance fluctuations may be calculated, and the results of various workers for these quantities will be discussed. In the last section the probability distribution of the irradiance fluctuations will be considered. The results discussed in this section will be for plane or spherical waves (see Chap. 5 for the beam wave case).

3.6.1 The Mean Field, $u(r)$

Equation (3.36) derived using the renormalization method, and (3.46) derived via the Markov random process assumption or the LMSM are identical, i.e.,

$$2ik\frac{\partial \langle u(\varrho, z)\rangle}{\partial z} + V_T^2\langle u\rangle + \frac{ik^3}{4} A_\varepsilon(0)\langle u(\varrho, z)\rangle = 0,$$

$$\langle u(\varrho, 0)\rangle = u(\varrho, 0).$$

(3.36)

Using the substitution

$$\langle u(r)\rangle = \exp\left[-\frac{k^2 A_\varepsilon(0)z}{8}\right] W(r),$$

(3.36) can be solved, and we get

$$\langle u(r)\rangle = \exp\left[-\frac{k^2 A_\varepsilon(0)z}{8}\right] u_0(r),$$

(3.115)

where $u_0(r)$ is the wave that would be propagated in free space. Equation (3.115) is the final result for the average field. It states that the average field is the same as the free space field, except that in the direction of propagation it is attenuated by an extra factor with an attenuation constant given by

$$\alpha = \frac{k^2 A_\varepsilon(0)}{8} = \frac{1}{8}k^2\sigma_\varepsilon^2 l_1.$$

(3.116)

If we use the spectrum for locally homogeneous isotropic turbulence suggested in Chapter 2, e.g. (2.38), we get [3.3]

$$\alpha = 0.391k^2 L_0^{5/3} \int_0^L C_n^2(z)dz.$$

(3.117)

Virtually the same results were derived using the diagram method by solving the resulting integral equations. In particular, *Bourret* [3.31, 32] and *Brown* [3.52, 53] kept the Green's function for the scalar wave equation instead of using the parabolic approximation. They concluded that all that is necessary to calculate the average field is to replace the free-space Green's function with an averaged Green's function given by

$$\langle G(r)\rangle = -\frac{\exp(ik_\varepsilon r)}{4\pi r},$$

(3.118)

where k_e is the effective wave number for the averaged field. *Bourret* concluded that k_e is given by

$$k_e \cong k - ik^2 \sigma_\varepsilon^2 l \quad \text{for} \quad kl \gg 1. \tag{3.119}$$

In Bourret's expression, l is the scale length of an assumed exponential type correlation function. Note that the imaginary part of k_e corresponds to the attenuation coefficient. *Brown* obtained a similar result, but l was an integral scale. A similar result was derived by *Keller* [3.78–80], but using a non-diagrammatic technique. Several authors [3.1, 51, 53] studied the question of the region of validity of the above expressions for the attenuation coefficient. They seem to be in agreement that the expressions are valid provided $\sigma_\varepsilon^2 k^2 l^2 \ll 1$. While this condition is undoubtedly met for many practical problems of interest, there may be cases of strong atmospheric turbulence in the ground layer in which it is violated.

3.6.2 The Mutual Coherence Function

The second quantity that has been carefully studied theoretically is the mutual coherence function,

$$\Gamma(\varrho_1, \varrho_2; z) = \langle u(\varrho_1, z) u^*(\varrho_2, z) \rangle.$$

The equation for this quantity was derived in Section 3.3.2, (3.53), and is given by

$$2ik \frac{\partial \Gamma}{\partial z} + (V_{T1}^2 - V_{T2}^2)\Gamma + \frac{ik^3}{2}[A_\varepsilon(0) - A_\varepsilon(\varrho_1 - \varrho_2)]\Gamma = 0,$$

$$\Gamma(\varrho_1, \varrho_2; 0) = u(\varrho_1, 0) u^*(\varrho_2, 0).$$

The coherence function is of great practical interest for the beam wave case because it gives the average intensity across the beam. Therefore, the solution of this equation is discussed in detail by Ishimaru in Section 5.2.1. For the plane wave case, $(V_{T1}^2 - V_{T2}^2)\Gamma = 0$, and the solution of the above equation is

$$\Gamma_p(\varrho_1, \varrho_2; z) = I_0 \exp\left\{-\frac{k^2 z}{4}[A_\varepsilon(0) - A_\varepsilon(\varrho_1 - \varrho_2)]\right\}$$

$$= I_0 \exp\left\{-\pi^2 k^2 z \int_0^\infty dK[1 - J_0(K|\varrho_1 - \varrho_2|)] K \Phi_\varepsilon(K)\right\}$$

$$= I_0 \exp\left\{-\frac{1}{2}[D_\chi(|\varrho_1 - \varrho_2|) + D_s(|\varrho_1 - \varrho_2|)]\right\}. \tag{3.120}$$

Equations for $D_\chi(\varrho)$ and $D_s(\varrho)$ are given in Chapter 2, i.e. (2.80–82).

In actual fact, this result can be derived from many different starting points. The result was derived by *Hufnagel* and *Stanley* [3.15] in 1964, by *Beran* [3.45] in 1966 based on an argument analogous to the LMSP, and by *Brown* [3.52] using the renormalization method and ladder approximation in the short wavelength limit. In addition, *Strohbehn* [3.25] showed that the result can be obtained quite easily from the method of smooth perturbations if it assumed that the log-amplitude and phase have a normal probability distribution.

In the case of a Kolmogorov spectrum, and using (2.85) for $D_s(\varrho)$, (3.120) becomes

$$\Gamma_p(\varrho;z) = I_0 \exp(-1.47 C_n^2 k^2 z \varrho^{5/3}), \quad (\lambda z)^{1/2} \ll \varrho \ll L_0. \tag{3.121}$$

Fante [3.3] included the effects of the outer scale and found

$$\Gamma_p(\varrho;z) = I_0 \exp\{-1.47 C_n^2 k^2 z \varrho^{5/3}[1 - 0.805(\varrho/L_0)^{1/3}]\}. \tag{3.122}$$

3.6.3 The Fourth-Order Coherence Function

As mentioned before, the fourth-order coherence function is of great practical interest because it gives information about the irradiance covariance function, and its Fourier transform gives the frequency spectrum of the irradiance fluctuations. In general, an integral equation for this function is too cumbersome to be useful, but a partial differential equation has been derived by *Shishov* [3.55], *Molyneux* [3.54], and *Brown* [3.81] using a diagram or renormalization approach, by *Tatarskii* and *Klyatskin* (Sec. 3.3) using the Markov random process approach, and by *Shishov* [3.42] (Sect. 3.4) and *Beran* and *Ho* [3.63] using the LMSP.

The equation for the fourth-order moment, (3.68), is a partial differential equation in nine variables. If the incident wave is plane, a coordinate transformation [3.9, 82] can reduce the problem to a partial differential equation in nine variables. Even the solution of this latter problem is a formidable task, no matter whether an analytical or a numerical technique is used.

Gochelashvily and *Shishov* [3.83–85] were the first to make significant progress on solving this problem, and an excellent summary of their work was given by *Prokhorov* et al. [3.2]. Recently results very similar to the Russian work were published by *Fante* [3.3, 86, 87]. For weak fluctuations, most workers have calculated the variance of the log-amplitude fluctuations, but for strong fluctuations it is most common to calculate the scintillation index, that is, the normalized variance of the irradiance fluctuations, which we shall designate by β^2,

$$\beta^2 \equiv \frac{\langle I^2 \rangle - \langle I \rangle^2}{\langle I \rangle^2}. \tag{3.123}$$

The classical theory (Chap. 2) gives the following expression for β^2 in the region of weak fluctuations:

$$\beta_0^2 = 1.23 C_n^2 k^{7/6} L^{11/6}. \tag{3.124}$$

Based on the partial differential equation for the fourth-order coherence function (3.68), $\beta^2 = \beta_0^2$ in the region where $D[\sqrt{(z/k)}] < 1$, where $D[\sqrt{(z/k)}]$ is the phase structure function evaluated for a separation equal to the radius of the first Fresnel zone. For $D[\sqrt{(z/k)}] > 1$, *Gochelashvily* and *Shishov* [3.85] got

$$\beta^2 = 1 + 0.85(\beta_0^2)^{-2/5}, \tag{3.125}$$

while *Fante* [3.3] got the same form but calculated a coefficient of 0.99 instead of 0.85. This expression assumes that $kl_0^2/z \ll 1$, which for normal wavelengths and turbulence conditions would require the path length to be greater than 1 km. Compare this with the result derived by *Clifford* et al. [3.64], which asymptotically goes as

$$\beta^2 \sim 0.92 + 1.44(\beta_0^2)^{-2/5}, \tag{3.126}$$

but note this was derived for the spherical wave case. *Tatarskii* [3.9] derived a form

$$\beta^2 \sim 1.36 - 0.907(\beta_0^2)^{-2/5}, \beta_0 \gg 1. \tag{3.127}$$

Note that while the functional form is the same for his expression, because of the minus sign in the constant it does not give the supersaturation effect observed experimentally or inferred by (3.125) and (3.126).

It is interesting to compare these results with the asymptotic form that can be obtained from the results of *Yura* [3.66]. For example, based on his equation for the variance, (3.110), we can examine the dependence on Q_0 by using (3.112). First of all, since $(\tan^{-1} x)/x$ is a monotonically decreasing function of x it can be shown that *Yura*'s expression for variance cannot exhibit supersaturation, i.e., as β_0^2 increases $(\beta_0^2 = Q_0^{5/3})$, his expression will approach a constant from below, not above. In addition while *Yura*'s expression depends on several parameters, as a function of β_0^2 it tends to go as $(\beta_0^2)^{-6/5}$, in contrast to the results of other workers who got $(\beta_0^2)^{-2/5}$.

Obtaining an expression for the covariance function is considerably more difficult. Gracheva et al. (Chap. 4) argue, based on similarity relations, that for a pure power law spectrum

$$\Phi_\varepsilon(k) = A C_\varepsilon^2 K^{-11/3},$$

there are two scales that describe the problem: a longitudinal scale $L_T = (\pi A C_\varepsilon^2 k^{7/6})^{-6/11}$ and a transverse scale $l_T = (L_T/k)^{1/2} = (\pi A C_\varepsilon^2 k^3)^{-3/11}$. The variance of the irradiance fluctuations, or the scintillation index, should be a function of z/L_T where z is the path length and L_T the longitudinal scale. Since $\beta_0^2 = 3(z/L_T)^{11/6}$, the dependencies discussed above agree with this argument. In addition, their results imply that the covariance function should be a function of the two scales L_T and l_T, though they do not give an expression for this function. Instead, they determine the function experimentally; their results are discussed in detail in the next chapter.

Gochelashvily and *Shishov* [3.2, 84, 85] obtained the following expression for the covariance function when $z \gg L_T$:

$$B_I(\varrho, z)/I_0^2 = \exp[-D(\varrho)] + 0.6\{D[\sqrt{(z/k)}]\}^{-2/5}$$
$$\cdot [b_1(q^*\varrho) + b_2(\varrho/\varrho^*)], \tag{3.128}$$

where

$$l_s = 1/q^* = \{D[\sqrt{(z/k)}]\}^{3/5} \sqrt{(z/k)} = (2.92)^{3/5}(C_n^2)^{3/5}k^{1/5}z^{8/5}, \tag{3.129}$$

$$\varrho^* = \{D[\sqrt{(z/k)}]\}^{-3/5} \sqrt{(z/k)} = (2.92)^{-3/5}(C_n^2)^{-3/5}k^{-6/5}z^{-3/5},$$
$$D(\varrho) = 2.92 C_n^2 k^2 \varrho^{5/3}. \tag{3.130}$$

Note that $D(\varrho)$ is the phase structure function calculated using geometrical optics. In terms of the similarity relations given by Gracheva et al.

$$l_s = 1/q^* = 3.225(z/L_T)^{8/5}l_T, \tag{3.131}$$

$$\varrho^* = 0.31(z/L_T)^{-3/5}l_T. \tag{3.132}$$

Note that the scales of *Gochelashvily* and *Shishov* are different from the scales of Gracheva et al., though they both obey the similarity constraints. However, the importance of this difference will be more apparent when we discuss the results of spectral measurements. The formulas for $b_1(\eta)$ and $b_2(\eta)$ are given in [3.85] as

$$b_1(\eta) = \tfrac{10}{9}\Gamma^{-1}(7/5) \, {}_2F_1[7/5, 2/3; 5/3; 5/8]$$
$$\cdot \int_0^1 d\zeta \zeta^2 \int_0^\infty dz z^{4/3} J_0(\eta) \exp[-z^{5/3}\zeta^{5/3}(1 - \tfrac{5}{8}\zeta)], \tag{3.133}$$

$$b_2(\eta) = \frac{10}{9\pi}\Gamma^{-1}(7/5) \, {}_2F_1[7/5, 2/3; 5/3; 5/8] (z/L_s)^{22/5}$$
$$\cdot \int_0^1 d\zeta \int dn \, n^{-11/3}\{1 - \cos[n \cdot (\eta + n\zeta)(z/L_s)^{-11/5}]\}$$
$$\cdot \exp\left[-|\eta + n\zeta|^{5/3}(1 - \zeta) - \zeta \int_0^1 ds|\eta + n\zeta s|^{5/3}\right], \tag{3.134}$$

where

$$L_s = (2.92 C_n^2 k^{7/6})^{-11/6} = 0.345 L_T.$$

Note that b_1 is only a function of the transverse scale η, while b_2 is a function of both η and z/L_s, and serves as a correction to the covariance function as z/L_s gets large. *Gochelashvily* and *Shishov* [3.85] calculated b_1 numerically but did not carry out the rather involved calculation for b_2. The function b_1 is equal to unity for $\varrho = 0$ but falls off to close to zero for $\varrho \geqq l_s$, while the function $b_2(\varrho)$ is equal to unity for $\varrho = 0$. In order to compare the theoretical results of *Gochelashvily* and *Shishov* [3.85] with the experimental results of Gracheva et al. (Chap. 4), a very rough approximation to $b_2(\eta)$ was found and used in (3.128) to calculate $B_I(\varrho, z)$. The results are shown in Fig. 3.14 for $\beta_0^2 = 4$ (a) and $\beta_0^2 = 36$ (b). In Fig. 3.14 the solid line represents the experimental data of Gracheva et al. (Chap. 4), and the short dashed line the approximate evaluations of the theoretical expressions of *Gochelashvily* and *Shishov* [3.85]. The other curves in the figure will be discussed later. While an exact correspondence is not found between the theory and the experiment, the correct general behaviour is shown. For example, as β_0^2 increases, the covariance function falls off much more rapidly, but has a longer tail. However, the long tail seems to be more pronounced in the experimental data. While the numerical calculation for $b_2(\eta)$ is very rough, it is felt that it would tend to overestimate the value of $b_2(\eta)$ and hence $B_1(\varrho)$. While the difference between the theory and experiment is larger than is desirable, given the complexities involved in the theory and the difficulties involved in carrying out good experiments, the agreement is quite remarkable.

Fante [3.3, 86–89] also derived an expression for the covariance function that is very similar to the expression of *Gochelashvily* and *Shishov* [3.85]. *Fante* [3.3, 86, 88] got

$$B_I(\varrho; z) = \exp(-11.2\beta_0^2 R^{5/3}) + \frac{1}{(\beta_0^2)^{2/5}}$$

$$\cdot \left\{ f\left[\frac{R}{(\beta_0^2)^{3/5}}\right] + g[(\beta_0^2)^{3/11}R] \right\} \tag{3.135}$$

where $R = \varrho/(\lambda z)^{1/2}$ and

$$f(s) = 1.43 \int_0^1 dy\, y^{-1/3} \int_0^\infty dt\, t^{2/5} \exp[-t(4.26 - 2.66y)]$$

$$\cdot J_0(3.54 t^{3/5} y^{-1} s), \tag{3.136a}$$

$$g(w) = 0.27 \int_0^\infty dt\, t^{-8/3}(1 - \cos t) \int_0^\infty ds\, \exp(-s)$$

$$\cdot J_0(2.43 t^{8/3} s^{-3/11} w). \tag{3.136b}$$

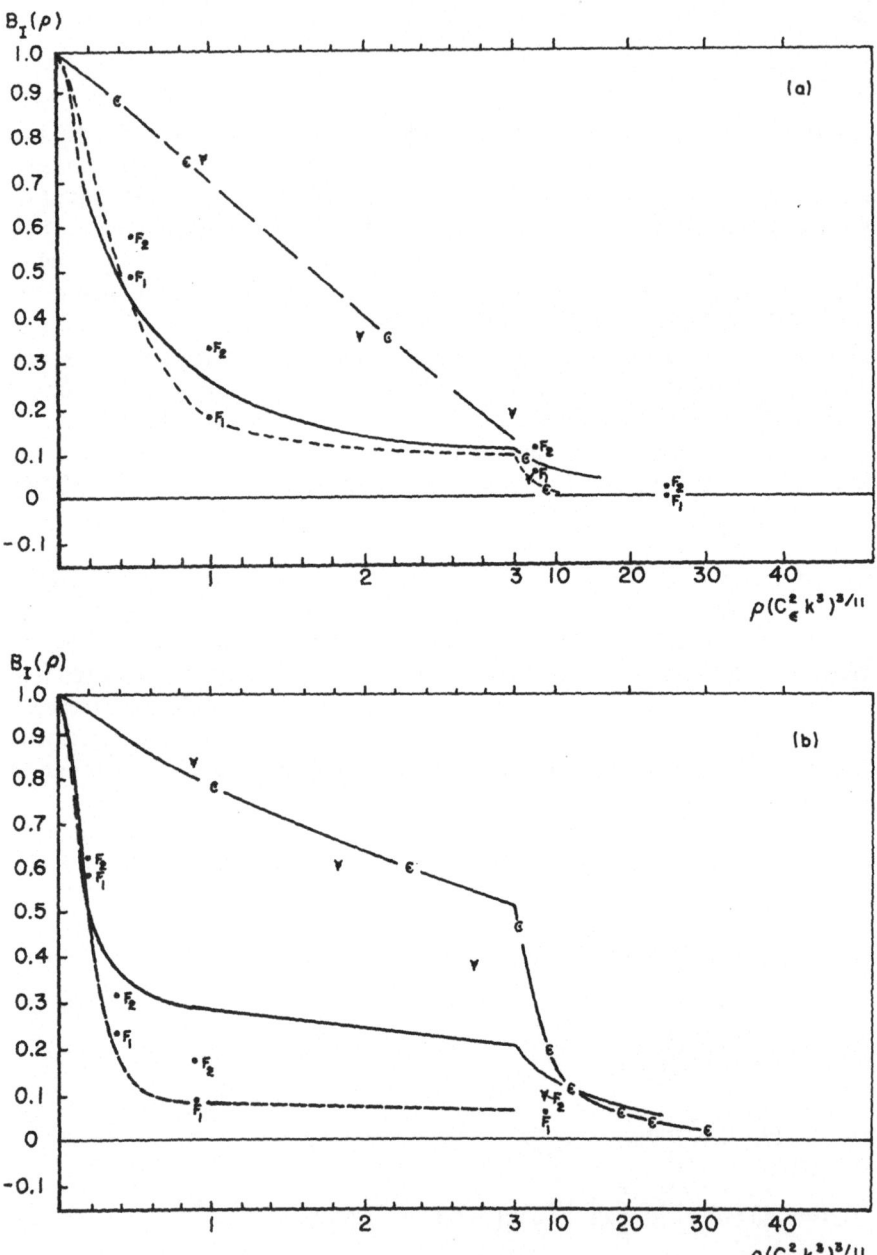

Fig. 3.14a and b. Comparison of the experimental data of Gracheva et al. (Chap. 4) (solid line) with the work of various theoretical groups: (a) $\beta_0^2 = 4$, (b) $\beta_0^2 = 36$

Fante [3.87] showed that his results for the first iteration are equivalent to those of *Gochelashvily* and *Shishov*, though the argument is based on their respective expressions for the spectrum. In particular, the exponential term and the term in $b_1(q^*\varrho)$ in (3.128) agree with the exponential term and the term in $f[R/(\beta_0^2)^{3/5}]$ in (3.135). In Fig. 3.14 are shown the results of *Fante*'s calculations [3.90] based on (3.135) and (3.136). The points marked F_1 are calculations based on what *Fante* calls his first iteration, and correspond to setting $g(\varrho)=0$ in (3.135). The points marked F_2 correspond to *Fante*'s second iteration, and include the g term in (3.135).

The expressions given by *Clifford* et al. [3.64, 65] for the covariance function (3.114) and *Yura* [3.66] (3.113) appear fundamentally different than those of *Gochelashvily* and *Shishov* [3.85] and the parametric relationship suggested by Gracheva et al. (Chap. 4) in that their expressions are a function of z/L_T and $\varrho/(\lambda L)^{1/2}$, not z/L_T and ϱ/l_T. However, remember that β_0^2 (σ_t^2 in *Clifford* et al.) is a dimensionless parameter, therefore, at least as far as similarity relations are concerned, $\varrho/(\lambda L)^{1/2}$ could be replaced by $\varrho(\lambda L)^{-1/2}(\beta_0^2)^\gamma$, where

$$\beta_0^2 \sim C_\varepsilon^2 k^{7/6} L^{11/6}.$$

If we let $\gamma = 3/11$,

$$\varrho(\lambda L)^{-1/2}(\beta_0^2)^\gamma \sim \varrho C_\varepsilon^{6/11} \lambda^{-9/11} \sim \varrho/l_T,$$

in agreement with the similarity relations.

In the paper by *Clifford* et al. they compared the general form of their theoretical expression with their experimental measurements taken over 500 and 1000 m paths. There is good qualitative agreement, but unfortunately these authors did not give quantitative values for their saturation parameters, so a direct comparison of their theoretical and experimental curves cannot be made. In addition, *Clifford* et al. calculated and measured the log-amplitude covariance function, while Gracheva et al. measured the irradiance covariance function, making a direct comparison between results difficult. In Fig. 3.14, the curves marked by the c's represent the curves of *Clifford* et al. for the log-amplitude covariance function. These curves appear to fall off much more slowly than the experimental data of Gracheva et al. However, as *Clifford* pointed out, if the assumption is made that the probability distribution of the amplitude fluctuations is log normal, the irradiance covariance function can be calculated from the log-amplitude covariance function. The resulting irradiance curves will fall off much more rapidly and be much closer to the experimental data and the calculations of *Gochelashvily* and *Shishov* [3.85] and *Fante* [3.90]. Finally, in Fig. 3.14 the points denoted by Y are based on Fig. 5 from *Yura* [3.66]. His values are also for the log-amplitude covariance function and are in agreement with *Clifford*. In summary, several authors now have theoretical expressions for the covariance function that seem to give reasonable agreement with experiment. However, parameter dependencies between different workers do not appear to be the same, and further comparison of experimental and theoretical work is needed to clarify these discrepancies.

Clearly the spectrum of the irradiance fluctuations and the covariance function are a Fourier transform pair. In actual fact, most workers first derived expressions for the spectral expansion and then calculated the covariance function. If we start with (3.68) for the fourth moment of $E(\varrho, z)$, but use the coordinate transformation [3.9],

$$\varrho_1 = R + \frac{r_1 + r_2}{2} + \tfrac{1}{4}\varrho \qquad R = \tfrac{1}{4}(\varrho_1 + \varrho_2 + \varrho_1' + \varrho_2')$$

$$\varrho_2 = R - \frac{r_1 + r_2}{2} + \tfrac{1}{4}\varrho \qquad r_1 = \tfrac{1}{2}(\varrho_1 - \varrho_2 + \varrho_1' - \varrho_2')$$

$$\varrho_1' = R + \frac{r_1 - r_2}{2} - \tfrac{1}{4}\varrho \qquad r_2 = \tfrac{1}{2}(\varrho_1 - \varrho_2 - \varrho_1' + \varrho_2') \tag{3.137}$$

$$\varrho_2' = R - \frac{r_1 - r_2}{2} - \tfrac{1}{4}\varrho \qquad \varrho = \varrho_1 + \varrho_2 - \varrho_1' - \varrho_2',$$

then $\Gamma_{2,2}(\varrho_1, \varrho_2, \varrho_1', \varrho_2'; z)$ is replaced by $\Gamma_4(R, r_1, r_2, \varrho; z)$, and (3.68) becomes

$$\frac{\partial \Gamma_4(R, r_1, r_2, \varrho; z)}{\partial z} = \frac{i}{k}(\nabla_R \nabla_\varrho + \nabla_{r_1} \nabla_{r_2})\Gamma_4$$

$$- \frac{k^2}{4} F(r_1, r_2, \varrho, z)\Gamma_4, \tag{3.138}$$

where

$$F(r_1, r_2, \varrho; z) = F(\varrho_1, \varrho_2, \varrho_1', \varrho_2'; z).$$

In terms of the spectral representation,

$$F(r_1, r_2, \varrho; z)$$
$$= 4\pi \iint\limits_{-\infty}^{\infty} d^2 K \Phi_\varepsilon(K, z)\left[1 + \cos(K \cdot r_1)\cos(K \cdot r_2)\right.$$

$$\left. - \cos(K \cdot r_1)\cos\left(\frac{K \cdot \varrho}{2}\right) - \cos(K \cdot r_2)\cos\left(\frac{K \cdot \varrho}{2}\right)\right]. \tag{3.139}$$

For an incident plane wave, the statistical homogeneity in the plane $z = \text{const.}$ implies $\nabla_R \Gamma_4 = 0$ and we may assume $R = 0$; therefore (3.138) becomes

$$\frac{\partial \Gamma_4}{\partial z} = \frac{i}{k}\nabla_{r_1}\nabla_{r_2}\Gamma_4 - \frac{k^2}{4} F(r_1, r_2, \varrho; z)\Gamma_4, \tag{3.140}$$

where

$$\Gamma_4(0, r_1, r_2, \varrho; 0) = |u_0|^4.$$

In addition, note that ϱ is just a parameter in (3.140) and may be set equal to zero, so that

$$\Gamma_4(0, r_1, r_2, 0; z)$$

$$= \left\langle E\left(\frac{r_1+r_2}{2}, z\right) E\left(-\frac{r_1+r_2}{2}, z\right) E^*\left(\frac{r_1-r_2}{2}, z\right) E^*\left(-\frac{r_1-r_2}{2}, z\right) \right\rangle.$$

Note that if we let $r_2 = 0$, then

$$\Gamma_4(0, r_1, 0, 0; z) = \left\langle I\left(\frac{r_1}{2}\right) I\left(-\frac{r_1}{2}\right) \right\rangle, \qquad (3.141)$$

which is the correlation function of the irradiance; clearly the covariance function is

$$B_I(r, z) = \langle I(r, z) I(-r, z) \rangle - \langle I(r, z) \rangle \langle I(-r, z) \rangle$$
$$= \Gamma_4(0, r, 0, 0; z) - \Gamma_2^2(0, 0, z). \qquad (3.142)$$

If we define a spectral expansion of Γ_4,

$$M(K, r_1; z) = \frac{1}{(2\pi)^2} \iint\limits_{-\infty}^{\infty} dr_2 \Gamma_4(0, r_1, r_2, 0; z) \exp(-iK \cdot r_2), \qquad (3.143)$$

and substitute (3.143) into (3.140), we get

$$\frac{\partial M(K, r; z)}{\partial z} + \frac{K}{k} V_r M(K, r; z) + \frac{\pi k^2}{2} H(r, z) M(K, r; z)$$

$$= \pi k^2 \iint\limits_{-\infty}^{\infty} dK' \Phi_\varepsilon(K', z) [1 - \cos(K' \cdot r)] M(K - K', r; z), \qquad (3.144)$$

where

$$H(r, z) = 2 \iint\limits_{-\infty}^{\infty} d^2 K \Phi_\varepsilon(K, z) [1 - \cos(K \cdot r)], \qquad (3.145)$$

and

$$M(K, r, 0) = |u_0|^4 \delta(K). \qquad (3.146)$$

Tatarskii [3.9] obtained the solution of this equation in the form

$$M(\boldsymbol{K}, \boldsymbol{r}, z) = M\left(\boldsymbol{K}, \boldsymbol{r} - \frac{\boldsymbol{K}z}{k}, 0\right) \exp\left\{-\frac{\pi k^2}{2} \int\limits_0^z H\left[\boldsymbol{r} - \frac{\boldsymbol{K}(z - \zeta)}{k}, \zeta\right] d\zeta\right\}$$

$$+ \int\limits_0^z \exp\left\{-\frac{\pi k^2}{2} \int\limits_{z'}^z H\left[\boldsymbol{r} - \frac{\boldsymbol{K}(z - \zeta)}{k}, \zeta\right] d\zeta\right\}$$

$$\cdot G\left[\boldsymbol{K}, \boldsymbol{r} - \frac{\boldsymbol{K}(z - z')}{k}, z'\right] dz', \tag{3.147}$$

where

$$G(\boldsymbol{K}, \boldsymbol{r}, z) = \pi k^2 \iint\limits_{-\infty}^{\infty} d^2 \boldsymbol{K}' \Phi_\varepsilon(\boldsymbol{K}', z) [1 - \cos(\boldsymbol{K}' \cdot \boldsymbol{r})] M(\boldsymbol{K} - \boldsymbol{K}', \boldsymbol{r}, z). \tag{3.148}$$

Equation (3.147) is an integral form of the radiative transfer equation. Note that since $G(\boldsymbol{K}, \boldsymbol{r}, z)$ is a function of $M(\boldsymbol{K}, \boldsymbol{r}, z)$, (3.148) is not a closed form solution.

The basic problem, then, is to find a solution of (3.140), (3.144), or (3.147), all of which represent the same problem. While (3.144) and (3.147) were derived by *Tatarskii* [3.9], an equation equivalent to (3.144) had also been derived by *Gochelashvily* and *Shishov* [3.84]. Since there are no analytical solutions presently known, and to date no good numerical techniques have been found, most authors have used an iterative technique in attempting to find a solution for the spectrum of the irradiance fluctuations. Attempts at solving (3.144) and (3.147) have been made by *Tatarskii* [3.9], *Gochelashvily* and *Shishov* [3.84, 85], and *Fante* [3.86]. The actual mathematics is based on an iterative solution to (3.144) or (3.147). *Tatarskii* [3.9] calculated the so-called single scattering approximation of the radiative transfer equation. *Gochelashvily* and *Shishov* [3.85] and *Fante* [3.86] carried the solution to a higher order, and, as shown by *Fante* [3.87], their solutions are equivalent to the first order at least. Following *Fante* [3.86, 87], and recalling that we are only considering the case of an initial plane wave,

$$\Phi(\boldsymbol{K}) \equiv M(\boldsymbol{K}; 0, z) = \delta(\boldsymbol{K}) + \pi k^2 \int\limits_0^z dz' \Phi_\varepsilon(\boldsymbol{K}, z') \left[1 - \cos\frac{K^2(z - z')}{k}\right]$$

$$\cdot \exp\left\{-\frac{\pi k^2}{2} \int\limits_{z'}^z H\left[\frac{\boldsymbol{K}(z - \zeta)}{k}, \zeta\right] d\zeta - \frac{\pi k^2}{2} \int\limits_0^{z'} H\left[\frac{\boldsymbol{K}(z - z')}{k}, \zeta\right] d\zeta\right\}$$

$$+ \frac{k^2}{16\pi} \int\limits_0^z dz' \exp\left\{-\frac{\pi k^2}{2} \int\limits_{z'}^z d\zeta H\left[\frac{\boldsymbol{K}(z - \zeta)}{k}, \zeta\right]\right\} Y(\boldsymbol{K}, z, z'), \tag{3.149}$$

where $H(\mathbf{r}, z)$ is given by (3.145), and

$$Y(\mathbf{K}, z, z') = \iint\limits_{-\infty}^{\infty} d^2 r_2 |\Gamma_2(\mathbf{r}_2, z')|^2 \exp(i\mathbf{K} \cdot \mathbf{r}_2)$$

$$\cdot \left\{ H\left[\mathbf{r}_2 - \frac{\mathbf{K}(z-z')}{k}, z'\right]\right.$$

$$\left. + H\left[\mathbf{r}_2 + \frac{\mathbf{K}(z-z')}{k}, z'\right] - 2H[\mathbf{r}_2, z']\right\}. \tag{3.150}$$

Based on a Kolmogrov spectrum, for $l_0 < r < L_0$,

$$H(\varrho) = 0.47 C_\varepsilon^2 \varrho^{5/3},$$

and if σ_1^2 is sufficiently large,

$$Y(\mathbf{K}, z, z') \simeq 2H\left[\frac{\mathbf{K}(z-z')}{k}, z\right] \iint\limits_{-\infty}^{\infty} d^2 r |\Gamma_2(\mathbf{r}, z')|^2 \exp(i\mathbf{K} \cdot \mathbf{r}),$$

where

$$|\Gamma_2(\mathbf{r}, z')| = \exp\left[-\frac{\pi k^2}{4} \int\limits_0^{z'} H(\mathbf{r}, \zeta) d\zeta\right].$$

Gochelashvili and *Shishov* [3.84, 85] and *Fante* [3.86, 87] showed that in the low-frequency region, for $K^2 z/k \ll 1$,

$$M_1(\mathbf{K}, z) \simeq \delta(\mathbf{K}) + \frac{K^4}{k^2 z} \Phi_s(\mathbf{K}, z) \int\limits_0^z dz'(z - z')^2$$

$$\cdot \exp\left[-\frac{\pi}{2} k^2 z H\left(\frac{\mathbf{K}z}{k}, z\right)\left(1 - \frac{z'}{z}\right)^{5/3}\left(\frac{5}{8}\frac{z'}{z} + \frac{3}{8}\right)\right], \tag{3.151}$$

where $\Phi_s(\mathbf{K}, z)$ is the spectrum of the phase fluctuations at a distance z computed in the geometrical optics approximation, i.e.,

$$\Phi_s(\mathbf{K}, z) = \frac{\pi k^2 z}{2} \Phi_\varepsilon(\mathbf{K}). \tag{3.152}$$

In the high-frequency region of the spectrum, they showed that

$$M_h(\mathbf{K}, z) = \frac{1}{(2\pi)^2} \iint\limits_{-\infty}^{\infty} d\mathbf{r} \exp\left[-\frac{\pi}{2} k^2 z H(\mathbf{r}, z)\right] \exp(i\mathbf{K} \cdot \mathbf{r}). \tag{3.153}$$

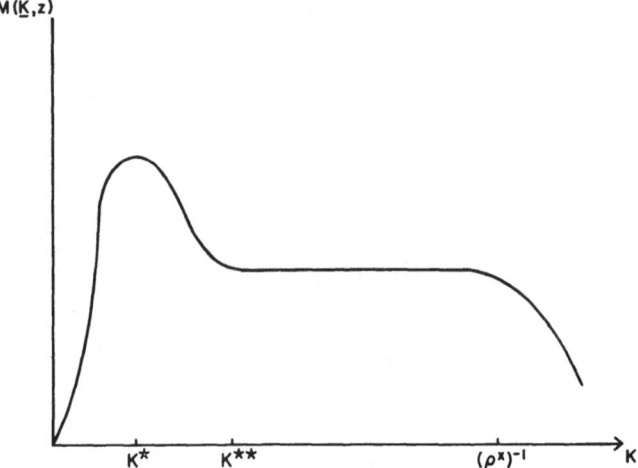

Fig. 3.15. A qualitative description of the two-dimensional spectrum of the irradiance scintillations in the saturation region [3.2, 85]

In Fig. 3.15 is shown a qualitative representation of $M(K, z)$ taken from *Prokhorov* et al. [3.2]. In this figure,

$$K^* = [(0.73)^{3/5} (C_\varepsilon^2)^{3/5} k^{1/5} z^{8/5}]^{-1},$$
$$K^{**} = K^* [\ln(0.43 C_\varepsilon^4 k^{7/3} z^{11/3})]^{3/5}, \tag{3.154}$$
$$(\varrho^*)^{-1} = 0.83 (C_\varepsilon^2)^{3/5} k^{6/5} z^{3/5}.$$

Fante also calculated the spectrum including his second iteration [3.90] and found three peaks in the spectrum at

$$K_1 \propto (C_\varepsilon^2)^{-3/5} k^{-1/5} z^{-8/5},$$
$$K_m = 0.41 (C_\varepsilon^2)^{3/11} k^{9/11},$$
$$K_h = 1.11 (C_\varepsilon^2)^{3/5} k^{6/5} z^{3/5}.$$

Note that *Fante's* upper and lower frequencies agree with *Prokhorov* et al.

Gracheva et al. [3.91] made a comparison of the effectiveness of the high-frequency scales $(\varrho^*)^{-1}$ or K_h proposed by *Prokhorov* et al. and *Fante* versus the scale l_T^{-1} proposed by Gracheva et al. (Chap. 4) and argued that the scale l_T^{-1} gives a much better fit to the experimental data. This result is demonstrated in Fig. 3.16. In Fig. 3.16a the abscissa is $\log\Omega_T$, where $\Omega_T = \omega l_T/v_\perp$ and $l_T = (0.033\pi C_\varepsilon^2 k^3)^{-3/11}$; in Fig. 3.16b the abscissa is $\log\Omega_c$, where $\Omega_c = \omega\varrho^*/v_\perp$. As can be seen from comparing the two figures, normalizing by l_T provides a much better fit than normalizing by ϱ^*. It is clear that further experimental and theoretical work is needed in this area in order to sort out the critical parameters. For a further discussion of this topic see [3.92].

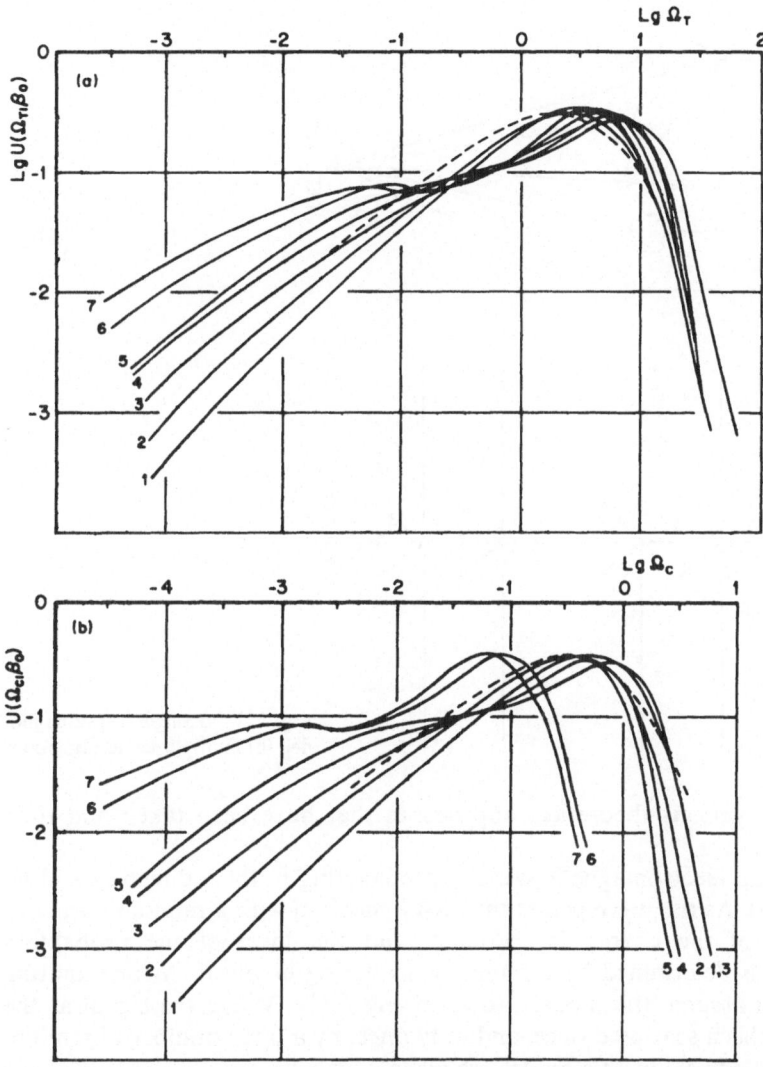

Fig. 3.16a and b. Experimental measurements of the frequency spectrum of irradiance fluctuations (see Chap. 4) as a function of the dimensionless frequency and β_0: (a) normalized by $\Omega_T = \omega l_T/v$; (b) normalized by $\Omega_c = \omega l_c/v_\perp$: $1\ \beta_0 = 0.95$, $2\ \beta_0 = 1.6$, $3\ \beta_0 = 2.8$, $4\ \beta_0 = 4.9$, $5\ \beta_0 = 7$, $6\ \beta_0 = 27$, $7\ \beta_0 = 35$

3.6.4 The Probability Distribution of the Irradiance

There has been considerable theoretical and experimental interest in the probability distribution of the irradiance fluctuations, particularly in the saturation region. *Strohbehn* et al. [3.93] recently reviewed the state of the art, and this section will closely follow part of that paper. First we shall discuss some physical arguments for various probability distributions, and then

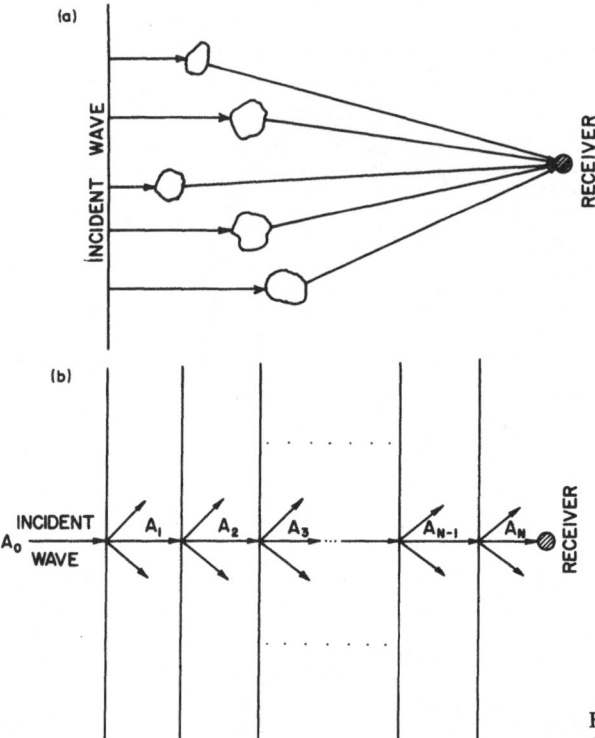

Fig. 3.17. (a) Single scattering model, (b) multiple scattering model

consider the various theoretical approaches that have been taken and their results.

Assume an electromagnetic wave is propagating in the z direction and let $E = u \exp(ikz)$. As the wave passes through a medium with a random refractive index, part of the energy is scattered, and the form of the probability distribution is determined by the type of scattering involved. As one limiting case, we can assume the model shown in Fig. 3.17a, where the signal at the receiver has been scattered once, and only once, by a large number of random scatterers, or eddies. In this model, we write

$$u = v_d + v_r + iv_i,\tag{3.155}$$

where v_d is the average field and v_r and v_i are random variables. If v_r and v_i are the components of the field resulting from the sum of a large number of random scatterings, then by the central limit theorem, v_r and v_i should be normal random variables. Furthermore, by writing down the Born expansion for u and assuming weak scattering, it can be shown that v_r and v_i are uncorrelated and have equal variances. Letting A represent the amplitude of the wave and assuming v_d is real,

$$A = |E| = (v_d^2 + 2v_d v_r + v_r^2 + v_i^2)^{1/2},\tag{3.156}$$

and it may be shown that A has a Rice-Nakagami probability distribution [3.94, 95], i.e.,

$$p(A) = \frac{2A}{\sigma^2} \exp\left(\frac{A^2 + v_d^2}{\sigma^2}\right) I_0\left(\frac{2Av_d}{\sigma^2}\right), \tag{3.157}$$

where $\sigma^2 = \langle v_r^2 \rangle + \langle v_i^2 \rangle = 2\langle v_r^2 \rangle$ and I_0 is a modified Bessel function. Furthermore,

$$\langle I \rangle = \langle A^2 \rangle = v_d^2 + \sigma^2 .$$

If $v_d = 0$, (3.157) reduces to a Rayleigh distribution,

$$p(A) = \frac{2A}{\sigma^2} \exp(-A^2/\sigma^2). \tag{3.158}$$

Since we normally work with irradiance I, the Rayleigh distribution in amplitude corresponds to an exponential distribution in intensity,

$$p(I) = \frac{1}{\sigma^2} \exp(-I/\sigma^2). \tag{3.159}$$

In Fig. 3.18 are shown plots of the Rice-Nakagami distribution for different values of $a = v_d^2/\sigma^2$. The coordinates in this figure are chosen to agree with the way much of the experimental data is displayed. The ordinate is $(\ln I - \langle \ln I \rangle)/\sigma_{\ln I}$, while the scale on the abscissa is chosen so that if the distribution of the irradiance follows the log normal, the cumulative probability distribution will be a straight line.

The above single scattering model is probably most appropriate where the receiver is located some distance from the scattering medium. When the receiver is in the scattering medium, which is the physical situation implicitly assumed throughout most of this chapter, the scattering is really multiplicative, i.e., the signal is first scattered from one eddy, then another, etc. This situation can be modeled by considering the turbulent medium as composed of a large number of independent slabs (see Fig. 3.17b). If we assume the slabs are oriented perpendicular to the direction of propagation, and that the thickness of each slab is large compared to the outer scale of turbulence, then the received field is the product of the incident field times a number of independent multiplicative terms. In this case

$$u = \prod_{i=1}^{n} M_i u_0 , \tag{3.160}$$

where the M_i are independent random variables. By taking the logarithm of (3.160) and applying the central limit theorem, we predict that u follows a log normal probability distribution [3.25]. Since the method of smooth per-

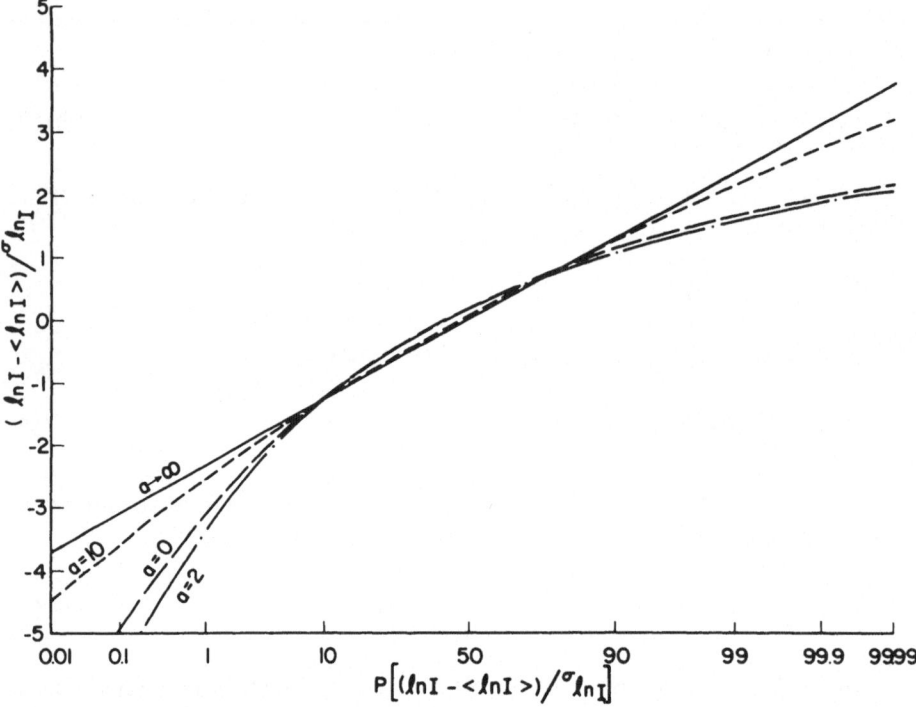

Fig. 3.18. Cumulative probability distribution of a Rice-Nakagami distribution

turbations (Rytov's method) is a perturbation technique applied to a logarithmic transformation of the original field, it includes multiple scattering effects, and is consistent with the assumptions necessary for a log normal distribution. In this case we let

$$u = \exp(\bar{\chi} + \chi_1 + i\bar{S} + S_1), \tag{3.161}$$

where χ and S denote the log amplitude and phase, respectively, $\bar{\chi}$ and $\bar{S}$ are their average values, and χ_1 and S_1 are random variables with zero mean. The log irradiance is

$$\ln I = 2\ln A = 2\bar{\chi} + 2\chi_1, \tag{3.162}$$

and the log normal distribution implies that χ_1 has a normal distribution. In this case

$$p(\ln I) = \frac{1}{(2\pi)^{1/2}\sigma_{\ln I}} \exp[-(\ln I - \overline{\ln I})^2/2\sigma_{\ln I}^2], \tag{3.163}$$

$$p(I) = \frac{1}{(2\pi)^{1/2}I\sigma_{\ln I}} \exp[-(\ln I - \overline{\ln I})^2/2\sigma_{\ln I}^2]. \tag{3.164}$$

For the case of plane wave propagation in a nonabsorbing medium with negligible backscatter, conservation of energy implies $\langle I \rangle = $ const, and the constant is usually set equal to unity. In this case

$$\overline{\ln I} = -\frac{1}{2}\sigma_{\ln I}^2$$

and   (3.165)

$$\bar{\chi} = -\sigma_\chi^2 .$$

Given the coordinates chosen in Fig. 3.18, the log normal distribution corresponds to the straight line denoted by $a \to \infty$, i.e., a log normal distribution corresponds to a Rice-Nakagami distribution where $a = v_d^2/\sigma^2 \to \infty$.

Wang and *Strohbehn* [3.96–98] showed that in the saturation region, for small and intermediate values of β_0^2, neither the log normal nor the Rayleigh distribution can be strictly true, since the assumption for either distribution leads to the prediction that the log irradiance variance will go negative. *Bissonnette* [3.99] also showed that asymptotically the irradiance variance diverges to infinity under the assumption of the log normal distribution. *Wang* and *Strohbehn* [3.98] proposed a distribution which is a combination of the log normal and Rice-Nakagami, and showed that this distribution is not inconsistent with the experimental results for the variance.

The above models are all heuristic and are based essentially on plausibility arguments. Mathematically there are several approaches that will lead to knowledge about the probability distribution of a propagating wave. We mention here applying the central limit theorem, calculating moments, finding the characteristic functional, and trial and error.

The central limit theorem was used above in justifying certain distributions by plausibility arguments. However, there have been a great number of statistical theorems developed governing when the central-limit theorem can be used. However, these theorems are difficult to apply to propagation problems and are not used much in practice. The plausible reasoning approach has been used in the region of weak fluctuations [3.8, 9] with apparently good results and also in the region of strong fluctuations [3.64, 100, 101], but with more conflicting results.

A more straightforward way to determine the probability distribution is to first calculate the moments of the distribution. In many cases, if the moments are known, the characteristic function $\Phi(\omega)$ can be determined from

$$\Phi(\omega) = \langle i\omega I \rangle = 1 + i\omega M_1 + \frac{(i\omega)^2}{2!}M_2 + \frac{(i\omega)^3}{3!}M_3 + \dots ,   (3.166)$$

where $m_n = \langle I^n \rangle$. Given $\Phi(\omega)$, $p(I)$ can be determined from the Fourier transform

$$p(I) = \frac{1}{2\pi}\int_{-\infty}^{\infty} \Phi(\omega)\exp(-i\omega I)d\omega .   (3.167)$$

However, for this process to work, the moments must satisfy certain convergence conditions. For example, for a Rice-Nakagami distribution the moments are

$$\langle I^n \rangle = \exp(-v_d^2/\sigma^2)\sigma^{2n}{}_1F_1 \quad (n+1, 1; z), \tag{3.168}$$

where $z = v_d^2/\sigma^2$. For a Rayleigh distribution this reduces to

$$\langle I^n \rangle = n!\sigma^{2n}. \tag{3.169}$$

For a log normal distribution the moments are

$$\langle I^n \rangle = \exp\left[n\left(\overline{\ln I} + \frac{1}{2} n\sigma_{\ln I}^2\right)\right],$$

which reduces to

$$\langle I^n \rangle = \exp\left[\frac{1}{2} n(n-1)\sigma_{\ln I}^2\right] \tag{3.170}$$

for a plane wave. *Heyde* [3.102] showed that the log normal distribution is not uniquely determined by its moments. If the moments satisfy the Carleman condition, [3.103], i.e., if

$$\sum_{n=1}^{\infty} \langle I^{2n} \rangle^{-\frac{1}{2n}} = \infty, \tag{3.171}$$

then the probability distribution is uniquely determined by its moments. Using the above expressions it is easy to show that the Rayleigh distribution satisfies the Carleman condition while the log normal does not.

Several authors have used the moment method to try to determine the probability distribution [3.100, 103–107]. *De Wolf* used the renormalization or diagram techniques discussed in Section 3.2 to try to calculate the moments. This approach is very difficult mathematically because a large number of approximations must be made. In one of his earliest papers [3.104] he concluded that the amplitude fluctuations obeyed the Rice-Nakagami distribution, but in his later papers he predicted a log normal distribution, not only in the region of weak fluctuations, but also in the saturation region, provided that

$$(L/kL_0^2)\sigma_1^2 \ll 1 \ll \sigma_1^2$$
$$\sigma_1^2 = K_m^{1/3}kLC_n^2, \tag{3.172}$$

while if

$$(L/kL_0^2)\sigma_1^2 \gg 1,$$

he predicted a Rayleigh distribution. Note his first result is directly opposite to that of *Wang* and *Strohbehn* [3.97], and his general trend does not seem to agree with the experimental observations of Gracheva et al. (Chap. 4). However, as *Wang* and *Strohbehn* pointed out, very small deviations from the log normal can be very important. *Klyatskin* [3.103] also tried to calculate the moments for large values of β_0^2 and found

$$\lim_{\beta_0^2 \to \infty} \langle I^n(L, \varrho) \rangle = n. \tag{3.173}$$

However, *Prokhorov* et al. [3.2] pointed out that this result is impossible, since

$$\sigma_{I^n}^2 = \langle I^{2n} \rangle - \langle I^n \rangle^2 = 2n - n^2, \tag{3.174}$$

which is negative for $n > 2$.

In conclusion then, we can state that while the use of moments is attractive from a theoretical point of view, it is very difficult mathematically, and has only led to results that are either wrong or controversial.

The third technique mentioned is the use of characteristic functionals. This technique was applied by *Furutsu* [3.48] in the wave propagation problem. In this approach, a characteristic functional is defined as

$$Z = \langle \exp\{\int dr[\bar{\eta}(r)E(r) + \bar{\eta}^*(r)E^*(r)]\} \rangle, \tag{3.175}$$

and the probability distribution can be found from (3.167) where

$$\langle \Phi(\omega) \rangle = \langle \exp[iE(r)E^*(r)] \rangle$$
$$= \exp\{it[\delta/\delta\bar{\eta}(r)][\delta/\delta\bar{\eta}^*(r)]\} Z|_{\bar{\eta} = \bar{\eta}^* = 0}. \tag{3.176}$$

Furutsu used this approach to derive an expression for the probability distribution assuming a collimated gaussian beam and a refractive index structure function, $D_n(r) \sim r^2$. His results were discussed in some detail by *Furuhama* et al. [3.108], but are not applicable to the plane wave case. Difficulties with their results were discussed by *Strohbehn* et al. [3.93].

The last technique, which is not very elegant mathematically, is trial and error. In essence, it is a method which takes a distribution based on plausible reasoning and then carries out certain calculations to see if the predicted distribution fits the experimental data. This technique was applied by *Wang* and *Strohbehn* [3.96–98] to the wave propagation problem. They showed that if the probability distribution is either log normal or Rice-Nakagami, then the fourth-order coherence function, which as we have shown is very difficult to determine in general, can be determined from the mean field $\langle u \rangle$, the mutual coherence function $\langle uu^* \rangle$, and the quantity $\langle uu \rangle$. Since equations for the first two quantities have been determined analytically, and the differential equation for $\langle uu \rangle$ can be solved numerically, it is possible to determine the variance of the log amplitude, i.e.,

$$\sigma_\chi^2(z) = \frac{1}{2} \mathrm{Re}[\ln\langle u(0,z)u^*(0,z)\rangle] - 2\mathrm{Re}[\ln\langle u(0,z)\rangle].$$

In the saturation region, this quantity goes negative, hence proving that the log normal distribution is not possible. *Klyatskin* [3.109] came to similar conclusions, based on a similar idea.

It seems fair to say that at this time the best models for the probability distributions are coming from the experimental data.

References

3.1   Yu.N.Barabanenkov, Yu.A.Kravtsov, S.M.Rytov, V.I.Tatarskii: Soviet Phys.-Usp. **13**, 551 (1971)
3.2   A.M.Prokhorov, F.V.Bunkin, K.S.Gochelashvily, V.I.Shishov: Proc. IEEE **63**, 790 (1975); also in Usp. Fiz. Nauk **114**, 415 (1974) [English transl.: Sov. Phys.-Usp. **17**, 826 (1975)]
3.3   R.L.Fante: Proc. IEEE **63**, 1669 (1975)
3.4   M.A.Plonus, H.H.Su, C.S.Gardner: IEEE Trans. AP-**20**, 801 (1972)
3.5   L.W.Pickering, R.E.McIntosh: IEEE Trans. AP-**20**, 528 (1972)
3.6   C.S.Gardner, M.A.Plonus: J. Opt. Soc. Am. **64**, 68 (1974)
3.7   S.M.Rytov: Izv. Akad. Nauk SSSR (Ser. Fiz.) **2**, 223 (1937)
3.8   V.I.Tatarskii: *Wave Propagation in a Turbulent Medium* (McGraw-Hill, New York 1961) p. 285
3.9   V.I.Tatarskii: *The Effects of the Turbulent Atmosphere on Wave Propagation* (National Technical Information Service, Springfield, Va. 1971) p. 472
3.10   T.A.Shirokova: Akustich Zh. **5**, 485 (1959) [English transl.: Sov. Phys.-Acoust. **5**, 498 (1960)]
3.11   V.V.Pisareva: Akustich Zh. **6**, 87 (1960) [English transl.: Sov. Phys.-Acoust. **5**, 81 (1960)]
3.12   Y.L.Feinberg: *Propagation of Radiowaves Along the Earth's Surface* (Academy of Sciences, Moscow USSR 1961)
3.13   V.V.Pisareva: Izv. VUZ, Radiofiz. **4**, 376 (1961)
3.14   V.I.Tatarskii: Izv. VUZ, Radiofiz. **5**, 490 (1962)
3.15   R.E.Hufnagel, N.R.Stanley: J. Opt. Soc. Am. **54**, 52 (1964)
3.16   D.A.de Wolf: J. Opt. Soc. Am. **55**, 812 (1965)
3.17   C.E.Coulman: J. Opt. Soc. Am. **56**, 1232 (1966)
3.18   W.P.Brown,Jr.: J. Opt. Soc. Am. **56**, 1045 (1966)
3.19   D.L.Fried: J. Opt. Soc. Am. **57**, 268 (1967)
3.20   L.S.Taylor: Radio Sci. **2**, 437 (1967)
3.21   D.A.de Wolf: J. Opt. Soc. Am. **57**, 1057 (1967)
3.22   G.R.Heidbreder: J. Opt. Soc. Am. **57**, 1477 (1967)
3.23   W.P.Brown,Jr.: J. Opt. Soc. Am. **57**, 1539 (1967)
3.24   J.W.Strohbehn: J. Opt. Soc. Am. **58**, 139 (1968)
3.25   J.W.Strohbehn: Proc. IEEE **56**, 1301 (1968)
3.26   H.T.Yura: J. Opt. Soc. Am. **59**, 111 (1969)
3.27   J.B.Keller: J. Opt. Soc. Am. **59**, 1003 (1969)
3.28   M.I.Sancer, A.D.Varvatsis: Proc. IEEE **58**, 140 (1970)
3.29   K.Mano: Proc. IEEE **58**, 1168 (1970)

3.30 K.Mano: Proc. IEEE **58**, 1405 (1970)
3.31 R.C.Bourret: Nuovo Cimento **26**, 1 (1962)
3.32 R.C.Bourret: Can. J. Phys. **40**, 782 (1962)
3.33 R.P.Feynman: Phys. Rev. **76**, 749 (1949)
3.34 R.P.Feynman: Phys. Rev. **76**, 769 (1949)
3.35 F.Dyson: Phys. Rev. **75**, 1736 (1949)
3.36 E.E.Salpeter, H.A.Bethe: Phys. Rev. **84**, 1232 (1951)
3.37 V.I.Tatarskii: Zh. Eksper. I. Teor. Fiz. **56**, 2106 (1969) [English transl.: Sov. Phys.-JETP **29**, 1133 (1969)]
3.38 V.I.Klyatskin: Zh. Eksper. I. Teor. Fiz. **57**, 952 (1969) [English transl.: Sov. Phys.-JETP **30**, 520 (1970)]
3.39 V.I.Klyatskin, V.I.Tatarskii: Zh. Eksper. I. Teor. Fiz. **58**, 618 (1970) [English transl.: Sov. Phys.-JETP **31**, 332 (1970)]
3.40 V.I.Klyatskin: Izv. VUZ, Radiofiz. **13**, 1069 (1970) [English transl.: Radiophys. Quant. Electron. **13**, 834 (1970)]
3.41 L.A.Chernov: "The Diffraction of Waves", Rec. 3rd All-Union Symp. (Nauka, Moscow, USSR 1964) p. 224
3.42 V.I.Shishov: Proc. P. N. Lebedev Phys. Inst. **38**, 171 (1967)
3.43 L.S.Dolyn: Izv. VUZ, Radiofiz. **11**, 840 (1968)
3.44 L.A.Chernov: Acoustich. Zh. **15**, 554 (1969)
3.45 M.J.Beran: J. Opt. Soc. Am. **56**, 1475 (1966)
3.46 M.J.Beran: IEEE Trans. AP-**15**, 66 (1967)
3.47 M.J.Beran: J. Opt. Soc. Am. **58**, 431 (1968)
3.48 K.Furutsu: J. Opt. Soc. Am. **62**, 240 (1972)
3.49 V.I.Klyatskin: Izv. VUZ, Radiofiz. **16**, 1629 (1973)
3.50 V.I.Tatarskii, M.E.Gertsenshtein: Zh. Eksper. I. Teor. Fiz. **44**, 676 (1963) [English transl.: Sov. Phys.-JETP **17**, 458 (1963)]
3.51 V.I.Tatarskii: Zh. Eksper. I. Teor. Fiz. **46**, 1399 (1964) [English transl.: Sov. Phys.-JETP **19**, 946 (1964)]
3.52 W.P.Brown,Jr.: IEEE Trans. AP-**15**, 81 (1967)
3.53 W.P.Brown,Jr.: In *Modern Optics*, ed. by J. Fox (Polytechnic Inst. Press, Brooklyn 1967) pp. 717–742
3.54 J.Molyneux: J. Opt. Soc. Am. **61**, 248 (1971)
3.55 V.I.Shishov: Izv. VUZ, Radiofiz. **11**, 866 (1968) [English transl.: Radiophys. Quant. Electron. **11**, 500 (1968)]
3.56 J.Molyneux: J. Opt. Soc. Am. **58**, 951 (1968)
3.57 V.I.Tatarskii: Izv. VUZ, Radiofiz. **17**, 570 (1974) [English transl.: Radiophys. Quant. Electron. **17**, 429 (1974)]
3.58 K.Furutsu: J. Res. Natl. Bur. Std. **67D**, 303 (1963)
3.59 E.A.Novikov: Zh. Eksper. I. Teor. Fiz. **47**, 1919 (1964) [English transl.: Sov. Phys.-JETP **20**, 1290 (1965)]
3.60 L.S.Dolin: Izv. VUZ, Radiofiz. **2**, 840, (1968)
3.61 V.I.Klyatskin, V.I.Tatarskii: Izv. VUZ, Radiofiz. **15**, 1437 (1972)
3.62 T.L.Ho, M.J.Beran: J. Opt. Soc. Am. **58**, 1335 (1968)
3.63 M.J.Beran, T.L.Ho: J. Opt. Soc. Am. **59**, 1134 (1969)
3.64 S.F.Clifford, G.R.Ochs, R.S.Lawrence: J. Opt. Soc. Am. **64**, 148 (1974)
3.65 S.F.Clifford, H.T.Yura: J. Opt. Soc. Am. **64**, 1641 (1974)
3.66 H.T.Yura: J. Opt. Soc. Am. **64**, 59 (1974)
3.67 H.T.Yura: J. Opt. Soc. Am. **64**, 1526 (1974)
3.68 H.T.Yura: J. Opt. Soc. Am. **64**, 357 (1974)
3.69 H.T.Yura: J. Opt. Soc. Am. **64**, 1211 (1974)
3.70 A.T.Young: J. Opt. Soc. Am. **60**, 1495 (1970)
3.71 V.I.Tatarskii: Izv. VUZ, Radiofiz. **10**, 1 (1967)
3.72 G.R.Ochs, R.R.Bergman, J.R.Snyder: J. Opt. Soc. Am. **59**, 231 (1969)

3.73 A.S.Gurvich, M.A.Kallistratova: Izv. VUZ, Radiofiz. **11**, 66 (1968) [English transl.: Radiophys. Quant. Electron. **11**, 37 (1968)]

3.74 G.K.Born, R.Bogenberger, K.D.Erben, F.Frank, F.Mohr, G.Sepp: Appl. Opt. **14**, 2857 (1975)

3.75 R.F.Lutomirski, H.T.Yura: Appl. Opt. **10**, 1652 (1971)

3.76 H.T.Yura: Appl. Opt. **10**, 2771 (1971)

3.77 D.L.Fried, H.T.Yura: J. Opt. Soc. Am. **62**, 600 (1972)

3.78 J.B.Keller: In "Proceedings of the 13th Symposium in Appl. Math." (American Mathematical Society, Providence 1962) p. 227

3.79 J.B.Keller: In "Proceedings of the 17th Symposium in Appl. Math." (American Mathematical Society, Providence 1964) p. 145

3.80 F.C.Karal,Jr., J.B.Keller: J. Math. Phys. **5**, 537 (1964)

3.81 W.P.Brown,Jr.: J. Opt. Soc. Am. **62**, 45 (1972)

3.82 W.P.Brown,Jr.: J. Opt. Soc. Am. **62**, 966 (1972)

3.83 K.S.Gochelashvily, V.I.Shishov: Opt. Acta **18**, 313 (1971)

3.84 K.S.Gochelashvily, V.I.Shishov: Opt. Acta **18**, 767 (1971)

3.85 K.S.Gochelashvily, V.I.Shishov: Zh. Eksper. I. Teor. Fiz. **66**, 1237 (1974) [English transl.: Sov. Phys.-JETP **39**, 605 (1974)]

3.86 R.L.Fante: Radio Sci. **10**, 77 (1975)

3.87 R.L.Fante: Radio Sci. **11**, 215 (1976)

3.88 R.L.Fante: J. Opt. Soc. Am. **65**, 548 (1975)

3.89 R.L.Fante: IEEE Trans. AP-**23**, 382 (1975)

3.90 R.L.Fante: Private communication

3.91 M.E.Gracheva, A.S.Gurvich, S.S.Kashkarov, V.V.Pokasov: Zh. Eksper. Teor. Fiz. **67**, 2035 (1974) [English transl.: Sov. Phys.-JETP **40**, 1011 (1975)]

3.92 A.S.Gurvich, V.I.Tatarskii: Radio Sci. **10**, 3 (1975)

3.93 J.W.Strohbehn, T.-i.Wang, J.P.Speck: Radio Sci. **10**, 59 (1975)

3.94 P.Beckmann: *Probability in Communication Engineering* (Harcourt, Brace, and World, New York 1967) p. 511

3.95 K.A.Norton, L.E.Vogler, W.V.Mansfield, P.J.Short: Proc. IRE **43**, 1354 (1955)

3.96 J.W.Strohbehn, T.-I.Wang: J. Opt. Soc. Am. **62**, 1061 (1972)

3.97 T.-I.Wang, J.W.Strohbehn: J. Opt. Soc. Am. **64**, 583 (1974)

3.98 T.-I.Wang, J.W.Strohbehn: J. Opt. Soc. Am. **64**, 994 (1974)

3.99 L.R.Bissonnette: 'Log-normal probability distribution of strong irradiance fluctuations: an asymptotic analysis", Paper 183 AGARD Conf. on Optical Propagation in the Atmosphere, Copenhagen (27–30 Oct., 1975)

3.100 D.A.de Wolf: J. Opt. Soc. Am. **64**, 360 (1974)

3.101 D.A.de Wolf: Proc. IEEE **62**, 1523 (1974)

3.102 C.C.Heyde: J. Roy. Statist. Soc. B **25**, 392 (1963)

3.103 V.I.Klyatskin: Zh. Eksper. I. Teor. Fiz. **60**, 1300 (1971) [English transl.: Sov. Phys.-JETP **33**, 703 (1971)]

3.104 D.A.de Wolf: J. Opt. Soc. Am. **58**, 461 (1968)

3.105 D.A.de Wolf: J. Opt. Soc. Am. **59**, 1455 (1969)

3.106 D.A.de Wolf: J. Opt. Soc. Am. **63**, 171 (1973)

3.107 D.A.de Wolf: J. Opt. Soc. Am. **63**, 1249 (1973)

3.108 Y.Furuhama, Y.Masuda, T.Shinozuka, M.Fukushima: Electron. Commun. Jap. **56**-B, 50 (1973)

3.109 V.I.Klyatskin: Izv. VUZ, Radiofiz. **4**, 540 (1972)

4. Similarity Relations and Their Experimental Verification for Strong Intensity Fluctuations of Laser Radiation

M. E. Gracheva, A. S. Gurvich, S. S. Kashkarov, and Vl. V. Pokasov
Translated and adapted by J. W. Strohbehn

With 13 Figures

The results of experimental investigations of the statistical characteristics of strong fluctuations of the irradiance of a laser beam when propagating through the turbulent atmosphere over paths of 250, 1750, and 8500 m are presented. Similarity relations for the fourth-order coherence function of a light wave are derived, and the results of verifying the similarity relations on the basis of experimental data are given.

The experimental data obtained in this work confirm that the covariance function and the frequency spectrum of the irradiance fluctuations appear to be a function of two dimensionless scales: a longitudinal scale $L_T = (\pi A C_\varepsilon^2 k^{7/6})^{-6/11}$ and a transverse scale $l_T = (L_T/k)^{1/2}$, while the probability distribution and variance are a function only of the scale L_T.

We also note that changes in the geometry of the initial laser beam have practically no effect on the statistical characteristics for strong fluctuations.

4.1 Background

Strong fluctuations in the irradiance of a laser beam, arising from turbulence on paths extending some hundreds of meters or more in the earth's atmosphere, are the object of both theoretical and experimental investigations. At the present time it has been reliably established experimentally that on sufficiently long paths there exists a "saturation" region for the irradiance fluctuations [4.1, 2]. Questions about the form of the probability distribution of the fluctuations and their correlation properties appear more complex. In the theory of "strong fluctuations" an equation for the fourth-order coherence function of a light wave propagating in a turbulent medium is obtained [4.3–6]. However, an analytical solution of this equation in the region of strong fluctuations has not been found, and therefore the necessity arises of finding similarity relations, which would permit generalizing the experimental data and obtaining results suitable for practical estimates.

In this work similarity relations are deduced and experimental data is presented along with the results of verifying the relationships.

4.2 Derivation of the Similarity Formulas

4.2.1 The Fourth-Order Coherence Function

Consider the fourth-order coherence function,

$$\Gamma_4(\varrho_1, \varrho_2, \varrho_3, \varrho_4, z) = \langle u(\varrho_1, z) u(\varrho_2, z) u^*(\varrho_3, z) u^*(\varrho_4, z) \rangle,$$

of a plane wave $u(\varrho, z)$, where z is the coordinate in the direction of propagation of the unperturbed wave u_0, and ϱ_j lies in the plane perpendicular to the z-axis. In [4.3–6] an equation is obtained, which for the case when

$$\varrho_3 - \varrho_1 = \varrho_2 - \varrho_4 = r_1 \quad \text{and} \quad \varrho_4 - \varrho_1 = \varrho_2 - \varrho_3 = r_2$$

reduces to the form

$$\frac{\partial \Gamma_4}{\partial z} = \frac{i}{k} \nabla_{r_1} \nabla_{r_2} \Gamma_4 - \frac{\pi k^2}{4} F(r_1, r_2, z) \Gamma_4, \tag{4.1}$$

where $k = 2\pi/\lambda$, λ is the wavelength, and the function $F(r_1, r_2, z)$ is expressed in terms of the three-dimensional spectrum of the fluctuations of the dielectric constant $\Phi_\varepsilon(\kappa, z)$.

$$F(r_1, r_2, z) = 4 \int_{-\infty}^{\infty} \Phi_\varepsilon(\kappa, z) [1 - \cos(\kappa \cdot r_1)] [1 - \cos(\kappa \cdot r_2)] d^2\kappa. \tag{4.2}$$

In order to derive (4.1) it is assumed that the turbulence is locally isotropic. If for $\Phi_\varepsilon(\kappa)$ we use the Kolmogorov spectrum,

$$\Phi_\varepsilon(\kappa) = AC_\varepsilon^2 |\kappa|^{-11/3}, \tag{4.3}$$

where C_ε^2 is the structure coefficient for the dielectric constant fluctuations and $A = 0.033$ then (4.1) may be written in the form

$$\frac{\partial \Gamma_4}{\partial z} = \frac{i}{k} \nabla_{r_1} \nabla_{r_2} \Gamma_4$$

$$- \left[\pi A k^2 C_\varepsilon^2 \int_{-\infty}^{\infty} |\kappa|^{-11/3} (1 - \cos\kappa \cdot r_1)(1 - \cos\kappa \cdot r_2) d^2\kappa \right] \Gamma_4. \tag{4.4}$$

Despite the fact that an analytical solution of (4.4) in explicit form has not been obtained at the present time, we may raise the following question: Do there not exist some characteristic scales of this problem which may be determined directly from the above equation? Having determined such scales and processed suitable experimental data, we may verify empirically the similarity relations for the fourth-order coherence function $\Gamma_4(r_1, r_2, z)$.

4.2.2 The Similarity Relations

From the same statement of the problem for the derivation of (4.1), based on
- using the parabolic equation of Leontovich
- assuming a locally isotropic turbulent medium

the boundary condition $\Gamma_4(r_1, r_2, 0) = |u_0|^4$,
corresponding to a plane wave, it is reasonable to assume that there must exist, at least, two characteristic scales: a longitudinal scale L_T and a transverse scale l_T. Let us determine these scales from (4.4) by introducing the new variables

$$z = L_T \xi, r_1 = l_T \eta_1, r_2 = l_T \eta_2, \tag{4.5}$$

and transforming (4.4) to dimensionless coordinates,

$$\frac{\partial \Gamma_4}{L_T \partial \xi} = \frac{i}{k l_T^2} \nabla_{\eta_1} \nabla_{\eta_2} \Gamma_4 - \pi A k^2 C_\varepsilon^2 l_T^{5/3}$$

$$\cdot \left[\int_{-\infty}^{\infty} |\kappa|^{-11/3} (1 - \cos\kappa \cdot \eta_1)(1 - \cos\kappa \cdot \eta_2) d^2\kappa \right] \Gamma_4. \tag{4.6}$$

If we use the relations

$$l_T = (L_T/k)^{1/2}, l_T = (\pi A C_\varepsilon^2 k^3)^{-3/11},$$
$$L_T = (\pi A C_\varepsilon^2 k^{7/6})^{-6/11}, \tag{4.7}$$

then (4.6) becomes

$$\frac{\partial \Gamma_4}{\partial \xi} = i \nabla_{\eta_1} \nabla_{\eta_2} \Gamma_4 - \left[\int_{-\infty}^{\infty} |\kappa'|^{-11/3} (1 - \cos\kappa' \cdot \eta_1)(1 - \cos\kappa' \cdot \eta_2) d^2\kappa' \right] \Gamma_4, \tag{4.8}$$

which does not contain any dimensional parameters and its solution will be a universal function,

$$\Gamma_4(\eta_1, \eta_2, \xi) = \Gamma_4 \left(\frac{r_1}{l_T}, \frac{r_2}{l_T}, \frac{z}{L_T} \right). \tag{4.9}$$

4.2.3 Physical Meaning of L_T and l_T

Notice that the scales L_T and l_T have a clear physical meaning. Setting $z = L_T$ in the expression for the variance of the normalized irradiance fluctuations calculated by the method of smooth perturbations (MSP)[1] [4.7],

$$\beta_0^2 = 0.31 C_\varepsilon^2 k^{7/6} z^{11/6}, \tag{4.10}$$

---

[1] MSP also called Rytov's method in the Western Literature.

we find that the scale L_T corresponds to the distance for which $\beta_0^2 \simeq 0.75$, i.e., a distance for which the calculation based on the MSP is definitely not valid [4.7], as has been confirmed experimentally [4.1,2].

The transverse scale l_T corresponds to a quantity of the order of the radius of the first Fresnel zone for a distance L_T. A calculation using the approximation of a Markov random process [4.3, 8, 9] shows that for such a scale the mutual coherence function $\Gamma_2(\varrho, z) = \langle u(0, z)u^*(\varrho, z)\rangle$ has appreciably fallen off $[\Gamma_2(l_T, L_T)/|u_0|^2 \simeq 0.03]$ and therefore a purely geometrical definition of a Fresnel zone specified for free space loses meaning for a turbulent medium with $(z/L_T) \gg 1$.

4.2.4 Comparison with Experiment

Equation (4.9) is the basis for the experimental verification of the similarity of the scales L_T and l_T. We notice that (4.1) was derived for describing strong fluctuations, when perturbation methods are not applicable, and consequently it is advantageous to use the scales L_T and l_T for solving the problem of "strong fluctuations", when $\beta_0^2 \gg 1$ or $z/L_T = \beta^{6/11} \gg 1$.

In the present work we obtained from the experiments values of the normalized variance of the fluctuations[2], β^2:

$$\beta^2 = \frac{\Gamma_4(0,0,z) - [\Gamma_2(0,z)]^2}{[\Gamma_2(0,z)]^2} = \frac{\Gamma_4(0,0,z)}{u_0^4} - 1 \tag{4.11}$$

and the covariance function of the irradiance

$$b_I(\varrho) = \frac{\Gamma_4(\varrho,0,z) - [\Gamma_2(0,z)]^2}{\Gamma_4(0,0,z) - [\Gamma_2(0,z)]^2}. \tag{4.12}$$

The frequency spectrum $U(\omega) = \omega W(\omega)$ under the assumption of "frozen turbulence" is related to the spatial spectrum $\kappa V(\kappa)$ by

$$\omega W(\omega) = \kappa V(\kappa), \tag{4.13}$$

[$W(\omega)$ is the spectral density of the irradiance fluctuations]. In turn the spatial spectrum $\kappa V(\kappa)$ is determined through the covariance function $b_I(\varrho)$,

$$\kappa V(\kappa) = \frac{\kappa}{\pi} \int_0^\infty b(\varrho') \cos(\kappa \varrho')d\varrho'. \tag{4.14}$$

From (4.8, 11–14) it is reasonable to expect

$$\beta^2 = f_\beta(z/L_T), \tag{4.15}$$

$$b_I(\varrho) = f_b(\varrho/l_T, z/L_T), \tag{4.16}$$

$$U(\omega) = f_u(\omega l_T/v, z/L_T),$$

---

[2] Often called the scintillation index.

where the functions f_β, f_b, f_u are to be determined from experimental data. We note that for comparison with previous experiments sometimes it is more convenient to use the parameter $\beta_0^2 = 3(z/L_T)^{11/6}$ instead of the dimensionless length z/L_T and take into account the obvious connection between the scales l_T and $(z/k)^{1/2}$,

$$l_T = \left(\frac{z}{k}\frac{L_T}{z}\right)^{1/2} = 1.35\beta_0^{-6/11}(z/k)^{1/2}. \tag{4.17}$$

It also should be noticed that the scale l_T was obtained in [4.3] in an approximate calculation of the spectrum of the irradiance fluctuations.

4.2.5 Similarity Relations for the Probability Distribution and Moments

For a Kolmogorov spectrum (4.3) and a plane wave $u(\varrho,0)=u_0$, it is possible to obtain a more general result for the form of the coherence function,

$$\Gamma_{2n} = M_{n,n} = \langle u(\varrho_1,z)u(\varrho_2,z)...u(\varrho_n,z)u^*(\varrho_1',z)u^*(\varrho_2',z)...u^*(\varrho_n',z)\rangle. \tag{4.18}$$

If we use the equation for $M_{n,n}$ obtained in [4.3], then

$$\frac{\partial M_{n,n}}{\partial z} = \frac{i}{2k}(\Delta_1 + \Delta_2 + ... + \Delta_n - \Delta_1' - \Delta_2' - ... - \Delta_n')M_{n,n}$$

$$- \frac{k^2}{8}Q_{n,n}M_{n,n}, \tag{4.19}$$

where Δ_j and Δ_l' are the transverse Laplacians in the coordinates ϱ_j and ϱ_l', respectively, and

$$Q_{n,n} = \pi \sum_{j=1}^{n}\sum_{l=1}^{n}[H(\varrho_j - \varrho_l) - H(\varrho_j - \varrho_l') + H(\varrho_j' - \varrho_l') - H(\varrho_l - \varrho_j')],$$

$$H(\varrho) = 2\int_{-\infty}^{\infty}(1 - \cos\boldsymbol{\kappa}\cdot\varrho)\Phi_\varepsilon(\boldsymbol{\kappa})d^2\kappa.$$

If the new variables

$$z = L_T\xi, \varrho_j = l_T\eta_j, \varrho_l = l_T\eta_l$$

are introduced, then (4.19) will not contain any dimensional parameters. Furthermore, for the case when all points are set equal, $\varrho_j = \varrho_l' = \varrho$, then

$$\Gamma_{2n} = \langle[u(\varrho,z)u^*(\varrho,z)]^n\rangle = \langle[I(\varrho,z)]^n\rangle,$$

where $I(\varrho,z)$ is the light irradiance.

For a plane wave and locally isotropic turbulence $\langle [I(\varrho, z)]^n \rangle$ does not depend on ϱ and consequently

$$\frac{\langle [I(\varrho, z)]^n \rangle}{[\langle I(z) \rangle]^n} = f_{I,n}(z/L_T); f_{I,1} = 1; f_{I,2} = 1 + f_\beta. \tag{4.20}$$

Since all moments for the normalized irradiance fluctuations depend only on the dimensionless length z/L_T, it follows that we should expect that the probability distribution for the random quantity $I/\langle I \rangle$ also is determined only by the single parameter z/L_T, or equivalently, β_0^2.

4.2.6 Limitations in Comparing Theory and Experiment

Formulas analogous to (4.9) and (4.15)–(4.17) may be obtained also for the spherical wave case. However, for the experiments that were carried out we did not realize, strictly speaking, the limiting cases of either a plane or a spherical wave. We are always limited in practice to a bounded beam. We shall not write down here in place of (4.1) the more complex equation, suitable for the bounded beam, but notice that the number of parameters on which the coherence functions depend includes in addition the characteristic radius of the beam, α_0, its divergence, θ, etc.

The study of such a multiparameter problem for bounded beams is extremely complicated. However, we may expect that for measurements close to the beam axis and with sufficiently wide, collimated beams, for which the Fresnel number $k\alpha_0^2/z \gg 1$, the results will not be strongly affected by the beam's finite size. For diverging and especially for focused beams it is necessary to carry out a special investigation.

In the experiments carried out in order to verify the similarity relations, measurements were performed primarily with a collimated beam for which the Fresnel number was not less than 26. In addition, a small number of experiments were carried out with a weakly diverging beam, $\theta = 1 \text{ mrad}$, and with a focused beam.

4.3 The Experimental Plan and Measurement Procedures

In Fig. 4.1 a block diagram for the fluctuation measurements is shown. The measurements were made at Tsimlyansk at two path lengths, $z = L_1 = 1750 \text{ m}$ and $z = L_2 = 8500 \text{ m}$, and also at Zvenigorod at a path length of 650 m. The beam from a helium-neon laser ($\lambda = 0.63 \text{ nm}$), operating in a single axial mode, was diverged through a collimator with an exit diameter of 0.5 m. Measurements were carried out with a collimated beam with an effective radius $\alpha_0 = 15 \text{ cm}$ at path lengths of 650, 1750, and 8500 m and also with a diverging

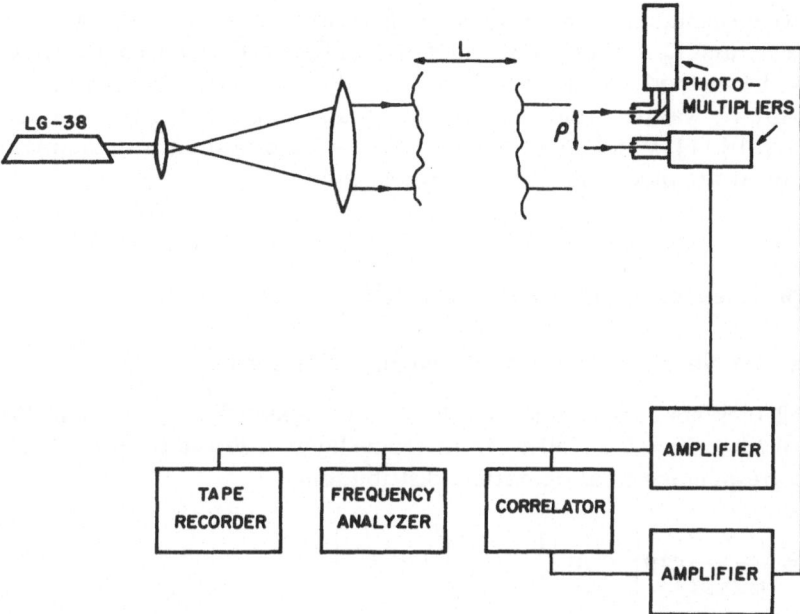

Fig. 4.1. Block diagram of the experiment

beam at the 1750 m path. The latter experiments were made with the same exit diameter but with a divergence angle of 1 mrad. The light receivers were photomultipliers in front of which stops with a diameter of 0.3 mm were placed. For protection from stray light during the daytime, lens hoods and interference light filters were used.

The photomultiplier current, after amplification, was recorded on magnetic tape for subsequent processing on a digital computer and simultaneously was fed to special-purpose analog equipment, a correlator and a frequency analyzer, which permitted obtaining directly under field conditions measurements of the values of the normalized variance β^2, the normalized covariance function $b_I(\varrho)$, and the frequency spectrum $U(\omega)$. In order to obtain stable results, the measurements were averaged over time intervals of 3.5 to 4 min.

The analog system for field measurements permitted operational monitoring of the results and introducing needed corrections in the experimental program. It should be noted that although it follows formally from (4.14) that measurements of $\omega W(\omega)$ and $b_I(\varrho)$ give identical results in the limit when the "frozen turbulence" hypothesis is valid, in actual fact, measurements of the frequency spectrum give much greater output when compared with measurements of the spatial covariance function. Actually, from a single time measurement (in our case 3.5 to 4 min) the frequency spectrum $U(\omega)$ is obtained—a complete set of results—while only one value of $b_I(\varrho)$ is found.

Processing of the magnetic records on a computer allowed extending the range of the spectral measurements on the low frequency side and obtaining the probability distribution of the irradiance. In order to independently calculate the scales L_T and l_T, C_ε^2 was determined from micrometeorological measurements by the method described in [4.10]. In 1972 for measurements on the 8500 m path, C_ε^2 was determined as well by an optical method, similar to that described in [4.11]. Wind speed and direction were measured with standard instruments at the time of the optical experiments.

4.4 Experimental Results and Their Discussion

4.4.1 The Log Normal vs Rayleigh Probability Distribution

Figure 4.2 presents an example of a histogram obtained by processing the records for $\beta_0^2 = 25(z = L = 1750 \text{ m})$. In this same figure is shown the probability density corresponding to a log normal distribution,

$$W(I) = \frac{1}{\sqrt{2\pi}\sigma I} \exp\left[-\left(\ln\frac{I}{\langle I\rangle} + \frac{\sigma^2}{2}\right)^2 (2\sigma^2)^{-1}\right], \qquad (4.21)$$

with mean and variance calculated from the experimental histogram. The straight line in Fig. 4.2 corresponds to a Rayleigh distribution for the amplitude of the light wave

$$W(I) = \frac{1}{\langle I\rangle} \exp(-I/\langle I\rangle). \qquad (4.22)$$

The distribution (4.22) was proposed in [4.12] for describing "strong" fluctuations. From the graph in Fig. 4.2 it is evident that the distribution (4.22) is in much poorer agreement with the experimental data than the distribution given by (4.21).

For a more detailed comparison of the experimental histogram with the distributions (4.21) and (4.22), the ratio m_n/m_1^n was calculated, where m_n is the nth order moment of the distribution,

$$m_n = \int I^n W(I)dI. \qquad (4.23)$$

The calculation of higher order moments from a finite length sample leads to errors [4.13] that increase as the order of the moment increases. The curves in Fig. 4.3, which represent the integrand in (4.23) and are obtained from the experimental data, indicate that because of the finite sample, the calculations of the moments from the experimental data are systematically underestimated, and for moments of order $n > 4$ the value, evidently, is questionable. This is related to the fact that the contribution to the moment of order n of rarely occurring but large overshoots increases as n increases. In Table 4.1 the ratio

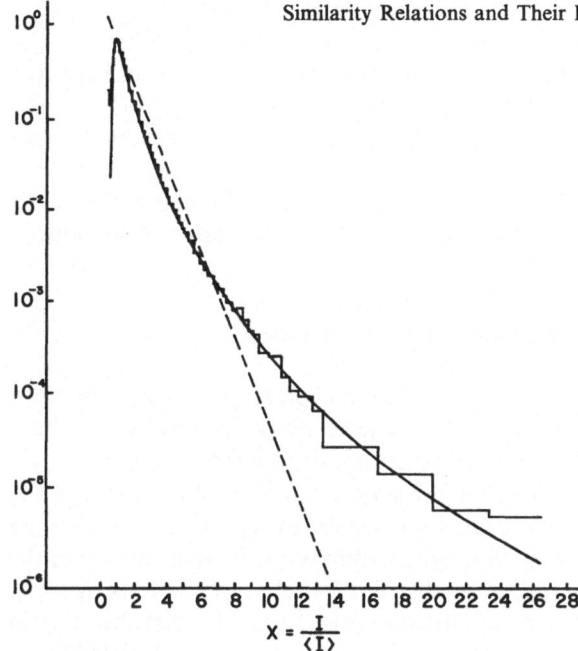

Fig. 4.2. The probability density W of the normalized fluctuations of irradiance, $I/\langle I \rangle$: ⌐ : the histogram obtained after processing the recorded signal; ——— log normal probability distribution; — — — Rayleigh probability distribution

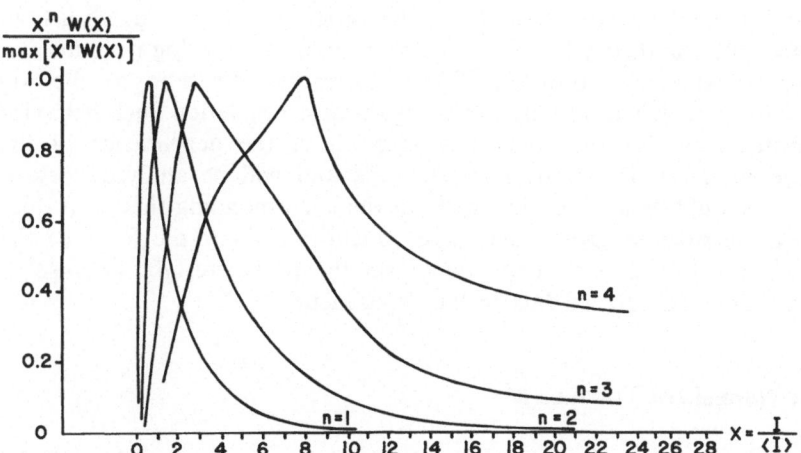

Fig. 4.3. The dependence of the integrand of expression (4.23) normalized by its maximum value as a function of $x = I/\langle I \rangle$

Table 4.1

n	$m_n/(m_1)^n$			$[m_n/(m_1)^n]\exp$
	Exp.	log normal	Rayleigh	$[m_n/(m_1)^n]\log normal$
2	2.18	2.18	2	1.0
3	9.76	10.33	6	0.94
4	79.32	106.71	24	0.74

$m_n/(m_1)^n$ is presented, for both the experimental data $(\beta_0^2 > 25)$ and for the probability laws (4.21) and (4.22).

From a comparison of the data presented in this table it is evident that the experimentally obtained distribution strongly diverges from (4.22), which was predicted in [4.12], and taking into account the remarks expressed when discussing the curves in Fig. 4.3, does not contradict a log normal distribution.

4.4.2 The Dependence of the Probability Distribution on β_0^2

In the composite curve in Fig. 4.4 the distribution is presented on a probability scale along the ordinate and on a logarithmic scale along the abscissa. In these coordinates the log normal distribution corresponds to a straight line.

Also in Fig. 4.4 the distributions are presented that were obtained in [4.14] for small values of $\beta_0^2 < 1$, for intermediate values of $\beta_0^2 \simeq 1$, and for strong fluctuations $\beta_0^2 \gg 1$. Measurements of the logarithm of the irradiance and of the irradiance itself permitted obtaining the distribution of the variations of $I/\langle I \rangle$ over a three-order range for "strong" fluctuations. From the distributions in Fig. 4.4 it is clear that the largest deviations from a log normal distribution occur for intermediate values of β_0^2 of the order of unity[3]. For $\beta_0^2 \ll 1$, in accordance with the theory [4.7], the distribution is close to log normal.

For $\beta_0^2 \gg 25$ and right up to $\beta_0^2 \simeq 100$ the distribution again approaches log normal, with some deviation only in the region of strong fades. Such behavior of the distribution function definitely tells about the dependence of the normalized variance β^2 on the parameter β_0^2 and will be analyzed below. However, it should be noticed that without the corresponding theory, it does not appear possible to give a definitive conclusion about the form of the distribution function and its dependence on the parameter β_0^2, because of analyzing a finite amount of data in the experiment.

4.4.3 The Normalized Variance β^2

In Fig. 4.5 we see the dependence of the root-mean-square value of the normalized irradiance fluctuations β on β_0, obtained in the experiments described above. The data presented in the figure were taken in the daytime, i.e., under conditions of fully developed turbulence, on paths of $z = L_1 = 1750$ m and $z = L_2 = 8500$ m. The largest value of β_0, which was obtained on the 8.5 km path, reached a value of $\beta_0 = 11$. The values of β obtained at a distance of 8500 m were in agreement with measurements at shorter distances. The values of β for small β_0, shown in Fig. 4.5, were obtained earlier in [4.2] for a path length $z = 250$ m.

---

[3] Deviations from the log normal distribution noticed in [4.15] were not connected with the propagation conditions.

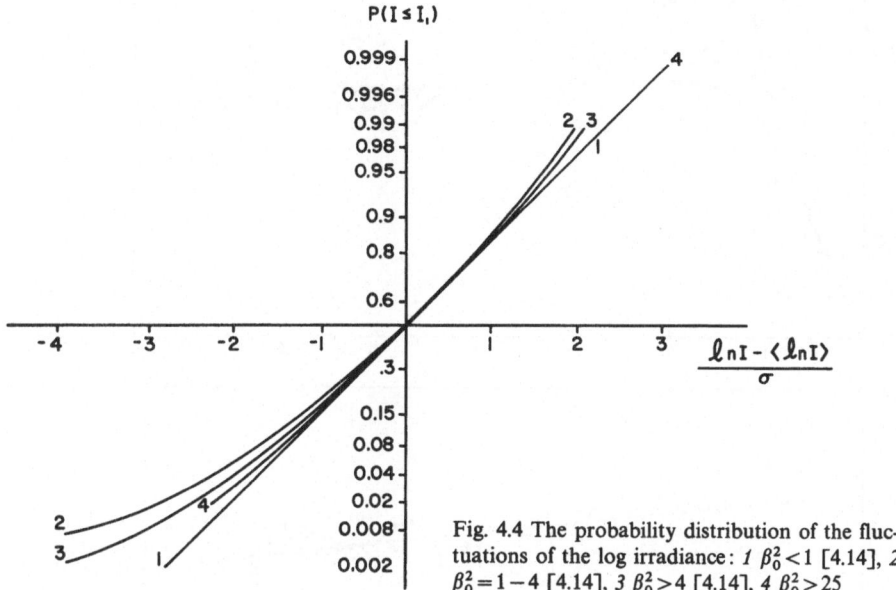

Fig. 4.4 The probability distribution of the fluctuations of the log irradiance: 1 $\beta_0^2 < 1$ [4.14], 2 $\beta_0^2 = 1 - 4$ [4.14], 3 $\beta_0^2 > 4$ [4.14], 4 $\beta_0^2 > 25$

The dependence of β on β_0 shown in Fig. 4.5 reflects the effects noticed earlier [4.1, 2, 16–18] of the saturation of the rms value of the normalized fluctuations β for $\beta_0 \gtrsim 1$ and the tendency for β to decrease with further increase in β_0. In addition, it is necessary to notice also the small difference in the measurements both for a collimated beam at two distances ($z = 1750$ m and $z = 8500$ m) (Curves 1 and 2 in Fig. 4.5), and also between a collimated and a diverging beam at the distance $z = 1750$ m (Curves 1 and 3).

Comparison of the results of measurements of β for a collimated beam with Fresnel numbers $k a_0^2/z$ equal to 130 and 26, for distances of $z = 1750$ m and $z = 8500$ m, respectively, and at the same time with the results of measurements for a diverging beam, indicates that for $\beta_0^2 \gg 1$ the geometrical factors of the laser beam do not play a determining role when estimating the normalized fluctuations on the beam axis. Further analysis of the geometrical factors, without the availability of a sufficiently developed theory, appears as a difficult problem because of the presence of many parameters, among which it would be necessary to include such characteristics of the turbulence as the microscale and the outer scale.

If we take into account that the similarity relations (4.15) were obtained for a plane wave, then the best estimate of the universal function $f_\beta(z/L_T)$ $= f_\beta[(\beta_0^2/0.75)^{3/11}]$ appears to be Curve 1, obtained with the largest Fresnel number. At the same time, Curves 2 and 3 may be useful for practical calculations. Notice that the above features of the probability density of the fluctuations permit a comparison between the dependence obtained above with the results of measurements of $\sigma^2 = \langle (\ln I - \langle \ln I \rangle)^2 \rangle$, published in [4.2, 16–19]. From similarity considerations it follows that $\sigma^2 = \sigma^2(\beta_0^2)$.

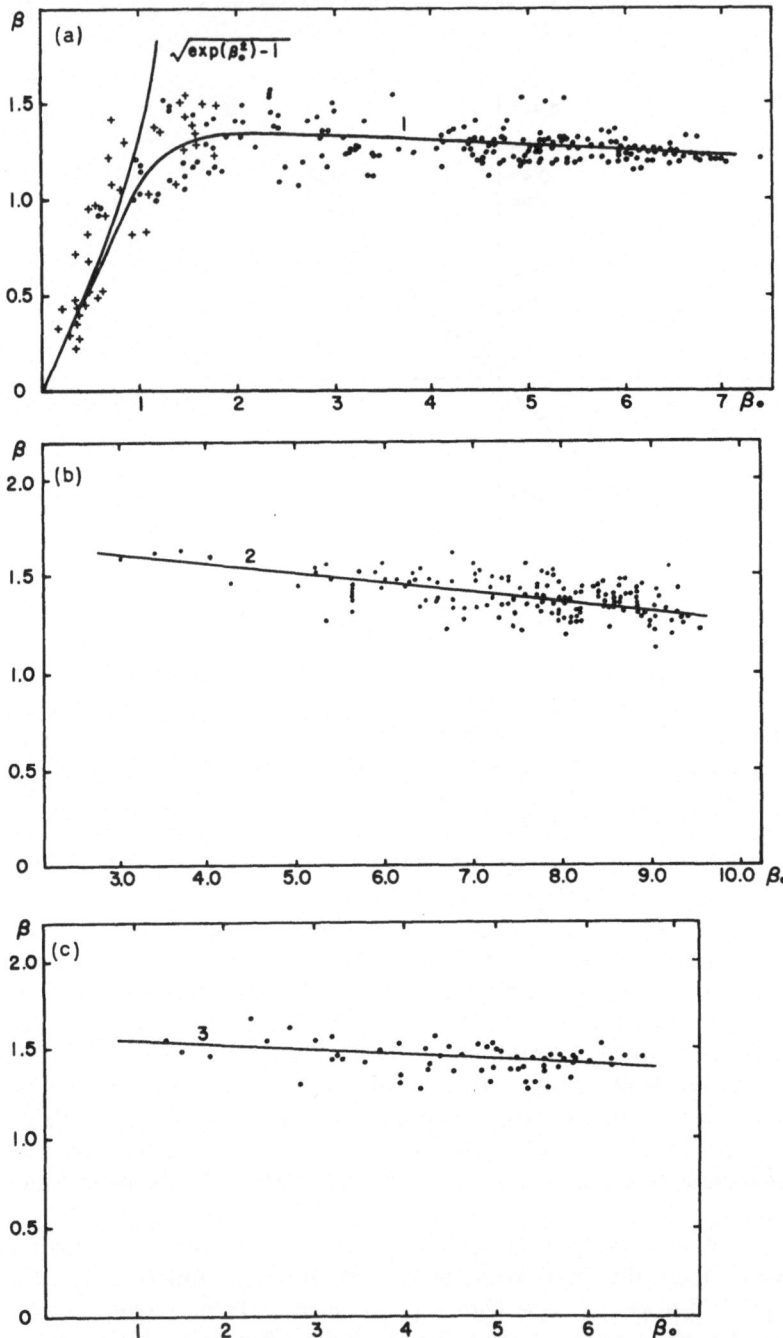

Fig. 4.5. (a) The dependence of β on β_0 for a collimated beam for $z = 250\,\mathrm{m}$ (+) and for $z = 1750\,\mathrm{m}$ ($\cdot$): Curve 1 is the average of the measurements; $[\exp(\beta_0^2) - 1]^{1/2}$ is calculated by the method of smooth perturbations. (b) The dependence of β on β_0 for a collimated beam for $z = 8500\,\mathrm{m}$: Curve 2 is the average of the measurements. (c) The dependence of β on β_0 for a diverging beam for $z = 1750\,\mathrm{m}$: Curve 3 is average of the measurements

The function $\sigma(\sigma_0^2)$ vividly demonstrates the maximum for intermediate values, $\sigma_0^2 \simeq 1$. This circumstance is connected with the fact that in the region of intermediate values of $\beta_0^2 \simeq 1$, the probability of a very deep fade increases (Fig. 4.4), which inherently causes a marked increase in the variance of the log irradiance but has significantly less effect on the variance of the irradiance itself. In an assumed log normal distribution the variance of the logarithm σ^2 is connected with β^2 by the formula

$$\sigma^2 = \ln(1 + \beta^2). \tag{4.24}$$

If in (4.24) we substitute β^2, as presented above, then for σ we obtain the value $\sigma = 0.96$, in excellent agreement with $\sigma = 0.95$ obtained in [4.20]. Notice that in [4.20], as finally has become clear, an underestimated value of $\sigma = 0.8$ was obtained from measurements with an incoherent source. This underestimate is connected, as is shown by the estimate based on the correlation function [4.21], with the weakening of the light fluctuations due to the averaging by the receiver's objective, which had a diameter of 5 mm [4.20].

4.4.4 The Normalized Covariance Function

In Fig. 4.6, experimental results are presented for the covariance function of the irradiance fluctuations as a function of the separation of the observation points. For clarity the scales along the abscissa are expanded for small values of ϱ. From the obtained data it is clear that the values of $b_I(\varrho)$ for strong fluctuations have a rather wide spread (Fig. 4.6b). At the same time for weak fluctuations the experimental data are concentrated in a relatively narrow band (Fig. 4.6a). Giving recognition to the theoretical estimates [4.7], and to all the previous

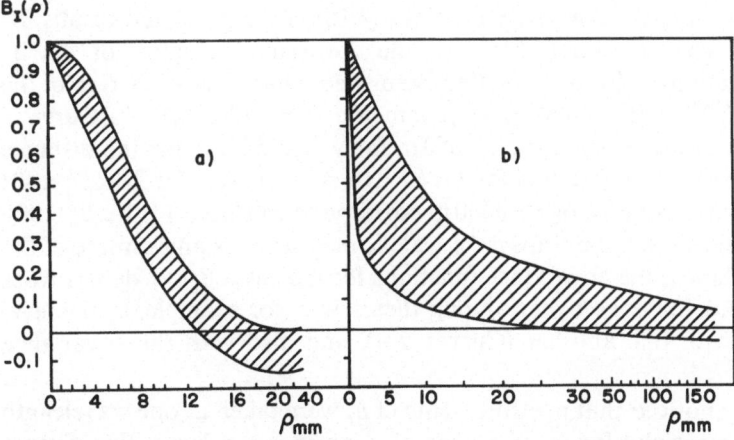

Fig. 4.6a and b. The covariance function of the irradiance fluctuations as a function of the separation of the observation points: (a) for the case of weak fluctuations, $z = 650$ m; (b) for the case of strong fluctuations, $z = 1750$ m

experimental results, we shall present the covariance function b_I as a function of the scale $\varrho/\sqrt{\lambda L}$ and its dependence on β_0^2.

The results of appropriate processing are presented in Fig. 4.7a for a collimated beam and in Fig. 4.7b for a diverging beam. From Fig. 4.7a it is clear that for $\beta_0^2 < 1$ the experimental values agree very well with the theoretical calculations [4.7]. The covariance function for the collimated and diverging beams have similar shapes: with increasing β_0^2 the correlation between two close observation points $(\varrho < 0.5\sqrt{\lambda L})$ decreases, but for large $\varrho(\varrho > \sqrt{\lambda L})$ the correlation increases. In general this agrees with the estimates obtained in [4.3], i.e., for $\beta_0^2 \gg 1$ the covariance function of the irradiance is characterized by two scales, one of which is significantly larger than $\sqrt{\lambda L}$, and at the same time the other is significantly smaller. The effect of large scale fluctuations for $\beta_0^2 \gg 1$ was noticed also in the measurements reported by *Gracheva* [4.20]. However, as has become clear from the results obtained in the present work, the decrease in the covariance function for $\varrho < 0.5\sqrt{\lambda L}$ in the region of "strong" fluctuations remained unnoticed in the measurements of [4.20] due to the fact that the receiving aperture was too large—its diameter was equal to 5 mm, which corresponded to $\sqrt{\lambda L}/6$.

The averaging of the small scale fluctuations by such an aperture, as is shown by a numerical estimate based on the results of the present work, introduces a considerable distortion of the value of b_I in the region $\varrho < 0.5\sqrt{\lambda L}$.

In Fig. 4.7b is presented also the covariance function for fluctuations of the log irradiance, obtained by *Ochs* et al. [4.23] with a diverging beam at a distance of 45 km and also pertaining to the region of strong fluctuations. From Fig. 4.7b it is possible to judge about the adequate qualitative agreement of the results of measurements of b_I for strong fluctuations at distances differing by factors of 25.

In order to verify the similarity relations (4.16) of the presented results for the transverse scale $l_T = (\pi A C_\varepsilon^2 k^3)^{-3/11}$, the covariance function of the irradiance fluctuations b_I for a collimated beam were plotted as a function of the quantity $\varrho(C_\varepsilon^2 k^3)^{3/11}$ (Fig. 4.8). The parameter is $\beta_0^2 = 3(z/L_T)^{11/6}$. Curve *1* corresponds to measurements for $z = 650$ m for which $\beta_0^2 < 1$; for the distance $z = 1750$ m, Curve *2* was obtained for $1 < \beta_0^2 < 10$, and Curve *3* for $24 < \beta_0^2 < 48$; measurements at a distance of $z = 8500$ m correspond to Curve *4* $(11 < \beta_0^2 < 90)$. As is clear from Fig. 4.8, the transverse scale l_T may serve as an estimate of the correlation distance; the covariance functions for the cases $\beta_0^2 \gg 1$ depart from the ordinate and converge better among themselves (for example, at the level $b_I = 0.5$) both with one another (Curves *2–4*), and also with the covariance function for $\beta_0^2 < 1$.

It should be noticed that measurements of b_I were taken at one wavelength and the data, especially for small values of ϱ, are not too large. For a more thorough verification of the similarity of the covariance function with the scale l_T, additional measurements are necessary, especially with two wavelengths.

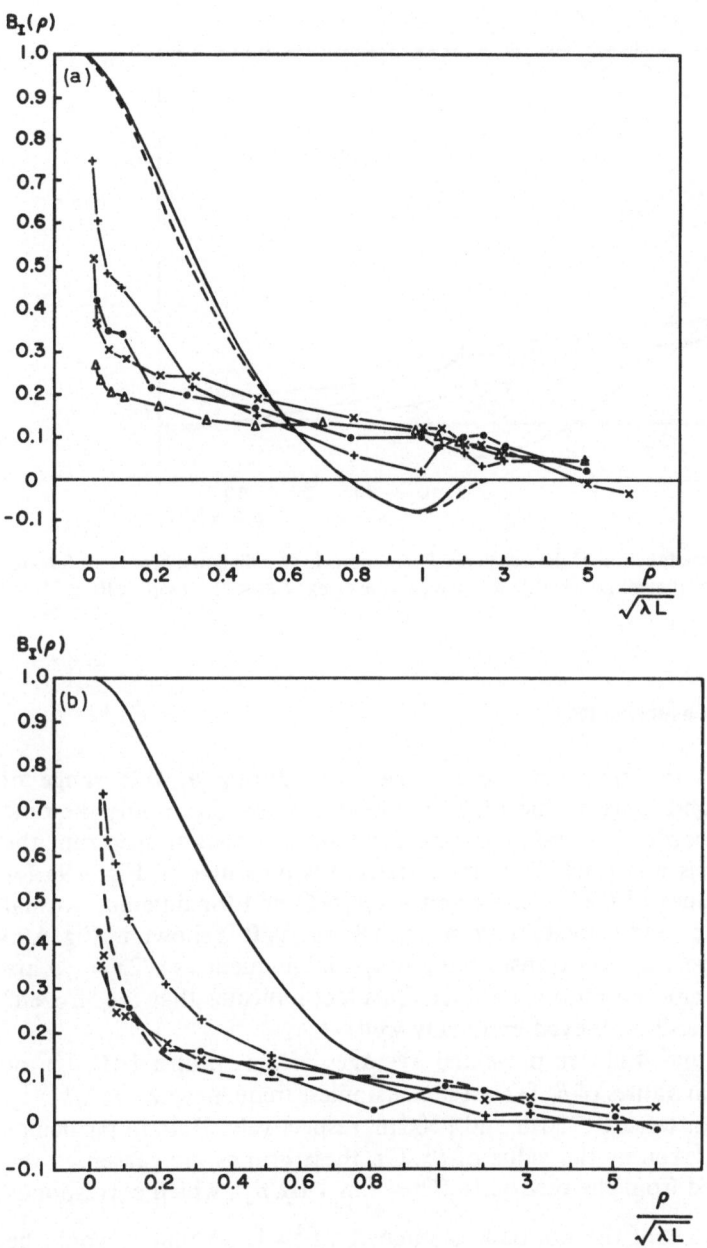

Fig. 4.7. (a) The covariance function of the irradiance fluctuations for a collimated beam: (————) theoretical for $\beta_0^2 \ll 1$; (– – – –) experimental for $\beta_0^2 < 1$ ($x=650\,\mathrm{m}$); (+) $1 < \beta_0^2 < 5.3$; (·) $7.9 < \beta_0^2 < 27$; (×) $\beta_0^2 > 27$ ($z=1750\,\mathrm{m}$); (△) $11 < \beta_0^2 < 90$ ($z=8500\,\mathrm{m}$). (b) The covariance function of the irradiance fluctuations for a diverging beam: (+) $1 < \beta_0^2 < 5.3$; (·) $7.9 < \beta_0^2 < 27$; (×) $\beta_0^2 > 27$ ($z=1750\,\mathrm{m}$); (————) theoretical for a spherical wave and $\beta_0^2 \ll 1$ [4.22]; (– – – –) the average of four curves from [4.23] and $\beta_0^2 > 1$

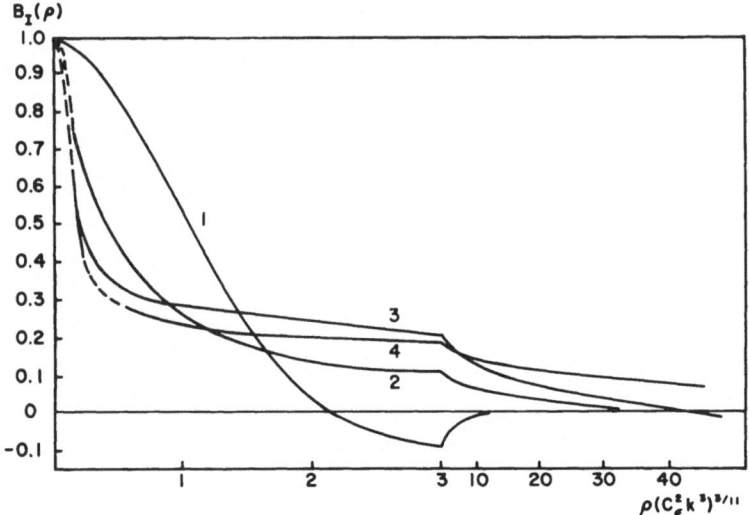

Fig. 4.8. The covariance function of the irradiance fluctuations for a collimated beam as a function of the dimensionless argument $\varrho(C_\varepsilon^2 k^3)^{3/11}$ and the parameter β_0^2: *1* $\beta_0^2 < 1$; *2* $1 < \beta_0^2 < 10$; *3* $24 < \beta_0^2 < 48$; *4* $11 < \beta_0^2 < 90$

4.4.5 Spectral Measurements

Measurements of the spectrum were carried out during a wide range of meteorological conditions on the 1750 and 8500 m paths. Since only the time spectrum was recorded, in order to transform it to a spatial spectrum the "frozen" hypothesis was used. The time spectrum is presented in Fig. 4.9a for two groups of values of $\beta_0 (6 > \beta_0 > 4$ and $8 > \beta_0 > 6)$ and for different normal components of the wind velocity (v from 1.5 to $8\,\mathrm{m\,s^{-1}}$). It is shown in Fig. 4.9b that these same spectra, after transforming to spatial frequencies $\kappa/2\pi = f/v$, are practically superimposed on one another. This fact indicates that the "frozen" turbulence hypothesis is obeyed extremely well.

In Figs. 4.10 and 4.11 are presented averages of the spectra $U(\Omega, \beta_0^2)$ for groups with similar values of β_0 [Ω is a dimensionless frequency, $\Omega = \kappa(L/k)^{1/2}$], obtained for the distances 1750 m and 8500 m, respectively. Here in particular the parameter is taken as the value of β_0^2. On these graphs are presented the function calculated from the theory, $U_0(\Omega) = \lim_{\beta_0^2 \to 0} U(\Omega, \beta_0^2)$, which corresponds to small fluctuations of the normalized intensity, $\beta_0^2 \ll 1$. So that it would be possible during further analysis to obtain an idea of the range of the spatial scales in which the spectrum of the irradiance fluctuations were studied, in Fig. 4.10 the wavenumber scale $\kappa/2\pi = f/v$ is shown. On this scale notice the condition for the lower boundary of the inertial range $\kappa_z = 2\pi/z$ (c), corresponding to the outer scale of turbulence. In this same figure are shown also the limits over which the upper boundary of the inertial range $\kappa_\eta = 2\pi/8/\eta_\kappa$ varies (a to b),

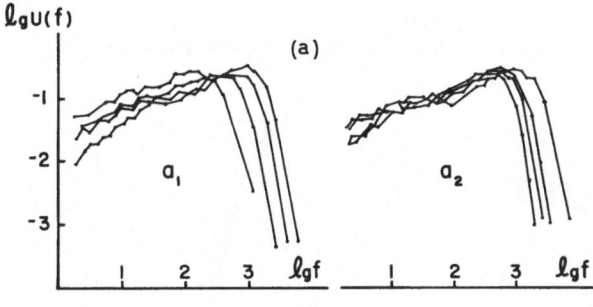

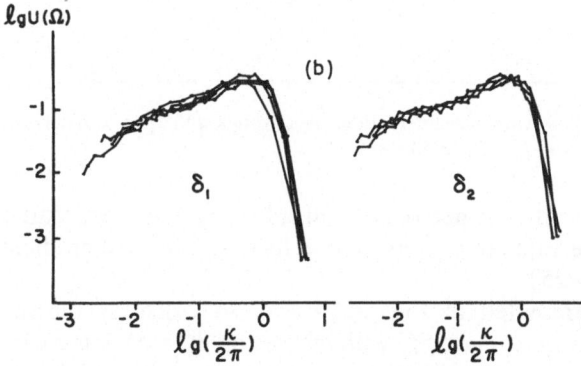

Fig. 4.9a and b. The spectra of the irradiance fluctuations: (a) for different wind speeds for groups with the values $16 < \beta_0^2 < 36$ (1), and $\beta_0^2 > 36$ (2); (b) the same spectra after transforming to wave numbers $\kappa/2\pi = f/v$

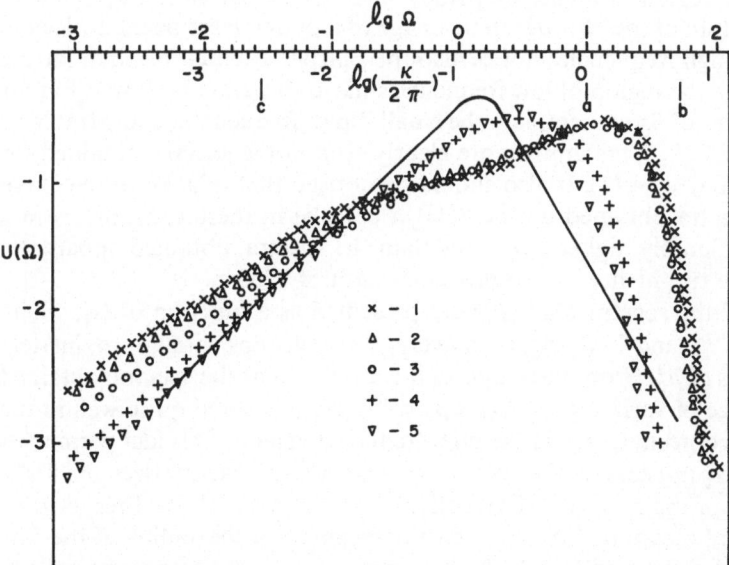

Fig. 4.10. Averages of the spectra of the irradiance fluctuations over groups with approximately the same value of β_0^2 for $z = 1750$ m: 1 $\beta_0^2 > 36$; 2 $36 > \beta_0^2 > 16$; 3 $16 > \beta_0^2 > 4$; 4 $4 > \beta_0^2 > 1.7$; 5 $1.7 > \beta_0^2 > 0.5$

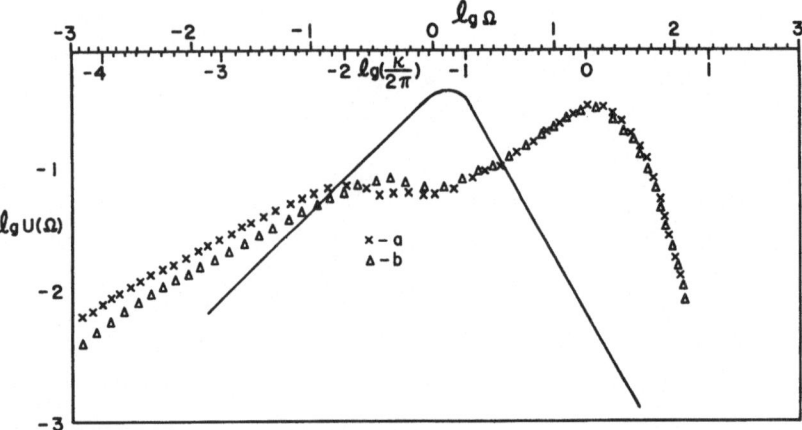

Fig. 4.11. Averages of the spectra of the irradiance fluctuations over groups with approximately the same value of β_0^2 for $z = 8500 \, \text{m}$: a $90 > \beta_0^2 > 50$; b $50 > \beta_0^2 > 35$

where η_κ is the Kolmogorov turbulence scale, multiplied by 8 in accordance with the work in [4.24]. The value of η_κ is estimated from the vertical gradient of wind and temperature [4.25].

The spectra $U(\Omega, \beta_0^2)$, presented in Figs. 4.10 and 4.11, display the dependence of the form of the spectrum on β_0^2: with increasing β_0^2, associated with the growth of either the turbulence or with distance, the spectra widen and their peaks shift to a region of higher frequency. The latter observation agrees with the small structure of the photographic image of a pulsed laser beam noticed in [4.26]. At the high frequencies notice also the sharper attenuation than for the spectrum U_0. In the region of low frequencies the attenuation is slower, but for increasing values of β_0^2 contemplate the small dip at frequencies comparable to the maximum of U_0. This dip is more clearly seen in the spectra obtained for the 8500 m path (Fig. 4.11). It also should be noticed that relative to the curve for U_0, the spectra obtained on the 8500 m path lie in these coordinates in a region of considerably higher frequency than the spectra, obtained apparently under the same turbulence conditions, on a path of 1750 m.

In Fig. 4.12 the spectra $U(\Omega_T, \beta_0^2)$ are presented as a function of Ω_T, where $\Omega_T = \kappa (C_\varepsilon^2 k^3)^{-3/11}$, and $U(\Omega_T, \beta_0^2)$ is an average over groups with approximately the same values of β_0^2. From these figures it is evident that the spectra, obtained for a wide range of variation of $\beta_0 (1 < \beta_0 < 40)$, are presented quite well in the dimensionless coordinates Ω_T in the high-frequency region. This fact permits us to assert that Ω_T appears as the transverse scale which characterizes the high-frequency part of the spectrum. Evidently, for the case $\beta_0^2 > 1$ the Fresnel zone loses its original meaning, but takes on the meaning of the radius of the first Fresnel zone for a distance L_T, which also corresponds to the transverse scale l_T.

Measurements of the spectrum for a diverging and a focused beam under conditions of "strong" fluctuations show that in the high-frequency region they do not register important differences compared to the data obtained from a

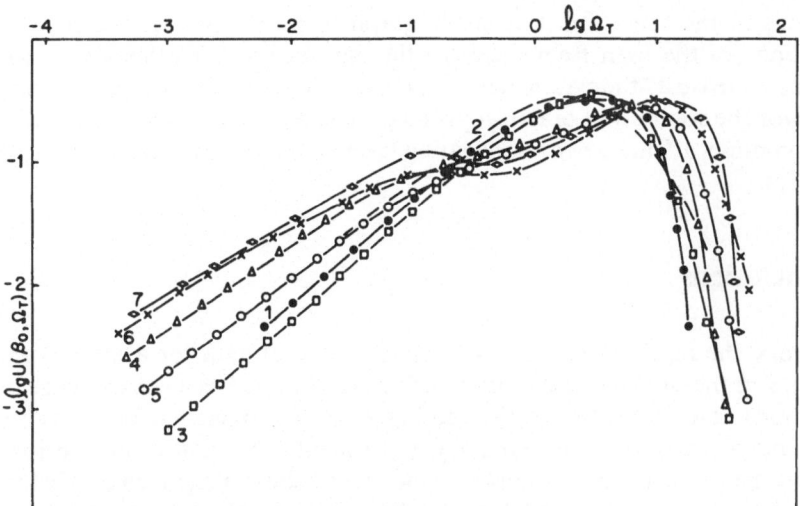

Fig. 4.12. The spectra $U(\Omega_T)$ as a function of the dimensionless frequency $\Omega_T = \kappa(C_\epsilon^2 k^3)^{-3/11}$; the notation for the curves is the same as in Figs. 4.10 and 4.11

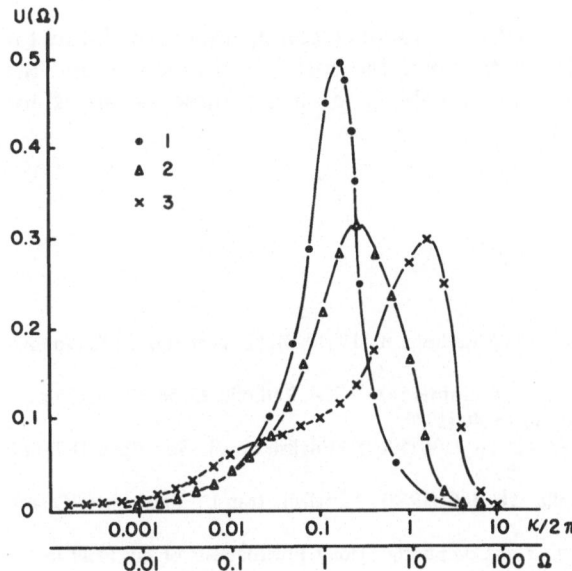

Fig. 4.13. The spectral distribution of the power in the fluctuations for different values of β_0^2: 1 $\beta_0^2 \to 0$; 2 $1.7 < \beta_0^2 < 4$; 3 $\beta_0^2 > 16$

collimated beam. The similar results for beams with different geometries permit us to make the supposition that the distortion of the high-frequency part of the spectrum, connected with the boundedness of the beam, investigated in [4.27], has not appeared in the present work.

Figure 4.13, in which the spectra U are presented in semi-logarithmic coordinates, allows a clear representation of the distribution of the power of the fluctuations in the spectrum. From this graph it is evident that the relative contributions of the high frequencies to the variance under otherwise equal conditions increase as the parameter β_0^2 increases. The curves of Fig. 4.13 may be useful for the solution of practical problems concerned with the value of the fluctuation noise, arising when propagating laser radiation through a turbulent atmosphere.

4.5 Conclusions

In this work the results are presented of experimental data for a laser beam which confirm the similarity relations which were obtained as a consequence of the equations describing the fourth- and higher order coherence functions for an unbounded plane wave. Preliminary experimental estimates indicate that changes in the geometry of the initial beam do not show a significant effect in the statistical characteristics for strong fluctuations. Nonetheless, without question there is interest in further experimental investigations of the effects of geometrical factors, also the effects of the characteristic scales and the form of the turbulent spectrum, and finally verification of the relationships with wavelength.

The empirical dependencies that were obtained can be used in the future for estimating the variance of the fluctuations, the correlation function, and the frequency spectrum, and give the possibility of using these estimates for analyzing the fluctuation noise.

References

4.1   M. E. Gracheva, A. S. Gurvich: Izv. VUZ, Radiofiz. **8**, 717 (1965); [English transl.: Radiophys. Quant. Electron **8**, 511 (1965)]

4.2   M. E. Gracheva, A. S. Gurvich, M. A. Kallistratova: Izv. VUZ, Radiofiz. **13**, 56 (1970); [English transl.: Radiophys. Quant. Electron **13**, 40 (1970)]

4.3   V. I. Tatarskii: Zh. Eksper. I. Teor. Fiz. **56**, 2106 (1969); [English transl.: Sov. Phys.-JETP **29**, 1133 (1969)]; also Ref. [4.5], Chap. 5

4.4   V. I. Shishov: Izv. VUZ, Radiofiz. **11**, 866 (1968); [English transl.: Radiophys. Quant. Electron **11**, 500 (1968)]

4.5   L. A. Chernov: "The Local Method of Computing Strong Fluctuations of the Field in the Problem of Wave Propagation in a Medium with Random Inhomogeneities", in Russian, presented at the 6th All-Union Acoustical Conf. (Moscow, USSR 1968)

4.6   W. P. Brown, Jr.: J. Opt. Soc. Am. **62**, 45 (1972)

4.7   V. I. Tatarskii: *The Effects of the Turbulent Atmosphere on Wave Propagation* (National Technical Information Service, Springfield, Va. 1971) p. 472; *Rasprostranenie Volv v Turbulentnoi Atmosfere* (Nauka, Moscow, USSR 1967) p. 548

4.8   L. S. Dolin: Izv. VUZ, Radiofiz. **11**, 840 (1968); [English transl.: Radiophys. Quant. Electron **11**, 486 (1968)]

4.9   W. P. Brown, Jr.: J. Opt. Soc. Am. **61**, 1051 (1971)

4.10 M. A. Kallistratova, D. F. Timanovskiy: Izv. AN SSSR FAO **7**, 73 (1971); [English transl.: Akad. Nauk SSSR Izv. Atmos. Ocean. Phys. **7**, 46 (1971)]
4.11 G. A. Andreyev, V. M. Kuznetsov, V. Tseytlin: Izv. AN SSSR FAO **7**, 987 (1971); [English transl.: Akad. Nauk SSSR Izv. Atmos. Ocean. Phys. **7**, 653 (1971)]
4.12 D. A. DeWolf: J. Opt. Soc. Am. **58**, 461 (1968)
4.13 V. L. Alekseev: Izv. AN SSSR Tekhnicheskaya Kibernetika **6**, 171 (1970)
4.14 A. S. Gurvich, M. A. Kallistratova, N. S. Time: Izv. VUZ, Radiofiz. **11**, 1360 (1968); [English transl.: Radiophys. Quant. Electron **11**, 771 (1968)]
4.15 D. H. Höhn: Appl. Opt. **5**, 1427 (1966)
4.16 P. H. Deitz, N. J. Wright: J. Opt. Soc. Am. **59**, 527 (1969)
4.17 J. R. Kerr: J. Opt. Soc. Am. **62**, 1040 (1972)
4.18 G. R. Ochs, R. S. Lawrence: J. Opt. Soc. Am. **59**, 226 (1969)
4.19 R. H. Kleen, G. R. Ochs: J. Opt. Soc. Am. **60**, 1695 (1970)
4.20 M. E. Gracheva: Izv. VUZ, Radiofiz. **10**, 775 (1967); [English transl.: Radiophys. Quant. Electron **10**, 424 (1967)]
4.21 M. E. Gracheva, A. S. Gurvich, A. S. Khrupin: Izv. VUZ, Radiofiz. **17**, 155 (1974)
4.22 D. L. Fried: J. Opt. Soc. Am. **57**, 175 (1967)
4.23 G. R. Ochs, R. R. Bergman, J. R. Snyder: J. Opt. Soc. Am. **59**, 231 (1969)
4.24 H. L. Grant, R. W. Stewart, A. Moilliet: J. Fluid Mech. **12**, 241 (1962)
4.25 J. Lumley, H. Panoksky: *The Structure of Atmospheric Turbulence* (Interscience, New York, London, Sydney 1964) p. 239
4.26 V. Ya S'edin, S. S. Khmelevtsov, M. E. Nebol'sin: Izv. VUZ, Radiofiz. **13**, 44 (1970); [English transl.: Radiophys. Quant. Electron **13**, 32 (1970)]
4.27 N. S. Time: Izv. VUZ, Radiofiz. **14**, 1195 (1971); [English transl.: Radiophys. Quant. Electron **14**, 935 (1971)]

5. The Beam Wave Case and Remote Sensing

A. Ishimaru

With 8 Figures

The propagation characteristics of an optical beam in the earth's atmosphere are greatly affected by random fluctuations of the index of refraction caused by the turbulent motion of air. These effects include beam spreading and a corresponding decrease in the intensity, scintillations of the received intensity, a decrease in the spatial and temporal coherence, and beam wander.

Interaction of light beams with a turbulent atmosphere is important in astronomy in the study of twinkling of stars and image resolution. It also affects the coherence bandwidth, the coherence time, and other characteristics of optical communication in the earth's atmosphere. This interaction of a light beam with the medium in which it propagates can also be used for remote sensing of the structure and dynamics of the atmosphere. In this chapter we examine recent advances in the study of beam wave propagation and the use of optical beams for remote sensing.

An optical beam propagating in the earth's atmosphere often can be approximated by a plane or spherical wave. For example, a laser beam transmitted from a spacecraft behaves almost as a plane wave when it reaches the earth's atmosphere, and a laser beam emitted from a small aperture in the earth's atmosphere behaves as a spherical wave. Numerous investigations have been reported on the fluctuation characteristics of plane and spherical waves in a turbulent atmosphere. They include studies of amplitude and phase variances, correlation functions, structure functions, temporal frequency spectra, and probability densities. All these quantities depend only upon the distance, the wavelength, and the turbulence characteristics.

In many practical optical beam problems, however, these plane and spherical wave approximations are not sufficient to characterize the properties of wave propagation, particularly when the aperture is large. It is necessary to take into account the beam size and the focusing or diverging characteristics of the beam. In describing a beam, we assume that the amplitude distribution is gaussian and therefore may be expressed by a single parameter W_0, the beam size. In addition, we assume the focussing or diverging characteristics may be described by a parabolic phase front with a focal distance R_0. In this chapter, we use the two parameters W_0 and R_0 to characterize the beam. We should note, however, that in practice we always have a finite aperture size rather than the ideal gaussian distribution and that the phase front may not be strictly parabolic.

Considerable theoretical and experimental studies have been reported in recent years on the effects of the parameters W_0 and R_0 on the fluctuation characteristics. A beam may be collimated $(R_0 \to \infty)$ or focused $(R_0 > 0)$ or diverging $(R_0 < 0)$. It has been shown that while diverging and collimated beams exhibit characteristics similar to those of spherical and plane waves, a focused beam is significantly different, and theoretical and experimental studies conducted to date have revealed considerable disparity among various investigators.

Theoretical studies on beam wave propagation may be classified into two general areas: the weak fluctuation and strong fluctuation cases. The weak fluctuation theory generally makes use of the Rytov method, which is also called the smooth perturbation method or the SPM. It yields relatively simple mathematical formulas for a variety of quantities such as variances, correlation functions, structure functions, and spectra. Since this method is applicable only to the weak fluctuation case, it has an obvious limitation that the variance should not exceed a certain level. This is normally stated in terms of the log amplitude χ, i.e., that the variance σ_χ^2 should be considerably smaller than $0.2 \sim 0.5$. While this limitation appears to be adequate for collimated and diverging beams, theoretical and experimental studies on focused beams have revealed that this condition is necessary, but not sufficient. The range of validity of the focused beam formulation is a point of active contention at present, and in this chapter we hope to shed some light on this question.

Strong fluctuation theories include the following three approaches: 1) the Dyson and the Bethe-Salpeter integral equations, 2) the Huygens-Fresnel principle, and 3) the parabolic equation. These three are closely related to each other, and for the second moment, they can be shown to be equivalent to each other under certain conditions. It should be noted that even though a formal solution for the second moment has been obtained in an integral form, it is often not possible to obtain reasonably simple expressions for various statistical quantities and one needs to resort to numerical solutions. For the fourth moment, which is necessary to analyze the intensity fluctuations, available solutions are much less satisfactory than for the second moment.

Recently some careful experiments have been reported both in the U.S. and in the U.S.S.R. In some cases they support the existing theories, while in others they show a considerable disagreement, pointing to a need for modification of the theories.

We note here that in addition to well-known texts [5.1, 2], some important survey papers have appeared recently, and the reader is urged to study them carefully [5.3–9].

Before we go into details of the effects of refractive index turbulence on a beam wave, let us describe a beam wave in free space so that we have a clear picture of the unperturbed situation.

Let us consider a beam wave $U_0(r)$ propagating in the z direction in free space. At the transmitting aperture $(z=0)$, the wave has a gaussian amplitude distribution with beam size W_0 and a curved phase front with radius of curvature R_0. The beam size W_0 is equal to the radius at which the field amplitude becomes e^{-1} of that on the beam axis, and positive (negative) R_0 corresponds to a converging (diverging) beam. Then

$$U_0(0, \varrho) = \exp\left[-\left(\frac{1}{W_0^2} + i\frac{k}{2R_0}\right)\varrho^2\right],$$ (5.1)

where $k = (2\pi)/\lambda$ is the free wave number, λ is the wavelength, and ϱ is the radial vector from the z axis. For convenience, we write [5.10, 11] (5.1) as

$$U_0(0, \varrho) = \exp[-(1/2)k\alpha\varrho^2],$$ (5.1a)

where

$$\alpha = \alpha_1 + i\alpha_2 = \frac{\lambda}{\pi W_0^2} + i\frac{1}{R_0}.$$ (5.1b)

At an arbitrary point (ϱ, z), the beam wave in free space is given by

$$U_0(\varrho, z) = \frac{1}{(1 + i\alpha z)} \exp\left[ikz - \left(\frac{k\alpha}{2}\right)\frac{\varrho^2}{1 + i\alpha z}\right].$$ (5.2)

This is valid within a distance z [5.10] such that

$$z \ll \frac{\pi^3 W_0^4}{\lambda^3},$$ (5.2a)

which covers almost all the laser propagation distances encountered in practice. The intensity $I_0(\varrho, z)$ in free space is given by

$$I_0(\varrho, z) = |U_0(\varrho, z)|^2 = \frac{W_0^2}{W^2} \exp\left(-\frac{2\varrho^2}{W^2}\right),$$ (5.3)

where the beam size W at z is

$$W = W_0[(1 - \alpha_2 z)^2 + (\alpha_1 z)^2]^{1/2}.$$

The total power transmitted should be independent of z and is given by

$$P_t = \int I_0(\varrho, z)d\varrho = \left(\frac{\pi}{2}\right)W_0^2.$$ (5.4)

We also note that for a collimated beam, $R_0 \to \infty$ and $\alpha_2 = 0$. If the beam is focused at the observation point, $z = R_0$ and therefore $\alpha_2 z = 1$.

Sections 5.1 and 5.2 are devoted to the weak and strong fluctuation theories, respectively, of the effects of atmospheric turbulence on the beam wave described above.

In Section 5.3, we discuss recent research activities on optical remote sensing. Remote sensing of atmospheric conditions has become increasingly important in recent years because it provides a useful tool for studying the structure and dynamics of the atmosphere, which are important for weather forecasting, weather modification, pollution studies, storm warning, and air traffic safety. In that section we present recent work on remote sensing of the turbulence strength, represented by the refractive index structure constant C_n, and the wind velocity. We note that what we probe is the structure constant of the refractive index fluctuations and that the structure constant of the turbulence velocity field itself can only be inferred from the refractive index fluctuations, even though these two are usually highly correlated.

We discuss two areas of remote sensing. One is to probe the average quantity over the propagation path and the other is to find the profile of these quantities along the path. Even though some successful theoretical and experimental studies have been reported, they are still in early stages of development and further work on new and more reliable techniques is needed. We also note that the studies reported thus far are almost all based entirely on weak fluctuation theory, and there is a definite need to employ strong fluctuation theory in order to cover longer distances or stronger turbulence.

5.1 Weak Fluctuation Theory

When the turbulence is weak or when the propagation distance is short, we expect that the wave fluctuations are small and a certain approximate formulation is possible. The beam wave theory for this case has been developed using an extension of the Rytov method, which was described in detail by *Tatarskii* [5.1, 2]. The first attempt in this direction was made by *Kon* and *Tatarskii* [5.12], who considered a collimated beam. *Schmeltzer* [5.13], *Fried* and *Seidman* [5.14], and *Kinoshita* et al. [5.15] made extensive studies of both focused beam and collimated beams. *Ishimaru* [5.10, 11] used spectral representations of the field and obtained correlation and structure functions for locally homogeneous and isotropic turbulence. *Gebhardt* and *Collins* [5.16] considered the mean of the log amplitude using the method of *Schmeltzer*. In this section, we outline the theory developed by *Ishimaru* [5]. It is a straightforward application of the spectral technique, and it can be shown to reduce to the results of *Schmeltzer*, *Kon* and *Tatarskii*, *Fried*, and *Seidman*.

We desire the Rytov solution of $U(r)$ satisfying the scalar wave equation

$$\{V^2 + k^2[1 + 2n_1(r, t)]\}U(r, t) = 0, \tag{5.5}$$

where $n_1(r, t)$ is the fluctuation of the index of refraction and assumed to be small compared to unity. The Rytov solution is given by

$$U(r, t) = U_0(r)\exp\psi_1(r, t)$$
$$= U_0(r)\exp[\chi(r, t) + iS_1(r, t)] \tag{5.6}$$

$$\psi_1(r, t) = \frac{k^2}{2\pi U_0(r)} \int_{V'} n_1(r, t)U_0(r)\frac{\exp(ik|r - r'|)}{|r - r'|}dV', \tag{5.7}$$

where the volume integration is over the space $0 \leq z' \leq z$. The real part χ and the imaginary part S_1 of ψ_1 are called the "log amplitude" and "phase" fluctuations, respectively.

Now, we introduce the transverse wind velocity $\bar{V}$ and assume that changes in the index of refraction field are primarily due to its transverse motion and not due to evolutions of the turbulence. Thus we have

$$n_1(r, t) = n_1(r - Vt, 0). \tag{5.8}$$

This assumption is called "Taylor's hypothesis" and is usually a good approximation for optical propagation in the atmosphere.

We assume that the refractive index fluctuations are locally homogeneous and isotropic and therefore can be expressed in the following spectral representation:

$$\begin{aligned} n_1(r, 0) &= n_1(\varrho, z, 0) \\ &= n_1(0, z, 0) + \int [1 - \exp(i\boldsymbol{\kappa} \cdot \boldsymbol{\varrho})] dv(\boldsymbol{\kappa}, z), \end{aligned} \tag{5.9}$$

where $\boldsymbol{\kappa}$ and $\boldsymbol{\varrho}$ are the wave number vector and the radial vector, respectively, in two dimensions and the random amplitude $dv(\boldsymbol{\kappa}, z)$ satisfies the condition

$$\langle dv(\boldsymbol{\kappa}, z) dv^*(\boldsymbol{\kappa}', z') \rangle = F_n(\boldsymbol{\kappa}, z - z') \delta(\boldsymbol{\kappa} - \boldsymbol{\kappa}') d\kappa d\kappa'.$$

The angle brackets represent an ensemble average. The quantity $F_n(\boldsymbol{\kappa}, z)$ is the two-dimensional spectrum of the refractive index variations and is related to the three-dimensional spectrum $\Phi_n(\boldsymbol{\kappa}, \kappa_3)$ through

$$\int_0^\infty F_n(\boldsymbol{\kappa}, z) dz = \pi \Phi_n(\boldsymbol{\kappa}, 0),$$

where κ_3 is the wave number in the z direction. Using the above relationships, we obtain the following correlation and structure functions for the log amplitude and phase fluctuations:

$$\begin{aligned} B_\chi &= B_\chi(\varrho_1, \varrho_2, z, \tau) = \langle \chi(\varrho_1, z, t) \chi(\varrho_2, z, t + \tau) \rangle \\ B_S &= B_S(\varrho_1, \varrho_2, z, \tau) = \langle S_1(\varrho_1, z, t) S_1(\varrho_2, z, t + \tau) \rangle \\ D_\chi &= D_\chi(\varrho_1, \varrho_2, z, \tau) = \langle [\chi(\varrho_1, z, t) - \chi(\varrho_2, z, t + \tau)]^2 \rangle \\ D_S &= D_S(\varrho_1, \varrho_2, z, \tau) = \langle [S_1(\varrho_1, z, t) - S_1(\varrho_2, z, t + \tau)]^2 \rangle. \end{aligned} \tag{5.10}$$

For homogeneous and isotropic turbulence, B_χ and B_S are given by

$$\begin{rcases} B_\chi \\ B_S \end{rcases} = 2\pi^2 \int_0^L d\eta \int_0^\infty d\kappa \kappa \operatorname{Re}[J_0(\kappa P)|H|^2 \pm J_0(\kappa Q)H^2]\Phi_n(\kappa, \eta). \tag{5.11a}$$

The variances of the log amplitude and the phase fluctuations are given by

$$\left.\begin{array}{c}\sigma_\chi^2\\\sigma_S^2\end{array}\right\} = 2\pi^2 \int\limits_0^L d\eta \int\limits_0^\infty d\kappa\kappa \, \text{Re}[I_0(2\gamma_i\kappa\varrho)|H|^2 \pm (H^2)]\Phi_n(\kappa,\eta), \tag{5.11b}$$

where $I_0(z)=J_0(iz)$ is a modified Bessel function.

For locally homogeneous and isotropic turbulence, we have

$$\left.\begin{array}{c}D_\chi\\D_S\end{array}\right\} = 4\pi^2 \int\limits_0^L d\eta \int\limits_0^\infty d\kappa\kappa \, \text{Re}\left\{\left[\frac{1}{2}I_0(2\gamma_i\kappa\varrho_1)+\frac{1}{2}I_0(2\gamma_i\kappa\varrho_2)\right.\right.$$

$$\left.\left. - J_0(\kappa P)\right]|H|^2 \pm [1-J_0(\kappa Q)]H^2\right\}\Phi_n(\kappa,\eta). \tag{5.12}$$

In the above expressions, the upper and lower signs correspond to the upper and lower functions, respectively, and we used the following notations:

$$|H|^2 = k^2 \exp\left[-\frac{\gamma_i(L-\eta)}{k}\kappa^2\right]$$

$$H^2 = -k^2 \exp\left[-i\frac{\gamma(L-\eta)}{k}\kappa^2\right]$$

$$\gamma = \frac{1+i\alpha\eta}{1+i\alpha L} = \gamma_r - i\gamma_i$$

$$\varrho_1 = (x_1^2+y_1^2)^{1/2} \quad \text{and} \quad \varrho_2 = (x_2^2+y_2^2)^{1/2}$$

$$P = [(\gamma x_1 - \gamma^* x_2 + V_x\tau)^2 + (\gamma y_1 - \gamma^* y_2 + V_y\tau)^2]^{1/2}$$

$$Q = \{[\gamma(x_1-x_2)+V_x\tau]^2 + [\gamma(y_1-y_2)+V_y\tau]^2\}^{1/2}. \tag{5.13}$$

For the plane wave case the range of validity of the Rytov solution of the variance of the log amplitude fluctuations is generally given by the following:

$$\sigma_\chi^2 < 0.2 \sim 0.5. \tag{5.14}$$

The Rytov solution for the phase variance σ_s^2 is considered to be valid even in the strong fluctuation region. We shall show in Section 5.1.3 that while (5.14) is an adequate condition for the variance of the log amplitude fluctuations for a collimated beam, it is not sufficient for a focused beam.

5.1.1 The Variance of the Log Amplitude Fluctuations

Let us assume that the atmospheric turbulence is characterized by the modified von Karman form of the Kolmogorov spectrum [see Chap. 2, Eq. (2.38)]

$$\Phi_n(\kappa,\eta)=0.033C_n^2(\eta)\left(\kappa^2+\frac{1}{L_0^2}\right)^{-11/6}\exp\left(-\frac{\kappa^2}{\kappa_m^2}\right), \tag{5.15}$$

where $\kappa_m = 5.92/l_0$, and L_0 and l_0 are the outer scale and the microscale of turbulence. The structure constant $C_n(\eta)$ may be a function of position along the propagation path.

The variance of the log amplitude fluctuations σ_χ^2 is obtained by substituting (5.15) into (5.12). We note that the integrand of (5.12) in the range of $\kappa < 2\pi/L_0$ is negligibly small as long as $\sqrt{\lambda L} < L_0$, and that the effect of the microscale l_0 is also relatively small. Therefore, we can let $L_0 \to \infty$ and $l_0 \to 0$. Also, we assume that the turbulence is uniform along the propagation path and, therefore, C_n is constant.

Under these assumptions, we obtain

$$
\sigma_\chi^2(L, \varrho) = \pi^2 (0.033 C_n^2) \left[-\Gamma\left(-\frac{5}{6}\right) \right] k^{7/6}
$$

$$
\cdot \operatorname{Re} \int_0^L d\eta \left\{ [i\gamma(L-\eta)]^{5/6} - [\gamma_i(L-\eta)]^{5/6} {}_1F_1\left(-\frac{5}{6}, 1; \frac{2\varrho^2}{W^2}\right) \right\}
$$

$$
= 2.176 C_n^2 k^{7/6} L^{11/6} \{ \operatorname{Re}[g_1(\alpha L)] - g_2(\alpha L, \varrho) \}, \tag{5.16}
$$

where

$$
g_1(\alpha L) = \frac{6}{11} i^{5/6} {}_2F_1\left(-\frac{5}{6}, \frac{11}{6}; \frac{17}{6}; \frac{i\alpha L}{1 + i\alpha L}\right)
$$

$$
g_2(\alpha L, \varrho) = \frac{3}{8} \left[\frac{\alpha_1 L}{(\alpha_1 L)^2 + (1 - \alpha_2 L)^2} \right]^{5/6} {}_1F_1\left(-\frac{5}{6}, 1; \frac{2\varrho^2}{W^2}\right)
$$

and ${}_2F_1(a, b; c; z)$ is a hypergeometric function and ${}_1F_1(a, b; z)$ is a Kummer function. We note that for both the plane wave ($\alpha_1 L \to 0$, $\alpha_2 L \to 0$) and spherical wave ($\alpha_1 L \to \infty$, $\alpha_2 L \to 0$) cases, g_2 reduces to zero and σ_χ^2 correctly reduces to the results given by *Tatarskii*, i.e.,

$$
\sigma_{\chi P}^2(L) = 0.307 C_n^2 k^{7/6} L^{11/6} \quad \text{plane wave}
$$

$$
\sigma_{\chi s}^2(L) = 0.124 C_n^2 k^{7/6} L^{11/6} \quad \text{spherical wave}. \tag{5.17}
$$

For a collimated beam ($\alpha_2 L \to 0$), the variance σ_χ^2 approaches $\sigma_{\chi P}^2$ as $\alpha_1 L \to 0$ and $\sigma_{\chi s}^2$ as $\alpha_1 L \to \infty$. When $\alpha_1 L \approx 1$, g_2 becomes significant and σ_χ^2 is slightly lower than $\sigma_{\chi s}^2$. This behavior was first shown by *Fried* and *Seidman* [5.14] and was experimentally verified by *Khmelevtsov* and *Tsvik* [5.17] (see Fig. 5.1). Note that *Khmelevtsov* and *Tsvik* used a reflective path, while the expression (5.16) is for a line-of-sight path.

For a focused beam, g_1 and g_2 in (5.16) tend to cancel each other and the variance σ_χ^2 becomes extremely small at the focal point ($\alpha_2 L = 1$), particularly when the transmitting aperture is large ($\alpha_1 L \ll 1$). This is also shown in Fig. 5.1. Several experimental studies have been conducted to verify this reduction of scintillation predicted by the theory [5.17, 18]. However, these experimental

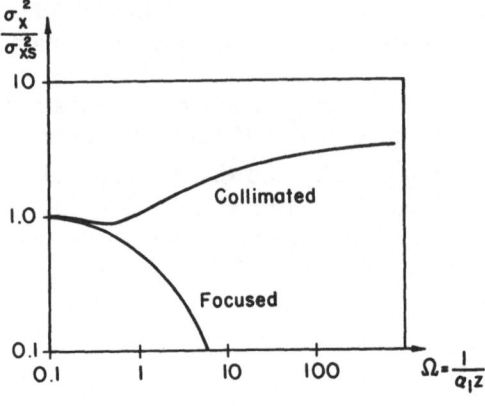

Fig. 5.1. Log amplitude variance σ_χ^2 as a function of $\Omega = 1/(\alpha_1 z) = \pi W_0^2/\lambda z$

results show that either this reduction does not exist or it is very difficult to achieve experimentally [5.19]. For example, the experimental variance σ_χ^2 for a focused beam appears to be closer to the case of a collimated beam than the value calculated by (5.16) [5.17]. However, *Kerr* and *Eiss* [5.19] discussed the difficulty in getting the receiver exactly at the focal point, which could explain the discrepancy between the theoretical prediction and the experiment. Despite this, at present, the validity of the weak fluctuation theory for a focused beam wave appears to be in serious doubt. We shall attempt to clarify the situation in the next section. Before we do this, however, we need to examine the general propagation characteristics of a focused beam.

5.1.2 Average Intensity and Beam Spread

Let us consider the average intensity $\langle I(L,\varrho)\rangle$ of a beam wave. Writing the field $U(r)$ as

$$U(r) = U_0(r)\exp(\chi_t + iS_t),$$ (5.18)

we have

$$\langle I(L,\varrho)\rangle = I_0(L,\varrho)\langle\exp(2\chi_t)\rangle,$$

where $I_0(L,\varrho) = |U_0(r)|^2$ is the intensity of the beam in free space and is given by (5.3). In (5.18), we used χ_t and S_t to denote the true log amplitude and phase fluctuation in contrast with the approximate Rytov solution χ and S_1 given in (5.7).

There appears to be reasonable experimental evidence that the intensity I is approximately log normally distributed at least under some conditions [5.17, 18]. Under this assumption we have

$$\langle I(L,\varrho)\rangle = I_0(L,\varrho)\exp[2\langle\chi_t\rangle + 2\langle(\chi_t - \langle\chi_t\rangle)^2\rangle].$$ (5.19)

The Rytov solution (5.7) is the first term of iteration and its average $\langle\chi\rangle+i\langle S_1\rangle$ is clearly zero and, therefore, the Rytov log amplitude χ must be approximately equal to $\chi_t-\langle\chi_t\rangle$. It is clear that the first Rytov solution (5.7) cannot yield the average $\langle\chi_t\rangle$. For the beam wave case, some attempt has been made to utilize higher order Rytov iterations to obtain $\langle\chi_t\rangle$ [5.7, 16, 20]. Let the Rytov solution for $\langle\chi_t\rangle$ and $\langle(\chi_t-\langle\chi_t\rangle)^2\rangle$ be $\langle\chi\rangle$ and σ_χ^2, respectively. Then we have the Rytov expression for the intensity,

$$\langle I(L,\varrho)\rangle\approx I_0(L,\varrho)\exp(2\langle\chi\rangle+2\sigma_\chi^2). \tag{5.20}$$

In addition it is reasonable to assume that the beam intensity has a gaussian profile,

$$\langle I(L,\varrho)\rangle=\frac{W_0^2}{W_b^2}\exp\left(-\frac{2\varrho^2}{W_b^2}\right), \tag{5.21}$$

where the intensity in free space is normalized so that $I_0(0,0)=1$ according to (5.1). Also, the total power as given by

$$P_t=\int\langle I(L,\varrho)\rangle d\varrho \tag{5.22}$$

should be equal to the incident power given by (5.4).

The beam broadening is then given by

$$\frac{W_b^2}{W_0^2}=\frac{1}{\langle I(L,0)\rangle}. \tag{5.23}$$

Using (5.19), we get

$$\frac{W_b^2}{W_0^2}=\exp[-2\langle\chi_t\rangle-2\langle(\chi_t-\langle\chi_t\rangle)^2\rangle]_{\varrho=0}. \tag{5.24a}$$

The Rytov expression for the beam broadening is given by (5.20)

$$\frac{W_b^2}{W_0^2}\approx\exp(-2\langle\chi\rangle-2\sigma_\chi^2)_{\varrho=0}. \tag{5.24b}$$

Bunkin and *Gochelashvily* [4.68] used Schmeltzer's results to calculate $\langle\chi\rangle$ and σ_χ^2 in (5.24b). *Gebhardt* and *Collins* [5.16] also made similar calculations of (5.24b) for collimated and focused beams. For example, *Bunkin* and *Gochelashvily* [5.20] obtained the following expression for the broadening of a collimated beam valid when $\sigma_\chi^2(\text{spherical})\ll1$:

$$a_{eff}^2=a_0^2+1.6C_n^2k^{1/3}a_0^{1/3}L^{8/3}, \qquad \Omega\gg1$$
$$a_{eff}^2=(z/ka_0)^2+1.6C_n^2a_0^{-1/3}L^3, \qquad \Omega\ll1$$

where

$$2a_{eff}^2 = W_b^2, \quad 2a_0^2 = W_0^2 \quad \text{and} \quad \Omega = (\alpha_1 L)^{-1}. \tag{5.25}$$

In the above, it is tacitly assumed that the Rytov solutions for $\langle \chi_t \rangle$ and $\langle (\chi_t - \langle \chi_t \rangle)^2 \rangle$ are valid for any $\varrho \neq 0$, that they satisfy the conservation of power given by (5.22), and that their values at $\varrho = 0$ determine the beam broadening W_b^2/W_0^2. Actually the range of validity of the Rytov solution for a collimated and a diverging beam is almost the same as that of a plane and spherical wave, but the range of validity for a focused beam is much more restricted. This is seen from the following considerations.

The range of validity of the Rytov solution for log amplitude fluctuations in the plane and spherical wave cases is that the log amplitude variance σ_χ^2 be small compared with unity, and no higher than about $0.2 \sim 0.5$, i.e.,

$$\sigma_{\chi P}^2 = 0.307 C_n^2 k^{7/6} L^{11/6} \ll 1. \tag{5.26}$$

It is reasonable to assume that for a beam wave, the range of validity of the Rytov solution for the log amplitude fluctuation is also given by

$$\sigma_\chi^2(L, \varrho) \ll 1, \tag{5.27}$$

where $\sigma_\chi^2(L, \varrho)$ is given in (5.16).

Let us consider the situation when the radial distance $\varrho \gtrsim W$. Then $_1F_1(a, b; z)$ in (5.16) may be approximated by $1 + (a/b)z$ and we obtain

$$\sigma_\chi^2(L, \varrho) \approx \sigma_\chi^2(L, 0) + f \frac{\varrho^2}{W^2}, \tag{5.28}$$

where

$$f = 1.36 C_n^2 k^{7/6} L^{11/6} \left[\frac{\alpha_1 L}{(\alpha_1 L)^2 + (1 - \alpha_2 L)^2} \right]^{5/6}.$$

For a collimated beam $\alpha_2 = 0$ and for a diverging beam $\alpha_2 < 0$, and hence f is always no greater than $\sigma_{\chi P}^2$ and therefore for $\varrho \gtrsim W$, (5.27) is always satisfied as long as (5.26) is satisfied.

The situation is quite different for the focused beam case ($\alpha_2 L = 1$). Here, at the focal point, f in (5.28) becomes

$$f = 1.36 C_n^2 k^{7/6} L^{11/6} (\alpha_1 L)^{-5/6} = 0.428 C_n^2 k^2 L W_0^{5/3}. \tag{5.29}$$

Therefore, the condition (5.27) becomes

$$\sigma_\chi^2(L, 0) + (0.428 C_n^2 k^2 L W_0^{5/3}) \frac{\varrho^2}{W^2} \ll 1. \tag{5.30}$$

It is well known that at the focal plane, $\sigma_\chi^2(L,0)$ is extremely small, but the second term in (5.30) can be quite large even at a distance for which (5.26) is satisfied. Therefore we conclude that for a focused beam, the range of validity of the Rytov solution should be given by

$$f = 0.428 C_n^2 k^2 L W_0^{5/3} \ll 1. \tag{5.31}$$

We conclude from (5.20), (5.22), (5.24b), and (5.28) that the average intensity for a beam wave is given by

$$\langle I(L,\varrho) \rangle = \frac{W_0^2}{W_b^2} \exp\left(-\frac{2\varrho^2}{W_b^2}\right) \tag{5.32}$$

$$W_b^2 = \frac{W^2}{1-f}, \qquad W^2 = W_0^2[(\alpha_1 L)^2 + (1-\alpha_2 L)^2],$$

where f is approximately given by (5.28). A similar result was obtained by *Ishimaru* [5.10]. It was noted by *Khmelevtsov* [5.7] that (5.32) is consistent with (5.25) for the collimated beam case.

It will be shown in (5.66b) that the general characteristics given in (5.32) are consistent with a solution obtained from the mutual coherence function formulation in the weak fluctuation limit except for a minor difference in the constant for f and, therefore, (5.32) may be considered as a valid approximation for the beam characteristics in weak turbulence.

Buck [5.21] reported on some experimental measurements for a beam focused at the observation point. He showed that at a fixed distance and a given frequency, the beam size W_b at the focal point at first decreases as the transmitter aperture size W_0 increases, reaches a minimum, and then increases. This can be easily explained by (5.32). Note that at the focal point, $\alpha_2 L = 1$, and as W_0 increases, W decreases as W_0^{-1}, but $(1-f)^{-1}$ increases. *Buck* also noted that the beam size W_b increases with the distance L at a rate slightly greater than linear. This can also be seen from (5.32).

Recently, careful experimental work has been reported by *Kerr* and *Dunphy* [5.18] on a focused beam in the weak fluctuation region. They observed that in the focal plane, the beam size is almost equal to that in free space, but the beam wanders around. Since the theoretical beam spread discussed above includes the effects of both the beam wander and the scintillations, the theoretical beam spread W_b should be considered to be comparable to the range of the radius of the beam wander. Also, the experiment indicates that the correlation distance of the intensity in the focal plane is comparable to the beam size in free space. Since the correlation distance may be considered as the instantaneous size of the beam, this is a manifestation of the fact that the beam preserves its size in weak turbulence. In contrast with this, for the strong fluctuation case, the beam is broken up into many patches. This aspect will be discussed further in the next section.

5.1.3 Angle of Arrival

Kon and *Tatarskii* [5.12] obtained expressions for the angle of arrival $\Delta\alpha$ in the weak fluctuation case. If a wave is propagating at an angle $\Delta\alpha$ with respect to the z axis, then the phase difference ΔS_1 at two points separated by ϱ in the z plane is given approximately by $\Delta S_1 \approx (k\varrho)\Delta\alpha$. Therefore, the variance of the fluctuation of the angle of arrival is given by

$$\langle (\Delta\alpha)^2 \rangle = \lim_{\varrho \to 0} \left[\frac{D_s(\varrho)}{k^2 \varrho^2} \right], \tag{5.33}$$

where $D_s(\varrho) = \langle (\Delta S_1)^2 \rangle$ is the phase structure function. The theoretical results given by *Kon* and *Tatarskii* were verified experimentally by *Borisov* et al. [5.22].

5.1.4 Temporal Frequency Spectra

Temporal frequency spectra of wave fluctuations for plane and spherical waves were obtained previously by *Tatarskii* [5.1, 2] and *Clifford* [5.23]. Extension of this technique to a beam wave case is straightforward and was obtained by *Ishimaru* [5.24]. Let us consider the temporal frequency spectrum of the log amplitude fluctuations $W_\chi(\omega)$ on the beam axis. This is given by

$$W_\chi(\omega) = 4 \int_0^\infty B_\chi(\tau) \cos \omega \tau \, d\tau, \tag{5.34}$$

where $B_\chi(\tau)$ is given in (5.10) with $\varrho_1 = \varrho_2 = 0$. We then obtain

$$W_\chi(\omega) = \frac{8\pi^2 k^2}{V} \int_0^\infty f_\chi(\kappa) \Phi_n(\kappa) d\kappa', \tag{5.35}$$

where

$$f_\chi(\kappa) = \int_0^L d\eta \, \text{Re}(|h|^2 - h^2)$$

$$h = \exp\left[-i \frac{\gamma(L-\eta)}{2k} \kappa^2 \right]$$

$$\kappa = \left(\kappa'^2 + \frac{\omega^2}{V^2} \right)^{1/2}.$$

The phase spectrum $W_s(\omega)$ is given by the expression (5.35) also except that f_χ is replaced by f_s

$$f_s(\kappa) = \int_0^L d\eta \, \text{Re}(|h|^2 + h^2). \tag{5.36}$$

The cross spectrum $W_{\chi S}$ between the log amplitude fluctuations and the phase fluctuations is given by (5.35) with $f_{\chi S}$ in place of f_χ, i.e.,

$$f_{\chi S}(\kappa) = \int_0^L d\eta \, \mathrm{Im}(-|h|^2 - h^2).$$
(5.37)

Using the modified von Karman spectrum (5.15) for $\Phi_n(\kappa)$, we get

$$\left.\begin{array}{c} W_\chi(\omega) \\ W_S(\omega) \end{array}\right\} = \frac{8\pi^2 k^2}{V}(0.033 C_n^2) \int_0^L d\eta \, \mathrm{Re}(G_1 \mp G_2)$$

$$W_{\chi S}(\omega) = \frac{8\pi^2 k^2}{V}(0.033 C_n^2) \int_0^L d\eta \, \mathrm{Im}(-G_1 - G_2)$$
(5.38)

where

$$G_1 = \frac{\sqrt{\pi}}{2} C^{-8/3} \psi\left(\frac{1}{2}, -\frac{1}{3}; AC^2\right) e^{-A\frac{\omega^2}{V^2}}$$

$$G_2 = \frac{\sqrt{\pi}}{2} C^{-8/3} \psi\left(\frac{1}{2}, -\frac{1}{3}; BC^2\right) e^{-A\frac{\omega^2}{V^2}}$$

$$A = \frac{\gamma_i(L-\eta)}{k}, \qquad B = \frac{i\gamma(L-\eta)}{k}, \qquad C = \left(\frac{\omega^2}{V^2} + \frac{1}{L_0^2}\right)^{1/2}$$

and $\psi(a, b; z)$ is the confluent hypergeometric function which is independent of the Kummer function $_1F_1(a, b; z)$. Detailed theoretical and experimental investigations of the frequency spectrum for an optical beam have not been reported to date.

5.2 Strong Fluctuation Theory

As discussed in Section 5.1.3, the weak fluctuation theory is valid only when the log amplitude variance satisfies (5.26). When this limit is exceeded, the fluctuations are considered "strong". Several methods have been proposed to deal with the strong fluctuation problem. They are the diagram method [5.2, 6, 25, 26], the integral equation representations of the diagram results including the Dyson and the Bethe-Salpeter equations [5.26–28], the parabolic equation method [5.2, 29–34] and the local method of small perturbations [5.8, 20]. A phenomenological approach using the extended Huygens-Fresnel method [5.35–42] has also been used for weak as well as strong fluctuation problems. These have already been discussed in Chapter 3 and therefore this section is devoted only to those problems directly related to the beam wave problem.

The extended Huygens-Fresnel technique is based on an assumption which is valid through terms of second order in the refractive index fluctuations. This is equivalent to the assumption of a gaussian distribution for the log amplitude and phase fluctuations [5.41]. It has been shown [5.43] that for the second moment all the techniques indicated above are equivalent to each other. For the fourth moment, the Huygens-Fresnel method has recently been used to obtain a relatively simple solution [5.39, 41]. At present, several asymptotic solutions of the parabolic equation for the fourth-order moment are available [5.8, 9, 44]. Some numerical solutions and comparison with experimental data have also been reported [5.28, 45].

5.2.1 Strong Fluctuation Theory for the Coherence Function

As noted in the above, various techniques of obtaining the second moment in the strong fluctuation region are equivalent [5.43], and therefore in this section, we outline the solution based on the parabolic equation method. The validity of the parabolic equation technique has been examined recently.

Let us consider a wave equation for a scalar field $U(r)$

$$\{\nabla^2 + k^2[1 + \varepsilon_1(r)]\} U(r) = 0, \tag{5.39}$$

where $\varepsilon_1(r)$ is the fluctuation of the relative dielectric constant. It is assumed to be small, and therefore, it is approximately equal to $2n_1(r)$.

We first note that the wave propagates in the z direction and, writing

$$U(r) = u(r)\exp(i\,kz), \tag{5.40}$$

and noting that $u(r)$ is slowly varying in z, we get a simpler differential equation for $u(r)$,

$$2ik\frac{\partial u(r)}{\partial z} + \nabla_t^2 u(r) + k^2\varepsilon_1(r)u(r) = 0, \tag{5.41}$$

where ∇_t^2 is the transverse Laplacian.

Assuming that $\varepsilon_1(r)$ is a gaussian random field, we get the differential equation for the average field [5.2]

$$\left[2ik\frac{\partial}{\partial z} + \nabla_t^2 + \frac{ik^3}{4}A(0)\right]\langle u(z,\varrho)\rangle = 0, \tag{5.42}$$

where

$$A(\varrho) = 16\pi^2 \int_0^\infty J_0(\kappa\varrho)\phi_n(\kappa)\kappa d\kappa. \tag{5.43}$$

The solution of (5.42) is easily obtained;

$$\langle u(\varrho, z)\rangle = u_0(\varrho, z)\exp(-\alpha_0 z), \tag{5.44}$$

where $u_0(\varrho, z)$ is the field in free space and for a beam wave it is given by (5.2),

$$u_0(\varrho, z) = \frac{1}{1+i\alpha z}\exp\left[-\left(\frac{k\alpha}{2}\right)\frac{\varrho^2}{1+i\alpha z}\right], \tag{5.45}$$

and α_0 is given by

$$\alpha_0 = 2\pi^2 k^2 \int\limits_0^\infty \Phi_n(\kappa)\kappa d\kappa. \tag{5.46}$$

For the Kolmogorov spectrum given in (5.15) we get

$$\alpha_0 = 0.033\pi^2 C_n^2 k^2 L_0^{5/3}\psi\left[1,\frac{1}{6};(\kappa_m L_0)^{-2}\right], \tag{5.47}$$

where $\psi(a, b; z)$ is a confluent hypergeometric function which is independent of the Kummer function. Since $\kappa_m L_0 = (5.92\, L_0)/l_0$, $(\kappa_m L_0)^{-2}$ is much smaller than unity and, therefore, we can use

$$\psi(a, b; z) \to \frac{\Gamma(1-b)}{\Gamma(1+a-b)} \quad \text{for} \quad \text{Re}\, b < 1 \quad \text{and} \quad |z| \ll 1$$

and obtain

$$\alpha_0 = 0.391\, C_n^2 k^2 L_0^{5/3}. \tag{5.48}$$

Therefore the coherent intensity is given by

$$|\langle U(r)\rangle|^2 = \frac{W_0^2}{W^2}\exp\left(-\frac{2\varrho^2}{W^2} - 2\alpha_0 z\right), \tag{5.49}$$

indicating that it has exactly the same transverse dependence as the free space beam wave. The attenuation $\exp(-2\alpha_0 z)$ depends largely upon the outer scale L_0. We note that $2\alpha_0$ as given in (5.46) is equal to the total scattering cross section per unit volume of the turbulence. In a turbulent atmosphere, the attenuation constant $2\alpha_0$ is sufficiently large so that the coherent intensity as given by (5.49) disappears within a few hundred meters. Therefore the intensity in a turbulent atmosphere is almost totally incoherent.

The differential equation for the mutual coherence function $\Gamma(\varrho_1, \varrho_2, z)$ defined by

$$\Gamma(\varrho_1, \varrho_2, z) = \langle U(\varrho_1, z)U^*(\varrho_2, z)\rangle \tag{5.50}$$

is given by

$$\left\{ 2ik\frac{\partial}{\partial z} + (V_{t1}^2 - V_{t2}^2) + \frac{ik^3}{2}[A(0) - A(\varrho_1 - \varrho_2)] \right\} \Gamma(\varrho_1, \varrho_2, z) = 0. \qquad (5.51)$$

The exact solution of this equation has been obtained [5.2]

$$\Gamma(\varrho_c, \varrho_d, z) = \int M\left(\kappa_d, \varrho_d - \frac{\kappa_d z}{k}, 0 \right) \exp\left[i\kappa_d \cdot \kappa_c - H(\kappa_d, \varrho_d, z) \right] d\kappa_d, \qquad (5.52)$$

where $\varrho_c = \frac{1}{2}(\varrho_1 + \varrho_2)$ and $\varrho_d = \varrho_1 - \varrho_2$, and M is related to the mutual coherence function $\Gamma(\varrho_c, \varrho_d, 0)$ through a Fourier transform

$$M(\kappa_d, \varrho_d, 0) = \frac{1}{(2\pi)^2} \int \Gamma(\varrho_c, \varrho_d, 0) e^{-i\kappa_d \cdot \varrho_c} d\varrho_c. \qquad (5.53)$$

The function H is given by

$$H(\kappa_d, \varrho_d, z) = \frac{k^2}{4} \int_0^z \left[A(0) - A\left(\varrho_d - \frac{\kappa_d}{k} z' \right) \right] dz'. \qquad (5.54)$$

Alternatively, using $\varrho' = \varrho_d - (\kappa_d/k)z$, the solution (5.52) can be written as

$$\Gamma(\varrho_c, \varrho_d, z) = \left(\frac{k}{2\pi z} \right)^2 \int d\varrho_c' \int d\varrho_d' \Gamma(\varrho_c', \varrho_d', 0)$$

$$\cdot \exp\left[i\frac{k}{z}(\varrho_d - \varrho_d') \cdot (\varrho_c - \varrho_c') - H(\varrho_d, \varrho_d', z) \right], \qquad (5.55)$$

where

$$H(\varrho_d, \varrho_d', z) = \frac{k^2}{4} \int_0^z \left\{ A(0) - A\left[\varrho_d - (\varrho_d - \varrho_d')\frac{z'}{z} \right] \right\} dz'.$$

For a beam wave, we have

$$\Gamma(\varrho_c, \varrho_d, 0) = \exp\left(-\frac{k\alpha}{2}\varrho_1^2 - \frac{k\alpha^*}{2}\varrho_2^2 \right)$$

$$= \exp\left\{ -k\left[\alpha_1\left(\varrho_c^2 + \frac{1}{4}\varrho_d^2 \right) + i\alpha_2 \varrho_c \cdot \varrho_d \right] \right\} \qquad (5.56a)$$

and, therefore,

$$M(\kappa_d, \varrho_d, 0) = \frac{1}{4\pi k\alpha_1} \exp\left(-k\alpha_1 \frac{\varrho_d^2}{4} - \frac{|\kappa_d + k\alpha_2\varrho_d|^2}{4k\alpha_1} \right). \qquad (5.56b)$$

In order to evaluate the integrals in (5.52) or (5.55), we need to examine H in (5.54),

$$A(0) - A(\varrho) = 16\pi^2 \int_0^\infty [1 - J_0(\kappa\varrho)]\Phi_n(\kappa)\kappa d\kappa \,. \tag{5.57}$$

This integral cannot be evaluated for the Kolmogorov spectrum given in (5.15). However, we can write the spectrum in the following form:

$$\Phi_n(\kappa) = 0.033 \, C_n^2 \left(\kappa^2 + \frac{1}{L_0^2}\right)^{-11/6} \exp\left(-\frac{\kappa^2}{\kappa_m^2}\right)$$

$$= \Phi_{n0}(\kappa) - \Phi_1(\kappa, L_0) - \Phi_2(\kappa, L_0, l_0) \,, \tag{5.58}$$

where Φ_{n0} is independent of L_0 and l_0, Φ_1 represents the effect of L_0, and Φ_2 represents the effect of L_0 and l_0. These quantities are given by

$$\Phi_{n0}(\kappa) = 0.033 \, C_n^2 \kappa^{-11/3}$$

$$\Phi_1(\kappa, L_0) = 0.033 \, C_n^2 \left[k^{-11/3} - \left(\kappa^2 + \frac{1}{L_0^2}\right)^{-11/6} \right]$$

$$\Phi_2(\kappa, L_0, l_0) = 0.033 \, C_n^2 \left(\kappa^2 + \frac{1}{L_0^2}\right)^{-11/6} \left[1 - \exp\left(-\frac{\kappa^2}{\kappa_m^2}\right)\right] \,.$$

Now we note that Φ_2 is negligible for $\kappa < \kappa_m$ because of the factor $[1 - \exp(-\kappa^2/\kappa_m^2)]$ and furthermore L_0 has almost no effect on Φ_2. Therefore, we can approximate Φ_2 by

$$\Phi_2(\kappa, L_0, l_0) \approx \Phi_2(\kappa_2, l_0) = 0.033 \, C_n^2 \kappa^{-11/3} \left[1 - \exp\left(-\frac{\kappa^2}{\kappa_m^2}\right)\right] \,.$$

We can now obtain a closed form expression for $A(0) - A(\varrho)$. Corresponding to Φ_{n0}, Φ_1, and Φ_2, we have

$$16\pi^2 \int_0^\infty [1 - J_0(\kappa\varrho)]\Phi_n(\kappa)\kappa d\kappa = A_0(\varrho) - A_1(\varrho, L_0) - A_2(\varrho, l_0) \,, \tag{5.59}$$

where

$$A_0(\varrho) = 5.83 \, C_n^2 \varrho^{5/3} \,,$$

$$A_1(\varrho, L_0) = C_n^2 \left[5.83 \, \varrho^{5/3} - 3.127 \, L_0^{5/3} - 3.109(L_0\varrho)^{5/6} K_{-5/6}\left(\frac{\varrho}{L_0}\right) \right] \,,$$

$$A_2(\varrho, I_0) = C_n^2 \left\{ 5.83\varrho^{5/3} - 17.404 \, \kappa_m^{-5/3} \left[{}_1F_1\left(-\frac{5}{6}, 1; -\frac{\kappa_m^2\varrho^2}{4}\right) - 1 \right] \right\} \,.$$

Note that $A_1 \to 0$ as $L_0 \to \infty$ and A_1 can be expanded in a power series of $(1/L_0)$. Similarly, $A_2 \to 0$ as $l_0 \to 0$, and A_2 can be expanded in a power series of l_0. An examination of (5.59) reveals the following general characteristics:

$$A(0) - A(\varrho) = 6.56 \, C_n^2 \frac{\varrho^2}{l_0^{1/3}} \quad \text{for} \quad \varrho < l_0, \tag{5.60a}$$

$$= 5.83 \, C_n^2 \varrho^{5/3} \quad \text{for} \quad l_0 < \varrho < L_0, \tag{5.60b}$$

$$= 3.127 \, C_n^2 L_0^{5/3} \quad \text{for} \quad L_0 < \varrho. \tag{5.60c}$$

Since (5.52) contains the factor $\exp(-H)$, the significant contribution to the integral comes from the range $H \gtrsim 1$. Therefore we note that depending on the distance z, different appropriate approximations can be used. We divide the range into the following three cases [5.35]:

$$z \gg z_i, \tag{5.61a}$$

$$z_i \gg z \gg z_c, \tag{5.61b}$$

$$z_c \gg z, \tag{5.61c}$$

corresponding to (5.60a)–(5.60c), respectively, where

$$z_i = (0.39 \, C_n^2 k^2 l_0^{5/3})^{-1}$$
$$z_c = (0.39 \, C_n^2 k^2 L_0^{5/3})^{-1}.$$

We now examine each of the above three cases. We note, however, that for most optical propagation in the atmosphere, (5.61b) is applicable.

$z \gg z_i$ (5.61a)

This case does not occur under normal atmospheric conditions, since it represents the limiting case of an extremely large distance.

In this case, we can make use of (5.60a). Substituting this into (5.52) and (5.54) and making use of (5.55) and (5.56), we obtain the following solution:

$$\Gamma(\varrho_c, \varrho_d, z) = \left(\frac{W_0}{W}\right)^2 \exp\left(\frac{2\varrho_c^2}{W^2} - p\varrho_d^2 + i2q\varrho_d \cdot \varrho_c\right), \tag{5.62a}$$

where the beam size W is given by

$$W^2 = W_0^2[(\alpha_1 z)^2 + (1 - \alpha_2 z)^2] + 4.38 \, C_n^2 l_0^{-1/3} z^3$$

$$p = \frac{1}{2W_0^2}\left(1 + \frac{\alpha_2^2}{\alpha_1^2}\right) + \beta z - \frac{2}{W^2} F^2$$

$$q = \frac{2}{W^2} F, \quad F = \frac{(\alpha_1 z)}{2}\left(1 + \frac{\alpha_2^2}{\alpha_1^2}\right) - \frac{\alpha_2}{2\alpha_1} + \beta \frac{z^2}{k} \tag{5.62b}$$

$$\beta = 0.906 \, k^2 \kappa_m^{1/3} C_n^2.$$

Note that as $z \to \infty$ the beam size W expands as $z^{3/2}$ and the correlation distance, which is $p^{-1/2}$, decreases as $z^{-1/2}$.

Feizulin and *Kravtsov* [5.40] considered the broadening of a laser beam and obtained the following formula for the beam size:

$$\overline{\langle \varrho^2 \rangle} = a^2 + \frac{L^2}{k^2 a^2} + 2.19\, C_n^2 l_0^{-1/3} L^3 . \tag{5.62c}$$

It can be easily shown that for a collimated beam $(\alpha_2 L = 0)$, the beam size W^2 given in (5.62b) reduces exactly to (5.62c) by letting $W^2 = \overline{2\langle \varrho^2 \rangle}$ and $W_0^2 = 2a^2$.

$z_i \gg z \gg z_c$ (5.61b)

This is the case most often encountered in practice, and the expression for $A(0) - A(\varrho)$ is given by (5.60b). Substituting this into (5.54), we get

$$\Gamma(\varrho_c, \varrho_d, z) = \frac{W_0^2}{8\pi} \int d\kappa_d \exp(-a\varrho_d^2 - b\kappa_d^2 + c\varrho_d \cdot \kappa_d + i\kappa_d \cdot \varrho_c - H), \tag{5.63}$$

where

$$a = \frac{1}{2W_0^2}\left(1 + \frac{\alpha_2^2}{\alpha_1^2}\right)$$

$$b = \frac{W_0^2}{8}\left[(\alpha_1 z)^2 + (1 - \alpha_2 z)^2\right] = \frac{W^2}{8}$$

$$c = \frac{1}{2}\left[\alpha_1 z - \frac{\alpha_2}{\alpha_1}(1 - \alpha_2 z)\right]$$

$$H = 1.46 \int_0^z k^2 C_n^2 \left|\varrho_d - \frac{\kappa_d}{k} z'\right|^{5/3} dz' .$$

The average intensity $\langle I(\varrho_c, z) \rangle$ is, therefore, given by

$$\langle I(\varrho_c, z) \rangle = \Gamma(\varrho_c, z, 0) \tag{5.64a}$$

$$= \frac{W_0^2}{4} \int_0^\infty d\kappa_d \kappa_d J_0(\kappa_d \varrho_c) \exp(-b\kappa_d^2 - H)$$

$$H = 0.547\, C_n^2 k^{1/3} z^{8/3} \kappa_d^{5/3} .$$

If we neglect the $b\kappa_d^2$ term in comparison with H, then (5.64) can be evaluated for $\varrho_c = 0$ and yields the on-axis intensity given by *Lutomirski* and *Yura* [5.36]. We can express (5.64) in the following normalized form using $b\kappa_d^2 = t^2$:

$$\langle I(\varrho_c, z) \rangle = \frac{2W_0^2}{W^2} \int_0^\infty t\, dt\, J_0(t\varrho') \exp\left[-t^2 - (1/2)D_s t^{5/3}\right], \tag{5.64b}$$

where $\varrho' = \dfrac{2\sqrt{2}\varrho_c}{W}$ and

$$D_s = 6.2\,C_n^2 k^{1/3} z^{8/3} W^{-5/3}.$$

At the focal point, D_s becomes

$$D_s = 1.95\,C_n^2 k^2 z W_0^{5/3}.$$

This is consistent with (5.75b) given by *Banakh* et al. [5.41]. The correlation function on axis ($\varrho_c = 0$) can also be found from (5.63),

$$\Gamma(0, \varrho_d, z) = \frac{W_0^2}{8\pi} \int\limits_0^\infty dк_d к_d \int\limits_0^{2\pi} d\phi \exp\left(-a\varrho_d^2 - bк_d^2 + cк_d \varrho_d \cos\phi - H\right), \qquad (5.65)$$

where

$$H = 1.46 \int\limits_0^z k^2 C_n^2 \left[\varrho_d^2 + \left(\frac{z'}{k}к_d\right)^2 - 2\frac{z'}{k}к_d\varrho_d\cos\phi\right]^{5/6} dz'.$$

Equations (5.64) and (5.65) cannot be expressed in closed forms with known functions and must be evaluated either approximately or numerically. We note that in the above formulas, we used the approximation (5.60b) and, therefore, the effects of the inner and outer scales do not appear in the solution. To include these effects, we need to use (5.59).

Figure 5.2 shows the average intensity $\langle I(0, z)\rangle$ on the beam axis as a function of distance as calculated numerically from (5.64). It shows that in a turbulent medium, the average intensity for a focused beam approaches that of a collimated beam within a relatively short distance. Note that for this example, $(1/\alpha_1) = (\pi W_0^2)/\lambda = 13\,\text{km}$ and the log amplitude variance for a plane wave σ_{xp}^2 becomes 0.5 at a distance of 1.92 km.

Figure 5.3 shows the average intensity $\langle I(\varrho_c, z)\rangle$ as a function of ϱ_c for a collimated beam normalized to the value at the beam axis ($\varrho_c = 0$) as found from (5.64) and also the normalized mutual coherence function on the beam axis, $\varrho_c = 0$, as a function of separation ϱ_d, as given in (5.65). Again, these integrals are evaluated numerically. It shows that at a short distance $z = 0.5\,\text{km}$, the beam spread is negligible, but the correlation distance is reduced to almost (1/4) of the value for free space. In fact the distance at which the mutual coherence function reduces to e^{-1} becomes almost a half at $z = 100\,\text{m}$. At $z = 5\,\text{km}$, the beam spreads out and the e^{-1} distance is reduced considerably. The e^{-1} distance represents the size of the instantaneous speckle pattern of the wave amplitude (or intensity), and therefore a smaller correlation distance means broken-up patterns of the beam shape [5.18].

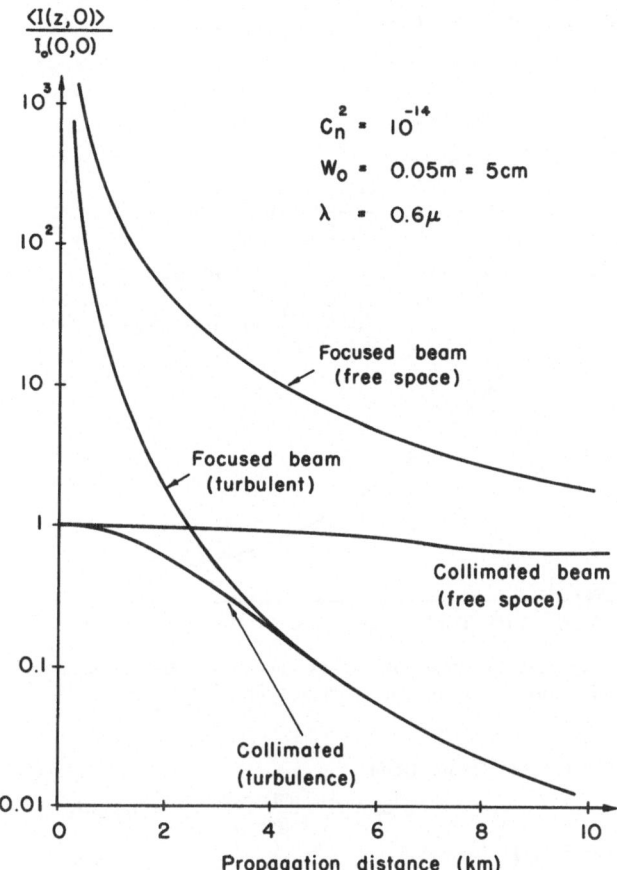

$$\frac{\langle I(z,0)\rangle}{I_e(0,0)}$$

$C_n^2 = 10^{-14}$

$W_0 = 0.05\,m = 5\,cm$

$\lambda = 0.6\mu$

Focused beam
(free space)

Focused beam
(turbulent)

Collimated beam
(free space)

Collimated
(turbulence)

Propagation distance (km)

Fig. 5.2. Average intensity $\langle I \rangle$ of collimated and focused beams on the beam axis as a function of distance

Figure 5.4, the focused beam case, it shows that at $z=500\,m$ the beam spreads slightly, but unlike the collimated beam the e^{-1} distance is not much different from that of the beam in free space, which is consistent with the experimental data [5.18]. At $z=5\,km$, the beam spread and the e^{-1} distance are almost the same as those of a collimated beam, indicating that practically no focusing is taking place at this distance, and that the speckle pattern of the focused beam at this distance is almost the same as that of the collimated beam.

We also note that the correlation distance is approximately equal to $\sqrt{\lambda z}$ for the plane wave case when the fluctuations are weak. For a collimated beam (see Fig. 5.3), the correlation distance is close to $\sqrt{\lambda z} = 0.0173$ at $z=0.5\,km$. However at $z=5\,km$, it is considerably smaller than $\sqrt{\lambda z} = 0.547$. For a focused beam, the correlation distance is considerably smaller than $\sqrt{\lambda z}$ even at $z=0.5\,km$.

It may be instructive to examine (5.63) in the weak fluctuation limit and compare it with the weak fluctuation theory (5.32). For weak fluctuations, we

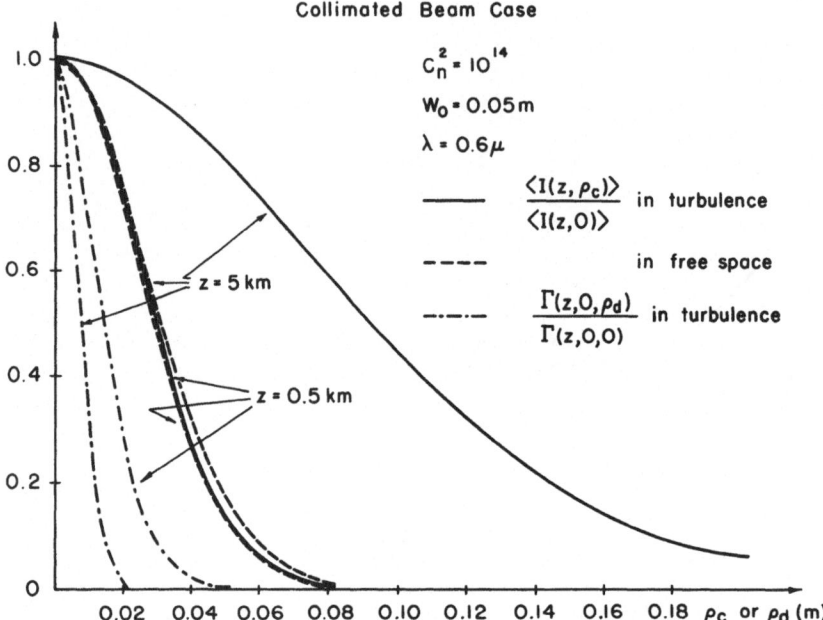

Fig. 5.3. Collimated beam case: Normalized average intensity as functions of transverse distance ϱ_c at $z=0.5$ km and $z=5$ km, and normalized mutual coherence function on beam axis as functions of separation ϱ_d

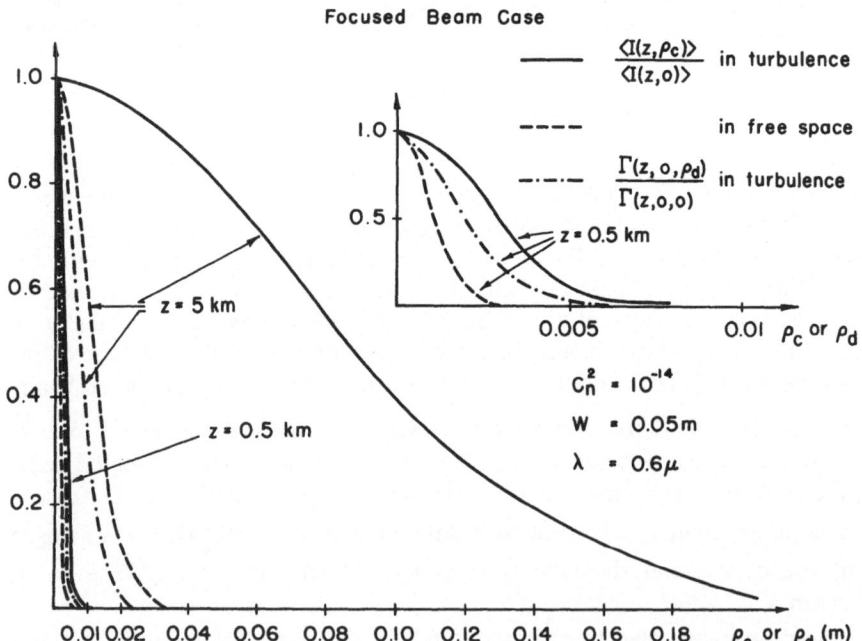

Fig. 5.4. Focused beam case: Normalized average intensity as functions of transverse distance ϱ_c and normalized mutual coherence function on beam axis as functions of separation ϱ_d

approximate $\exp(-H)\approx 1-H$ in (5.63). Now, we can easily perform the integration in (5.63) and obtain

$$\langle I(\varrho_c,z)\rangle = \frac{W_0^2}{8}\left[\frac{\exp\left(-\dfrac{2\varrho_c^2}{W^2}\right)}{b}-0.547\,C_n^2 k^{1/3} z^{8/3}\frac{\Gamma\left(\dfrac{11}{6}\right)}{\dfrac{11}{6}b^6}\,{}_1F_1\left(\frac{11}{6},1;-\frac{\varrho_c^2}{4b}\right)\right].$$

(5.66a)

Expanding ${}_1F_1$, keeping the first two terms, and equating them with an exponential function, we get

$$\langle I(\varrho_c,z)\rangle = \frac{W_0^2}{W^2}(1-f_s)\exp\left(-\frac{2\varrho_c^2}{W_b^2}\right),$$

where

$$f_s=1.63\,C_n^2 k^{7/6} z^{11/6}\left[\frac{\alpha_1 z}{(\alpha_1 z)^2+(1-\alpha_2 z)^2}\right]^{5/6}$$

$$W_b^2=\frac{W^2}{1-f}$$

(5.66b)

and f is given in (5.28). Comparing (5.66b) with (5.32), we note that both have the same beam spread W_b. The difference is that (5.66b) has $(1-f_s)$ while (5.26) has $(1-f)$ in magnitude at $\varrho_c=0$. This difference is minor as f_s is 1.2 times f.

$z_c\gg z$ (5.61c)

In this case, the effect of the inner scale is negligible and the effect of the outer scale is dominant, and we should use

$$A(0)-A(\varrho)=A_0(\varrho)-A_1(\varrho,L_0)$$

$$=C_n^2\left[3.127 L_0^{5/3}-3.109(L_0\varrho)^{5/6}K_{-5/6}\left(\frac{\varrho}{L_0}\right)\right].$$

(5.67)

Some detailed calculations for this distance were reported recently [5.46].

5.2.2 Temporal Frequency Spectra [5.46, 47]

If the wind with velocity V transverse to the propagation path is assumed "frozen" within the correlation time of observation, the parabolic equation (5.51) may be modified to give

$$\left\{2ik\frac{\partial}{\partial z}+(\nabla_{t_1}^2-\nabla_{t_2}^2)+\frac{ik^3}{2}[A(0)-A(\varrho_1-\varrho_2-V\tau)]\right\}$$

$$\Gamma(\varrho_1,\varrho_2,z,\tau)=0,$$

(5.68)

where

$$\Gamma(\varrho_1, \varrho_2, z, \tau) = \langle U(\varrho_1, z, t + \tau) U^*(\varrho_2, z, t) \rangle.$$

The solution to (5.68) is identical to (5.52) or (5.55) except that $A(\varrho)$ is replaced by $A(\varrho - V\tau)$. The frequency spectrum $W_f(z, \Omega)$ on the beam axis is given by

$$W_f(z, \Omega) = 2 \int_{-\infty}^{\infty} \Gamma(0, 0, z, \tau) e^{-i\Omega\tau} d\tau. \tag{5.69}$$

For the case $z_i \gg z \gg z_c$, we get

$$W_f(z, \Omega) = \frac{W_0^2}{4\pi} \int d\kappa_d \int_{-\infty}^{\infty} d\tau \exp(-b\kappa_d^2 - H - i\Omega\tau), \tag{5.70}$$

where

$$H = 1.46 \int_0^z k^2 C_n^2 \left| V\tau - \frac{\kappa_d}{k} z' \right|^{5/3} dz'.$$

The temporal frequency spectrum $W_f(z, \Omega)$ has not been studied in detail except for the plane wave case [5.46].

5.2.3 Two-Frequency Correlation Function

The correlation function between waves at two different operating frequencies $k_1 = \omega_1/c$ and $k_2 = \omega_2/c$ satisfies the following parabolic differential equation:

$$\left(2i \frac{\partial}{\partial z} + \frac{1}{k_1} \nabla_{t_1}^2 - \frac{1}{k_2} \nabla_{t_2}^2 + \frac{i}{4} \{ k_1^2 A_1(0) + k_2^2 A_2(0) \right.$$
$$\left. - k_1 k_2 [A_1(\varrho_d - V\tau) + A_2(\varrho_d - V\tau)] \} \right) \Gamma = 0, \tag{5.71}$$

where

$$\langle U(\varrho_1, z, t + \tau, k_1) U^*(\varrho_2, z, t, k_2) \rangle$$
$$= \Gamma(\varrho_1, \varrho_2, z, \tau, k_1, k_2) \exp[-i\omega_1(t + \tau) + i\omega_2 t],$$

and $A_1(\varrho)$ is given by

$$A_1(\varrho) = 16\pi^2 \int_0^{\infty} J_0(\kappa\varrho) \Phi_{n1}(\kappa) \kappa d\kappa, \tag{5.72}$$

where Φ_{n1} is the spectral density of the refractive index fluctuations at k_1, and $A_2(\varrho)$ is given by the same equation with Φ_{n1} replaced by Φ_{n2}, the spectral density at k_2.

The two-frequency correlation function is important for pulse propagation problems, particularly for a broad band pulse [5.48, 49].

5.2.4 Fourth-Order Moments

The parabolic equation for the higher order moments has been obtained by many investigators using a variety of techniques [5.2, 28, 32, 33, 50]. They all obtained the following equation:

$$\left[2ik\frac{\partial}{\partial z}+(\varDelta_1+\dots+\varDelta_n-\varDelta_1'-\dots\varDelta_m')+\frac{ik^3}{4}F_{nm}\right]\Gamma_{nm}=0,\tag{5.73}$$

where

$$\Gamma_{nm}=\langle U(\varrho,z)\dots U(\varrho_n,z)U^*(\varrho_1',z)\dots U^*(\varrho_m',z)\rangle,$$

and $\varDelta_1\dots\varDelta_n$, $\varDelta_1'\dots\varDelta_m'$ are the transverse Laplacian with respect to $\varrho_1\dots\varrho_n$, $\varrho_1'\dots\varrho_m'$, respectively. F_{nm} is given by

$$F_{nm}=\sum_{i=1}^{n}\sum_{j=1}^{n}A(\varrho_i-\varrho_j)-2\sum_{i=1}^{n}\sum_{k=1}^{m}A(\varrho_i-\varrho_k')$$

$$+\sum_{k=1}^{m}\sum_{l=1}^{m}A(\varrho_k'-\varrho_l').$$

To date, however, no general solution has been obtained even for the fourth-order moment except for approximate solutions [5.2, 8, 9, 20, 44].

The extended Huygens-Fresnel technique has been used to obtain the fourth-order moment. The solution for the focused beam case [5.31, 41] appears to give reasonable agreement with experiments.

The field $U(\varrho,z)$ at the observation point is expressed in terms of the field $U_0(\varrho',z)$ at the transmitting aperture, by means of the following extension of the Huygens-Fresnel principle:

$$U(\varrho,z)=\frac{k}{2\pi iz}\int_{S_0}U_0(\varrho')\exp\left[ikz+\frac{ik|\varrho-\varrho'|^2}{2z}+\psi(\varrho,\varrho')\right]d\varrho',\tag{5.74a}$$

where S_0 is the surface of the transmitting aperture. $\psi(\varrho,\varrho')$ is the random part of the complex phase of a spherical wave propagating from $(0,\varrho')$ to (ϱ,z), and may be considered to correspond to the Rytov solution $\psi=\chi+iS_1$ for a spherical wave. The solution obtained in this manner appears to be reasonable even in the strong fluctuation region [5.41]. However, this is based on the use of the Rytov solution, and its range of validity has not been established. If the parabolic equation can be solved for a beam wave case, this should give a much more satisfactory solution. However, at present this has not been accomplished.

We form the second moment of the intensity $I(\varrho, z)$ using (5.74a)

$$\langle I(\varrho_1, z)I(\varrho_2, z)\rangle = \langle U(\varrho_1, z)U^*(\varrho_1, z)U(\varrho_2, z)U^*(\varrho_2, z)\rangle. \tag{5.74b}$$

In calculating (5.74b), use is made of the assumption that for a focused beam, the scintillations are mainly determined by the phase fluctuations

$$\psi = \chi + iS_1 \approx iS_1,$$

and that the phase fluctuation S_1 is a gaussian random variable. Under this assumption the normalized intensity variance of a beam focused at the observation point

$$\sigma_I^2 = \frac{\langle I^2 \rangle}{\langle I \rangle^2} - 1 \tag{5.75a}$$

has been calculated [5.41] and for a focused beam it is shown to depend only on D_s

$$D_s = D_s(W_0) = 1.95C_n^2 k^2 L W_0^{5/3} \tag{5.75b}$$

and σ_I has a maximum ($\sigma_{I\,\text{max}} = 1.4$) at $D_s \approx 50$.

Even though the equivalence of the Huygens-Fresnel technique and the parabolic equation technique has been established for the second moment, the equivalence for the fourth- and higher order moments has not. Further investigations should be directed to solutions of the fourth-order parabolic equation and to investigations of its equivalence with the Huygens-Fresnel formulations.

5.2.5 Short- and Long-Term Beam Spreads

If we place a photographic plate in the propagation path of a light beam and record the light beam spot, we observe the following. At a short distance the beam shape in the turbulence is substantially the same as that in free space, but the beam spot wanders around due to the random motion of the atmosphere. Therefore, if the exposure time is short compared with the wandering time Δt of the beam, which is of the order of (beam size)/(wind velocity), the beam size should be substantially the same as that in free space. This beam radius ϱ_s is called the "short-term beam spread". If the exposure time is much longer than Δt, the wander of the center of gravity of the beam ϱ_c in addition to the short-term spread ϱ_s contributes to the recorded picture of the beam. This is called the "long-term beam spread" ϱ_L and is related to ϱ_s and ϱ_c through

$$\langle \varrho_L^2 \rangle = \langle \varrho_s^2 \rangle + \langle \varrho_c^2 \rangle. \tag{5.76}$$

Mathematically, the long-term spread is given by

$$\langle \varrho_L^2 \rangle = \frac{\int \langle I(\varrho, z) \rangle \varrho^2 d\varrho}{\int \langle I(\varrho, z) \rangle d\varrho},$$

(5.77)

where $\langle I(\varrho, z) \rangle = \Gamma(\varrho, z, 0)$ is given in (5.64). The center of gravity ϱ_c is given by

$$\varrho_c = \frac{\int I(\varrho, z) \varrho d\varrho}{\int \langle I(\varrho, z) \rangle d\varrho}.$$

(5.78)

Therefore we have

$$\langle \varrho_c^2 \rangle = \frac{\int\int \langle I(\varrho_1, z) I(\varrho_2, z) \rangle (\varrho_1 \cdot \varrho_2) d\varrho_1 d\varrho_2}{[\int \langle I(\varrho, z) \rangle d\varrho]^2}.$$

(5.79)

Some approximate calculations of the above have been made. Note that W^2 in (5.62b) is equal to $2\langle \varrho_L^2 \rangle$ for the case $z \gg z_i$,

$$\langle \varrho_L^2 \rangle = \frac{W_0^2}{2}[(\alpha_1 z)^2 + (1 - \alpha_2 z)^2] + 2.2 C_n^2 l_0^{-1/3} z^3.$$

(5.80)

For $z \ll z_i$, it has been shown that

$$\langle \varrho_L^2 \rangle = \frac{W_0^2}{2}[(\alpha_1 z)^2 + (1 - \alpha_2 z)^2] + \frac{4z^2}{k^2 \varrho_0^2},$$

(5.81)

where ϱ_0 is given by

$$\varrho_0 = [1.46 k^2 z C_n^2 \cdot (3/8)]^{-3/5}.$$

(5.82)

For the short-term beam spread in various cases, see *Fante* [5.9].

At a large distance, the beam no longer wanders as much; but it breaks up into a multitude of stringlike spots at random locations. The long exposure picture is a blurred version of the above short exposure picture, but their radii are approximately the same. From (5.76), it is seen that $\langle \varrho_c^2 \rangle$ should be much smaller than $\langle \varrho_L^2 \rangle$ or $\langle \varrho_s^2 \rangle$ in the strong fluctuation region.

It has been shown experimentally [5.18] that the beam wanders with a time constant of the order of 1 s. This wandering results in fading in laser systems in the atmosphere, and therefore it is an important factor. It has been recognized that this wander may be cancelled by using a fast-tracking transmitter, and some experimental work [5.51] has been conducted.

5.3 Optical Remote Sensing

Remote sensing of various atmospheric parameters has become increasingly important in recent years because it provides a new tool in the study of the structure and dynamics of the atmosphere and may be applied to weather forecasting, weather modification, pollution studies, storm warning, and air traffic safety [5.52].

Remote sensing may be divided into two categories: passive and active. A passive sensing system is one in which the observing instrument simply receives the natural radiation from the environment, such as the radiation from gas and aerosol constituents of the atmosphere or the radiation from the sun, the moon, or the planets. An active sensing system is one in which a signal is sent out from the transmitter, interacts with the environment or the target, and after interaction, is observed and measured [5.53, 54]. Examples include radar, lidar, acoustic sounding, and line-of-sight propagation techniques.

Atmospheric turbulence causes fluctuations of the refractive index of air, which in turn produces fluctuations in amplitude and phase of an optical beam propagating through it. Therefore, it should be possible to deduce the characteristics of turbulence from observations of the fluctuations of the optical wave. The turbulence characteristics to be probed include the strength of turbulence as represented by the structure constant C_n and the wind velocity. Remote sensing of these quantities may be divided into two areas: one is to find averages of the quantities over the total path length, and the other is to obtain the profile of these quantities as functions of position along the path. We shall discuss these two aspects separately in the following sections.

At present, remote sensing studies have been based almost entirely on weak fluctuation theory, and therefore at optical frequencies, it is applicable only within a distance of a few kilometers or so. Beyond this distance, the strong fluctuation theory must be used. No serious study of remote sensing in the strong fluctuation region seems to have been reported in the literature.

5.3.1 Remote Sensing of the Average Structure Constant C_n Over the Path

In well-developed turbulence, velocity fluctuations are known to follow the Kolmogorov spectrum. Certain quantities such as potential temperature and mass of water vapor move with the velocity field in the turbulence without appreciable change and therefore their fluctuation characteristics also obey the Kolmogorov spectrum (see Chap. 2). At optical frequencies, humidity effects are negligible, and therefore the temperature field is known to obey the Kolmogorov spectrum. The index of refraction field n is related to the temperature field by

$$N = (n-1)10^6 = 79\frac{P}{T},$$

(5.83)

where P is the atmospheric pressure in mb and T is the temperature in K. The relationship between the pressure and temperature depends on the process of heat transfer in turbulence. Considering the short lifetime of turbulence, we may regard the process as adiabatic, and under this assumption we have

$$\frac{\delta P}{P} = \frac{\gamma}{\gamma-1} \frac{\delta T}{T}, \tag{5.84}$$

where γ is the ratio of specific heats ($=C_p/C_v=1.4$ for air). Therefore, the structure function of the index of refraction $D_n(r)$ is related to the temperature structure function $D_T(r)$ through

$$D_n(r) = C_n^2 r^{2/3}, \qquad D_T(r) = C_T^2 r^{2/3},$$

$$C_n = \left(\frac{79}{\gamma-1} \frac{P}{T^2} 10^{-6}\right) C_T. \tag{5.85}$$

It is important to note that remote sensing by an optical beam detects refractive index fluctuations in turbulence, and not the velocity field fluctuations. These two are usually closely related to each other as noted in the above. However, in some cases such as in a neutral atmosphere, strong velocity fluctuations may exist with little optical effect [5.55].

The structure constant C_n of turbulence can be obtained by measuring the variance of log irradiance fluctuations of an optical wave propagating through it, and using the formulas (5.17)

$$\sigma_{\ln I}^2 = 1.228 k^{7/6} L^{11/6} C_n^2 \quad \text{plane wave}$$
$$= 0.496 k^{7/6} L^{11/6} C_n^2 \quad \text{spherical wave} \tag{5.86}$$

where $k = 2\pi/\lambda$ is the wave number and L is the propagation distance. For a beam wave, we use

$$\sigma_{\ln I}^2 = 4\sigma_\chi^2(L, \varrho), \tag{5.87}$$

where $\sigma_\chi^2(L, \varrho)$ is given in (5.16).

The measurement of the log irradiance fluctuations must be made with an aperture small compared with the correlation distance of the wave, which is approximately equal to $\sqrt{\lambda L}$ for plane and spherical waves. For a collimated beam wave, the correlation distance is also of the order $\sqrt{\lambda L}$, but for a focused beam, it can be considerably smaller (see Figs. 5.3 and 5.4).

5.3.2 Remote Sensing of the Average Wind Velocity Across the Path

Remote sensing of the average wind across a wave propagation path has been studied in recent years [5.24, 55–59]. In this section, we discuss the following three methods of remotely sensing the wind velocity: a) temporal frequency spectrum method. b) time delay method, and c) correlation slope method.

Temporal Frequency Spectrum Method

The scintillation pattern at the receiver drifts with the transverse wind, and the higher the wind velocity, the faster the drift of the scintillation. Therefore we expect that as the cross wind velocity increases, the frequency spectrum of amplitude and phase fluctuations contain higher frequency components. Consequently, it should be possible to obtain the average wind velocity by measuring the spectrum of the fluctuations. There may be three ways of obtaining the wind velocity: 1) spectrum shape, 2) ratio of spectra at two frequencies, and 3) coherence between two frequencies.

The theoretical spectral shape of the log amplitude fluctuations is that for low frequencies; the spectrum is relatively constant and for high frequencies, the spectrum tends to decrease as $f^{-8/3}$ (Fig. 5.5). For plane and spherical waves, the spectra are obtained from (5.38) and their asymptotic forms are

for plane wave

$$W_{\chi P} \to 0.8506 \frac{C_n^2}{V} k^{2/3} L^{7/3} \qquad \text{as} \quad f \to 0 \tag{5.88a}$$

$$\to 2.192 \frac{C_n^2}{V} k^{2/3} L^{7/3} \left(\frac{f}{f_0}\right)^{-8/3} \qquad \text{as} \quad f \to \infty \tag{5.88b}$$

for spherical wave

$$W_{\chi s} \to 0.1905 \frac{C_n^2}{V} k^{2/3} L^{7/3} \qquad \text{as} \quad f \to 0 \tag{5.88c}$$

$$\to 2.192 \frac{C_n^2}{V} k^{2/3} L^{7/3} \left(\frac{f}{f_0}\right)^{-8/3} \qquad \text{as} \quad f \to \infty \tag{5.88d}$$

where $f_0 = V(k/L)^{1/2}/(2\pi) = 0.4V/\sqrt{\lambda L}$ and V is the cross wind velocity. Equation (5.88) is applicable when the velocity is assumed constant. If V is substantially constant, V in (5.88) should be close to the average wind velocity. However if V is a function of position, we need to use a more general formula [5.24].

The shapes of the spectra obtained experimentally generally follow those in Fig. 5.5 and therefore the shape and particularly the frequency f_c at which the above two asymptotes meet give a good parameter from which the wind velocity can be obtained. Note that this frequency f_c is $1.43 f_0$ for plane waves and $2.60 f_0$ for spherical waves.

The above technique requires the measurement of fluctuations at a single operating frequency. In practice, however, the break-point frequency f_c is not easy to determine experimentally. If we can transmit two waves with different operating frequencies, then we have additional data from which to extract more information.

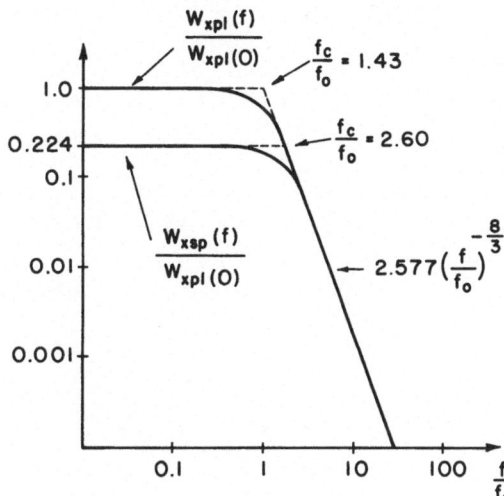

Fig. 5.5. Theoretical frequency spectra of log amplitude fluctuation

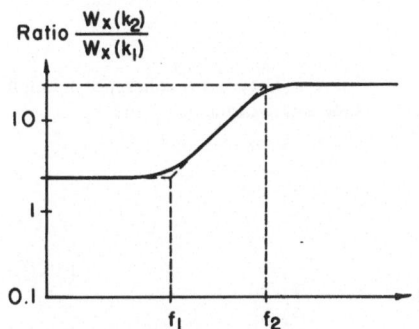

Fig. 5.6. Theoretical ratio of frequency spectra at two frequencies k_1 and k_2

We can compare the spectra at two different operating frequencies k_1 and k_2 and take their ratio. From (5.88a)–(5.88d), we note that as $f \to 0$,

$$\frac{W_\chi(k_2)}{W_\chi(k_1)} = \left(\frac{k_2}{k_1}\right)^{2/3}, \tag{5.89a}$$

and as $f \to \infty$

$$\frac{W_\chi(k_2)}{W_\chi(k_1)} = \left(\frac{k_2}{k_1}\right)^2. \tag{5.89b}$$

The general shape of this ratio is shown in Fig. 5.6. The break-point frequencies f_1 and f_2 are

$$f_i = 1.43 \left(\frac{V}{2\pi}\right)\left(\frac{k_i}{L}\right)^{1/2}, \quad i = 1 \text{ and } 2 \tag{5.90a}$$

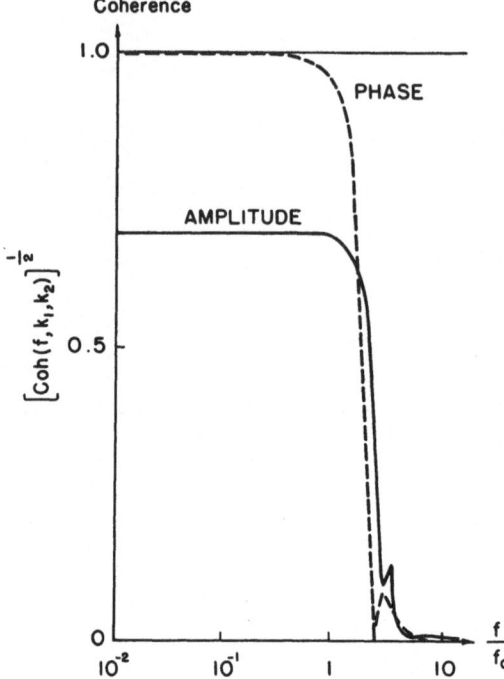

Fig. 5.7. Theoretical coherence of amplitude fluctuations at k_1 and k_2

for plane waves and

$$f_i = 2.60 \left(\frac{V}{2\pi}\right) \left(\frac{k_i}{L}\right)^{1/2}, \quad i = 1 \text{ and } 2 \tag{5.90b}$$

for spherical waves.

Therefore, it may be possible to obtain the wind velocity V by measuring the spectra at k_1 and k_2 and obtaining the ratio and the break-point frequencies f_i. However, at present, this technique has been used only at microwave frequencies [5.24].

It is also possible to make use of correlation characteristics of fluctuations at two operating frequencies in determining the wind velocity. The cross spectrum of log amplitude fluctuations at two different operating frequencies (k_1 and k_2) is given by an extension of (5.38) [5.24], i.e.,

$$W_\chi(\omega, k_1, k_2) = \left(\frac{8\pi^2 k_1 k_2}{V}\right)(0.033 C_n^2) k_1^{-4/3} L^{7/3} I_\chi, \tag{5.91}$$

where I_χ is given by Ref. [5.24], Eq. (15b).

The square of the cross spectrum $[W_\chi(\omega, k_1, k_2)]^2$ normalized to $W_\chi(\omega, k_1) W_\chi(\omega, k_2)$ approaches the following constant value as $\omega \to 0$:

$$\frac{[W_\chi(\omega, k_1, k_2)]^2}{W_\chi(\omega, k_1) W(\omega, k_2)} \to \left(\frac{k_2}{k_1}\right)^{4/3} \left[\left(\frac{1 + k_1/k_2}{2}\right)^{4/3} - \left(\frac{1 - k_1/k_2}{2}\right)^{4/3}\right]^2, \tag{5.92}$$

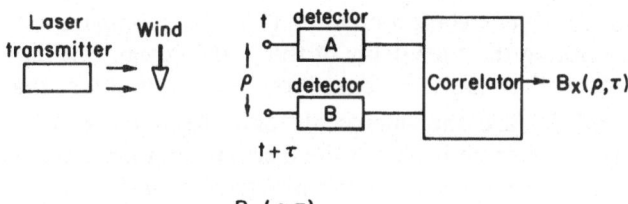

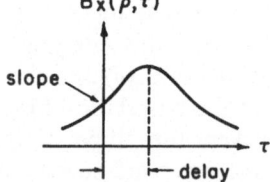

Fig. 5.8. Time delay and correlation slope method of wind velocity measurement

where $k_1 < k_2$. The square of the normalized cross spectrum

$$[W_\chi(\omega, k_1, k_2)]^2/[W(\omega, k_1)W_\chi(\omega, k_2)]$$

is called the "coherence" in analysis of random data [5.60].

The coherence is almost constant at the value given by (5.92) up to a certain frequency determined by the wind velocity and then it drops to a negligible value beyond this point [5.61]. A typical behavior is shown in Fig. 5.7. Therefore, it should be possible to obtain the wind velocity from the measurement of the cross spectrum.

Time Delay Method

If two detectors are placed perpendicular to the propagation path, parallel to the direction of wind velocity (Fig. 5.8), the fluctuation of a wave drifts with the wind. The fluctuation at receiver A at t becomes the fluctuation at the receiver B at a delayed time $t + \tau$. Therefore, we expect that at a certain delay time τ, the fluctuations at A and B are strongly correlated. This is shown in Fig. 5.8. The delay time τ of the peak of the correlation is related to the wind velocity and is approximately equal to the correlation distance of the field divided by the wind velocity. The correlation distance is approximately equal to $\sqrt{\lambda L}$.

A general expression for the correlation function of the log amplitude fluctuations χ is given in (5.11a). For a spherical wave it reduces to

$$B_\chi(\varrho, \tau) = 8\pi^2 k^2 \int_0^L d\eta \int_0^\infty \kappa d\kappa \, J_0\left(\kappa \left|\frac{\varrho\eta}{L} - V\tau\right|\right)$$

$$\cdot \sin^2\left[\frac{\eta(L-\eta)}{2kL}\kappa^2\right]\Phi_n(\kappa, \eta). \tag{5.93}$$

Using the Kolmogorov spectrum

$$\Phi_n(\kappa, \eta) = 0.033 C_n^2(\eta)\kappa^{-11/3}, \tag{5.94}$$

we can calculate the shape of the correlation function (5.93) and compare with the experimental data. By noting the time delay of the peak, we can calculate the wind velocity. We note that (5.94) is based on the assumption that $l_0 \ll \sqrt{\lambda L} \ll L$ where l_0 and L_0 are the microscale and the outer scale of turbulence, and therefore the spectrum is that of the inertial subrange and that the intensity of the turbulence is represented by the structure constant C_n. Also we note that the correlation function B_χ is in fact the covariance function because the average $\langle \chi \rangle$ is assumed to be zero in this analysis.

It has been found [5.56] that the variations in $C_n(\eta)$ and the wind velocity $V(\eta)$ cause severe distortions in the shape of the correlation function and thus contribute to a considerable error. However, the slope of the correlation function at $\tau = 0$ is also dependent on the average wind speed and seems to be relatively insensitive to these variations. Therefore, utilizing a measurement of the slope may be preferable in determining the average wind.

Correlation Slope Method

The slope of the correlation function at $\tau = 0$ is given by

$$\frac{\partial B_\chi}{\partial \tau}\bigg|_{\tau=0} = 8\pi^2 k^2 \int_0^L d\eta \int_0^\infty d\kappa \kappa^2 V J_1\left(\kappa \frac{\varrho\eta}{L}\right)$$

$$\cdot \sin^2\left[\frac{\eta(L-\eta)}{2KL}\kappa^2\right] \Phi_n(\kappa, \eta). \tag{5.95}$$

It has been found [5.56] that it is preferable to use a value of $\varrho = 0.33\sqrt{\lambda L}$ because it gives the most uniform weighting of C_n and V along the propagation path. Optically measured average wind velocity have been compared with the average of several anemometer readings along the path and good agreements have been obtained [5.56].

5.3.3 Remote Sensing of the Profile of the Structure Constant and Wind Velocity Along the Propagation Path

Up to this point, we have considered remote sensing of the average structure constant and the average wind velocity over the propagation path. Suppose that we wish to find the profile of the structure constant as a function of position along the path by means of the measurement of wave fluctuations. It should be obvious that finding the profile, rather than the average, requires more data taking. Typically, the number of detectors should be increased. To illustrate this point, let us consider the correlation function given in (5.93). Using (5.94), we write

$$B_\chi(\varrho, \tau) = \int_0^L d\eta K(\varrho, \tau, \eta) C_n^2(\eta), \tag{5.96}$$

where

$$K(\varrho,\tau,\eta)=8\pi^2k^2(0.033)\int_0^\infty \kappa^{-8/3}\,\mathrm{J}_0\left(\kappa|\frac{\varrho\eta}{L}-V\tau|\right)\sin^2\left[\frac{\eta(L-\eta)}{2kL}\kappa^2\right]d\kappa.$$

We now make a series of measurements and obtain M different values of the correlation function $B_\chi(\varrho_i,\tau_i)$ where $i=1,2,\ldots M$. We also approximate the integral in (5.96) by a series and obtain

$$B_\chi(\varrho_i,\tau_i)=\sum_{j=1}^N a(\varrho_i,\tau_i,\eta_j)C_n^2(\eta_j),\qquad(5.97)$$

where $a(\varrho_i,\tau_i,\eta_j)=W_jK(\varrho_i,\tau_i,\eta_j)$ and W_j is an appropriate weighting function depending upon the type of quadrature formula used to obtain the series (5.97) from the integral (5.96).

Let us write (5.97) in the following matrix form:

$$g=Af,\qquad(5.98)$$

where $g=(g_i)$ is a $M\times 1$ column matrix, $A=(a_{ij})$ is $M\times N$ rectangular matrix, $f=(f_j)$ is a $N\times 1$ column matrix, and

$$g_i=B_\chi(\varrho_i,\tau_i),\qquad a_{ij}=W_jK(\varrho_i,\tau_i,\eta_j),$$

and

$$f_j=C_n^2(\eta_j).$$

In (5.98), g_i $(i=1,\ldots,M)$ are the data obtained from M different measurements, (a_{ij}) are known elements of a matrix, and f_j $(j=1,\ldots,N)$ are unknown quantities. There are several methods to obtain the unknown f from the measured data g. They include a) least square estimation, b) smoothing method, c) statistical inversion method, and d) Backus-Gilbert inversion technique. These methods will be outlined below.

Least Square Estimation

In (5.98), the number of measurements M must be at least equal to the number of unknowns N, and in general $M>N$.

The least square estimation procedure consists of finding f such that the square of the difference $(g-Af)$ is minimized, i.e.,

$$|g-Af|^2=\text{minimum}.\qquad(5.99)$$

A solution to (5.99) is well-known [5.62]

$$f=(A^+A)^{-1}A^+g,\qquad(5.100)$$

where A^+ is the complex conjugate of the transpose of A. If $M = N$, (5.100) reduces to the usual inversion

$$f = A^{-1}g. \tag{5.101}$$

For many problems in remote sensing, as it will be shown below, (5.100) or (5.101) does not yield a useful solution because a small error in the measured data g causes an extremely large error in the unknown. To illustrate this point, consider (5.100) and assume a certain error δg in g causes a resulting error δf in f. We then have

$$\delta f = (A^+A)^{-1}A^+\delta g. \tag{5.102}$$

As a percentage error Δ_f in f, we take

$$\Delta_f = \frac{\|\delta f\|}{\|f\|}, \tag{5.103}$$

where $\|f\|$ is the norm of f defined by

$$\|f\| = \max |f_j|. \tag{5.104}$$

Using (5.102), we get

$$\Delta_f = \frac{\|\delta f\|}{\|f\|} < \|(A^+A)^{-1}\|\,\|A^+\|\frac{\|\delta_g\|}{\|g\|}, \tag{5.105}$$

where $\|A\|$ is the norm of a matrix A defined by

$$\|A\| = \max_i \sum_i |a_{ij}|, \tag{5.106}$$

and therefore the maximum percentage error $\Delta_{f\max}$ is given by

$$\Delta_{f\max} = \|(A^+A)^{-1}\|\,\|A^+\|\Delta_g, \tag{5.107}$$

where $\Delta_g = $ percentage error in $g = \|\delta g\|/\|g\|$. In many remote sensing problems, the elements a_{ij} are not very different from each other and therefore the norm $\|(A^+A)^{-1}\|$ can be extremely high, and often of order of magnitude of tens of powers of ten. This situation is called "unstable" and the problem is "ill posed" [5.63].

This instability problem has been experienced by *Shen* [5.59] who used only a limited number of unknowns to avoid this problem. In a more general case, however, it is necessary to devise a method by which a stable inversion of (5.98) can be accomplished. This will be discussed in the next section.

Statistical Inversion Method

The inversion of an ill-posed problem (5.98) has been studied by several methods including the "smoothing (regularization) method", the "statistical method", and the "Backus-Gilbert technique".

The regularization method has been studied by *Phillips, Twomey, Tikhonov* and others [5.64–68] with some success. The method is based on imposing certain additional conditions which ensure a "smooth" or "stable" solution. This usually takes the form of constructing a solution which depends upon a parameter (such as a Tikhonov's "regularization parameter") and choosing this parameter judiciously. The problems associated with this technique are summarized by *Turchin* et al. [5.69].

In many remote sensing problems, errors encountered are statistical, and it is more natural to consider the inversion problem by taking into account the statistical nature of the experimental errors and other statistical information. In this section, we discuss an element of the statistical inversion technique [5.63, 70, 71]. We note that even though at present this technique has been applied only to microwave problems [5.70], the same technique should be applicable to the optical propagation problems.

Let us consider an ill-posed problem

$$g = Af. \tag{5.108}$$

In practice, the true g is never known because it always contains a certain experimental error n. The measured data g_d is therefore

$$g_d = g + n, \tag{5.109}$$

where the $M \times 1$ matrix g_d is the measured value and known, and the $M \times 1$ matrix n is the experimental error. We may know some statistical characteristics of n.

Substituting (5.109) into (5.108), we write

$$g_d = Af + n. \tag{5.110}$$

We regard (5.110) as a stochastic equation and g_d, f, and n as random variables with zero mean

$$\langle g_d \rangle = 0, \quad \langle f \rangle = 0, \quad \langle n \rangle = 0. \tag{5.111}$$

The actual values of g_d, f, and n for a particular experiment are then considered as one outcome (member) of the ensemble.

The inversion problem may be stated as follows: Find an $N \times M$ matrix B such that Bg_d is as close to the desired unknown f as possible. This closeness may be stated mathematically as minimizing the average of the scalar product of $(f - Bg_d)$ and an $N \times 1$ vector a_1 for any arbitrary vector a_1,

$$\langle |(f - Bg_d)^+ a_1|^2 \rangle = \text{minimum}. \tag{5.112}$$

A solution to this is given by [5.63, 70]

$$B=(R_{ff}A^+ +R_{fn})(AR_{ff}A^+ +R_{fn}^+A^+ +AR_{fn}+R_{nn})^{-1},\tag{5.113}$$

where $R_{ff}=\langle ff^+\rangle$ is an $N\times N$ matrix representing the covariance of f and is called the "covariance" matrix. Similarly, $R_{fn}=\langle fn^+\rangle$ is an $N\times M$ covariance matrix and $R_{nn}=\langle nn^+\rangle$ is an $M\times M$ covariance matrix.

In many practical problems, the measurement error n and the unknown quantity f are independent of each other and in this case we have $R_{fn}=0$ and (5.113) reduces to

$$B=(R_{ff}A^+)(AR_{ff}A^+ +R_{nn})^{-1}.\tag{5.114}$$

The final solution f_d for the measured data g_d is then given by

$$f_d=Bg_d.\tag{5.115}$$

The maximum percentage error $\Delta_{f\max}$ in the unknown f is related to the percentage error Δ_n in the measurement [5.63, 70] by

$$\Delta_{f\max}=\|A\|\|B\|\Delta_n,\tag{5.116}$$

where

$$\Delta_f=\frac{\|\delta f\|}{\|f\|}\quad\text{and}\quad\Delta_n=\frac{\|n\|}{\|g_d\|}.$$

The norm of B as given in (5.113) or (5.114) is usually quite small compared with the norm of A^{-1}, and therefore the error $\Delta_{f\max}$ is comparable in magnitude to the experimental error Δ_n, and thus this procedure yields a stable solution.

As can be seen from (5.114), the effectiveness of this procedure depends upon the choice of the covariance matrices R_{ff} and R_{nn}. It is obvious that if one knows something about the statistical properties of the unknown f and the experimental error n, we can make use of this information to construct the covariance matrices R_{ff} and R_{nn}. Clearly, the more we know about f and n, the better choice of R_{ff} and R_{nn} can be made, which should result in a better solution.

For example, we may represent the unknown $f(x)=C_n^2(x)$, $0\leq x\leq L$, in terms of a random Fourier series

$$f(x)=\bar{f}+\sum_{k=0}^{K} b_k\left(a_k\cos k\pi\frac{x}{L} +a_k'\sin k\pi\frac{x}{L}\right),\tag{5.117}$$

where $\bar{f}$ is the mean value of $f(x)$, the b_k are constants, and a_k and a_k' are independent random coefficients with zero mean and variance one. K limits the rate at which the fluctuations occur. From (5.117), we get the elements of the covariance matrix $R_{ff} = (r_{ij})$

$$r_{ij} = \bar{f}^2 + \sum_{k=0}^{K} b_k^2 \cos\left[k\pi \frac{(x_i - x_j)}{L}\right]$$

$$\langle a_k a_{k'} \rangle = \delta_{kk'}, \quad \langle a_k \rangle = 0.$$

(5.118)

The covariance matrix of the error may be chosen to be

$$R_{nn} = (\sigma_n^2 \delta_{ij}).$$

(5.119)

Appropriate choice of $\bar{f}$, b_k, and σ_n^2 may be made by experience or good estimate. Some examples of this technique are shown in [5.70] for microwave applications.

The profile of the wind velocity $V(\eta)$ as a function of position along the propagation path may be obtained by a technique similar to that described above. We make use of the derivative $(\partial/\partial\tau)B_\chi$ given in (5.95) and obtain a matrix equation

$$g' = A'V,$$

(5.120)

where g' is the derivative data $[(\partial/\partial\tau)B_\chi]$, V is the wind velocity, and A' is given by

$$A' = (a'_{ij})$$
$$a'_{ij} = W_j K'(\varrho_i, \tau_i, \eta_j) C_n^2(\eta_j),$$

and K' is the derivative of K with respect to τ. A technique of taking into account the error in C_n^2 as well as the measurement error in g' is discussed in [5.70].

Backus-Gilbert Inversion Technique

The above inversion techniques require some knowledge of the unknown function f. In 1970, *Backus* and *Gilbert* [5.72] proposed a method which does not require a priori knowledge of the unknown and gives a measure of the "resolution" and the "accuracy". Here the resolution means the spread over which the unknown function is averaged, and the accuracy is expressed by the variance of the error in the unknown. The technique provides a control of the trade-off between the spread and the variance. The Backus-Gilbert inversion technique was originally proposed for the estimation of the earth data and was subsequently used for the determination of particle size distribution by light scattering techniques [5.72–75]. This technique offers an advantage of giving a measure of the resolution and the accuracy. However, further studies need to be made on its application in remote sensing.

5.3.4 Other Remote Sensing Techniques

Recently, Clifford and Ochs of NOAA and Wang of CIRES in Boulder reported extension of the technique used by *Lawrence* et al. [5.56]. They discussed a technique which makes use of the naturally occurring ambient illumination of a scene such as a mountainside and clouds. Thus this technique does not require an active light source such as a laser or headlight.

Two other promising methods of remote sensing the profile of the structure constant and the wind velocity have been recently proposed. They are a) the crossed beam method and b) the spatially filtered aperture technique. In this section, we give a brief outline of the methods.

Crossed Beam Method

Wang et al. [5.76] proposed the use of two transmitters and two receivers so that two beams cross at a certain location in the path. By measuring the cross correlation between the outputs of two receivers, it is possible to obtain the turbulence characteristics at this particular beam crossing point. The crossed beam technique was previously proposed by *Fisher* and *Krause* [5.77].

Spatially Filtered Aperture Technique

Lee [5.78] proposed that if the transmitting and receiving apertures are appropriately weighted (or spatially filtered), then the output can be made sensitive to one particular spatial wavenumber which in turn selects one particular location in the path. He conducted simple experiments using a Fresnel lens of aperture 0.5 by 0.6 m across which were taped 11 mm wide vertical strips of black paper separated by 11 m, forming a spatial filter. Using a similar lens for receiving aperture, he predicted the wind velocity along the path and his results agreed well with anemometer measurements.

References

5.1   V.I.Tatarskii: *Wave Propagation in a Turbulent Medium*, English transl. (McGraw-Hill, New York 1961)
5.2   V.I.Tatarskii: *The Effects of the Turbulent Atmosphere on Wave Propagation*, English transl. (U.S. Dept. of Commerce, NTIS, Springfield, VA 1971)
5.3   J.W.Strohbehn: Proc. IEEE **56**, 1301 (1968)
5.4   R.S.Lawrence, J.W.Strohbehn: Proc. IEEE **58**, 1523 (1970)
5.5   J.W.Strohbehn: In *Progress in Optics*, Vol. IX, ed. by E.Wolf (North-Holland, Amsterdam 1971) p. 75
5.6   Y.N.Barabunenkov, Y.A.Kravtsov, S.M.Rytov, V.I.Tatarskii: Sov. Phys.-Usp. **13**, 551 (1971)

5.7   S.S.Khmelevtsov: Appl. Opt. **12**, 2421 (1973)
5.8   A.M.Prokhorov, F.V.Bunkin, K.S.Gochelashvily, V.I.Shishov: Proc. IEEE **63**, 790 (1975)
5.9   R.L.Fante: Proc. IEEE **63**, 1669 (1975)
5.10   A.Ishimaru: Radio Sci. **4**, 295 (1969)
5.11   A.Ishimaru: Proc. IEEE **57**, 407 (1969)
5.12   A.I.Kon, V.I.Tatarskii: Izv. VUZ, Radiofiz. **8**, 870, (1965)
5.13   R.A.Schmeltzer: Quart. Appl. Math. **24**, 339 (1967)
5.14   D.L.Fried, J.B.Seidman: J. Opt. Soc. Am. **57**, 181 (1967)
5.15   Y.Kinoshita, T.Sakura, M.Suzuki: J. Opt. Soc. Am. **58**, 798 (1968)
5.16   F.G.Gebhardt, S.A.Collins,Jr.: J. Opt. Soc. Am. **59**, 1139 (1969)
5.17   S.S.Khmelevtsov, R.S.Tsvik: Izv. VUZ Radiofiz. **13**, 146 (1970)
5.18   J.R.Kerr, J.R.Dunphy: J. Opt. Soc. Am. **63**, 1 (1973)
5.19   J.R.Kerr, R.Eiss: J. Opt. Soc. Am. **62**, 682 (1972)
5.20   F.V.Bunkin, K.S.Gochelashvily: Izv.VUZ, Radiofiz. **13**, 1039 (1970)
5.21   A.L.Buck: Appl. Opt. **6**, 703 (1967)
5.22   B.D.Borisov, V.M.Sazanovich, S.S.Khmelevtsov: Izv. VUZ, Fiz. **6**, 103 (1969)
5.23   S.F.Clifford: J. Opt. Soc. Am. **61**, 1285 (1971)
5.24   A.Ishimaru: IEEE Trans. AP-**20**, 10 (1972)
5.25   D.A.deWolf: J. Opt. Soc. Am. **58**, 461 (1968)
5.26   U.Frisch: "Wave Propagation in Random Media", in *Probabilistic Methods in Applied Mathematics*, ed. by A.T.Baracha-Reid (Academic Press, New York 1968)
5.27   W.P.Brown,Jr.: J. Opt. Soc. Am. **61**, 1061 (1971)
5.28   W.P.Brown,Jr.: J. Opt. Soc. Ám. **62**, 966 (1972)
5.29   T.L.Ho, M.J.Beran: J. Opt. Soc. Am. **58**, 1335 (1968)
5.30   M.J.Beran, A.M.Whitman: J. Opt. Soc. Am. **61**, 1044 (1971)
5.31   J.A.Dowling, P.M.Livingston: J. Opt. Soc. Am. **63**, 846 (1973)
5.32   K.Furutsu: J. Opt. Soc. Am. **62**, 240 (1972)
5.33   V.I.Shishov: Izv. VUZ, Radiofiz. **11**, 886 (1968)
5.34   A.M.Whitman, M.J.Beran: J. Opt. Soc. Am. **60**, 1595 (1970)
5.35   R.F.Lutomirski, H.T.Yura: J. Opt. Soc. Am. **61**, 482 (1971)
5.36   R.F.Lutomirski, H.T.Yura: Appl. Opt. **10**, 1652 (1971)
5.37   H.T.Yura: Appl. Opt. **11**, 1399 (1972)
5.38   A.I.Kon: Izv. VUZ, Radiofiz. **13**, 61 (1970)
5.39   S.F.Clifford, G.R.Ochs, R.S.Lawrence: J. Opt. Soc. Am. **64**, 148 (1974)
5.40   Z.I.Feizulin, Y.A.Kravtsov: Izv. VUZ, Radiofiz. **10**, 68 (1967)
5.41   V.A.Banakh, G.M.Krekov, V.L.Mironov, S.S.Khmelevtsov, R.S.Tsvik: J. Opt. Soc. Am. **64**, 516 (1974)
5.42   H.T.Yura: J. Opt. Soc. Am. **63**, 567 (1973)
5.43   H.T.Yura: J. Opt. Soc. Am. **62**, 889 (1972)
5.44   R.L.Fante: Radio Sci. **10**, 77 (1975)
5.45   A.S.Gurvich, V.I.Tatarskii: Radio Sci. **10**, 3, (1975)
5.46   R.L.Fante: J. Opt. Soc. Am. **64**, 592 (1974)
5.47   H.T.Yura: J. Opt. Soc. Am. **64**, 357 (1974)
5.48   A.Ishimaru, S.T.Hong: Radio Sci. **10**, 637 (1975)
5.49   S.T.Hong, A.Ishimaru: Radio Sci. **11**, 551 (1976)
5.50   M.J.Beran, T.L.Ho: J. Opt. Soc. Am. **59**, 1135 (1969)
5.51   J.R.Dunphy, J.R.Kerr: J. Opt. Soc. Am. **64**, 1015 (1974)
5.52   V.E.Derr: *Remote Sensing of the Troposphere* (U.S. Government Printing Office, Washington, DC 1972)
5.53   H.W.Yates: Appl. Opt. **9**, 1971 (1970)
5.54   V.E.Derr, C.G.Little: Appl. Opt. **9**, 1976 (1970)
5.55   R.S.Lawrence: *Remote Sensing by Optical line-of-Sight Propagation*, Ref. 4.45 (U.S. Government Printing Office, Washington, DC 1972) Chapt. 25
5.56   R.S.Lawrence, G.R.Ochs, S.F.Clifford: Appl. Opt. **11**, 239 (1972)

5.57 R.W.Lee, J.C.Harp: Proc. IEEE **57**, 375 (1969)
5.58 A.Ishimaru: Proc. IEEE **57**, 407 (1969)
5.59 L.Shen: IEEE Trans. AP-**18**, 493 (1970)
5.60 J.S.Bendat, A.G.Piersol: *Random Data* (Interscience, New York 1971)
5.61 P.A.Mandics, J.C.Harp, R.W.Lee, A.T.Waterman,Jr.: Radio Sci. **9**, 723 (1974)
5.62 J.N.Franklin: *Matrix Theory* Englewood Cliffs, NJ (Prentice-Hall, 1968)
5.63 J.N.Franklin: J. Math. Analysis and Applications **31**, 682 (1970)
5.64 D.L.Phillips: J. Ass. Comput. Mach. **9**, 84 (1962)
5.65 S.Twomey: J. Ass. Comput. Mach. **10**, 79 (1963)
5.66 A.N.Tihonov: Dokl. Akad. Nauk. SSSR **153**, 49 (1963)
5.67 F.W.Stallman: Numer. Math. **15**, 297 (1970)
5.68 G.A.Deschamps, H.S.Cabayan: IEEE Trans. AP-**20**, 268 (1972)
5.69 V.F.Turchin, V.P.Kozlov, M.S.Malkevich: Sov. Phys.-Usp. **13**, 681 (1971)
5.70 J.M.Heneghan, A.Ishimaru: IEEE Trans. AP-**22**, 457 (1974)
5.71 O.N.Strand, E.R.Westwater: J. Ass. Comput. Mach. **15**, 100 (1968)
5.72 G.Backus, F.Gilbert: Philos. Trans. Roy. Soc. (London) A-**266**, 123 (1970)
5.73 E.R.Westwater, A.Cohen: Appl. Opt. **12**, 1340 (1973)
5.74 L.C.Chow, C.L.Tien: Appl. Opt. **15**, 378 (1976)
5.75 M.J.Post: J. Opt. Soc. Am. **66**, 483 (1976)
5.76 T.Wang, S.F.Clifford, G.R.Ochs: Appl. Opt. **13**, 2602 (1974)
5.77 M.J.Fisher, F.R.Krause: J. Fluid Mech. **28** (4), 705 (1967)
5.78 R.W.Lee: J. Opt. Soc. Am. **64**, 1295 (1974)

6. Imaging and Optical Communication Through Atmospheric Turbulence

J. H. Shapiro

With 15 Figures

Microscale temperature fluctuations, which are due to turbulent mixing, cause the refractive index of the earth's atmosphere to be a random function of position and time (cf. Chap. 2). Although the resulting refractive index fluctuations, which we shall henceforth refer to as atmospheric turbulence, are only a few parts in 10^6, their effect on optical wave propagation in the atmosphere is profound. For many years the random behavior of the refractive index has been the bane of astronomers, restricting their "seeing" to a few seconds of arc. The rapid development of optical technology in the years following *Maiman*'s announcement of the first working laser [6.1] has been accompanied by increasing interest in the problem of optical wave propagation through the earth's turbulent atmosphere. Indeed, when it was suggested that lasers be used to extend radio-frequency atmospheric communication and radar techniques to the optical-frequency band, it was recognized that even under clear-weather conditions the earth's atmosphere is not a quiescent propagation channel [6.2–8]. In particular, the same turbulence-induced optical phase perturbations that limit astronomical seeing gradually destroy the spatial coherence of a laser beam as it propagates through the atmosphere. This loss of spatial coherence limits the extent to which laser beams may be collimated or focused in the atmosphere, and hence restricts the received power levels in optical communication and radar systems. Moreover, heterodyne-detection optical receivers are extremely sensitive to the loss of spatial coherence, and, furthermore, both heterodyne-detection and direct-detection systems exhibit severe temporal fading associated with the turbulence-induced optical amplitude fluctuations that astronomers had previously termed scintillation. Thus, a major impetus for recent optical propagation work has been the desire to quantify the performance limitations imposed by atmospheric turbulence on specific imaging and communication systems, and to develop system configurations that are immune to atmospheric fluctuations.

During the past decade, the propagation physics community has taken enormous strides toward achieving a comprehensive theory for wave propagation in the turbulent atmosphere. The status of this activity, which is still progressing, is reviewed elsewhere in this volume. The present chapter comprises a self-contained survey of the effects of atmospheric turbulence on imaging and optical communication systems. Its roots lie in the pioneering work of *Fried* [6.5–7, 9–11], *Hufnagel* and *Stanley* [6.12], and *Kennedy* and *Hoversten* [6.13], who, following the translation of *Tatarskii*'s monograph [6.14], were the first to use the weak-perturbation propagation theory to evaluate the performance of imaging and communication systems and to consider the design of optimized systems. These early studies relied upon the Rytov approximation theory for infinite plane wave propagation. Much of the recent applications literature utilizes the linear system (extended Huygens-Fresnel principle) propagation model [6.15–21], wherein the statistics of a field received from an arbitrary extended source are obtained, through a superposition integral, from the statistics of a spherical wave (point source). This approach to the propagation problem, which we shall adopt, is known to be equivalent to a number of other widely used theories [6.21–23], and it has the merit of permitting a unified presentation of atmospheric propagation effects in both imaging and communication.

The chapter is organized as follows. Section 6.1 contains a brief development of the extended Huygens-Fresnel principle and its ramifications. In Section 6.2 this propagation model is used to delineate the limitations imposed by atmospheric turbulence on imaging systems. System performance is investigated for both incoherently illuminated or self-luminous objects (the astronomical problem), and coherently illuminated objects (the imaging radar problem), beginning with the early work on long-exposure telescopic photography and concluding with recent efforts in interferometric and phase-compensated imaging. In Section 6.3, atmospheric optical communication systems are considered. A simple diffraction-limited communication system, which functions efficiently in vacuum, is analyzed and shown, on physical and mathematical grounds, to suffer an enormous performance degradation in the presence of turbulence. Systems which circumvent this performance degradation, by effectively exploiting spatial field diversity, are discussed and compared.

The emphasis throughout the chapter will be on elucidating those implications of the propagation physics that bear on the design and performance of imaging and optical communication systems. Because the performance limitations on optical systems that are currently operational are often set by the nature of available optical-frequency devices, rather than propagation phenomena, many of the results in Sections 6.2 and 6.3 must be viewed as benchmarks whose achievement will require the development of the devices necessary to translate the relevant mathematically defined signal processors into hardware. To the extent that there is a large differential between the performance achievable with available technology and the fundamental limits set by the propagation medium, there is a strong justification for undertaking the device research and development necessary to bridge the gap.

6.1 Propagation Model

In this section we shall develop the atmospheric propagation model that is based on the extended Huygens-Fresnel principle. We limit our considerations to the line-of-sight propagation geometry shown in Fig. 6.1. The problem is to

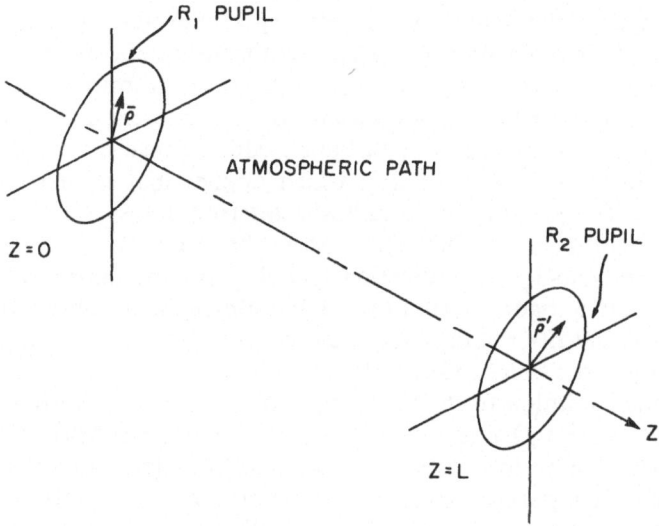

Fig. 6.1. Line-of-sight propagation geometry

statistically characterize the optical frequency field that is received over a planar pupil R_2 from a source that is located within another planar pupil R_1 a distance L meters away, when the earth's turbulent atmosphere comprises at least part of the intervening medium. For convenience we assume that the R_1 and R_2 pupils are parallel and circular, with diameters d_1 and d_2, respectively, and that their centers are located along a common perpendicular (which we take to be the z axis in a cartesian coordinate system) joining their respective planes. By using the appropriate turbulence-strength profile along the z axis, we may use this geometry to model propagation through free space, along a horizontal terrestrial path, or along a space-to-earth or an earth-to-space path. In this section we develop the propagation model for sources that are coherent in both space and time; results for incoherent sources will be presented in Section 6.2 as a prelude to the discussion of the astronomical imaging problem.

6.1.1 Extended Huygens-Fresnel Principle

Rigorous solution of the propagation problem we have posed requires vector diffraction theory. Specifically, we must solve Maxwell's equations for the source-free half-space $z > 0$, subject to the constitutive laws relating B to H and D to E in the atmosphere, and the appropriate boundary conditions on the $z = 0$ plane and the surface at infinity. To the extent that the source under consideration produces power densities that are low enough to ignore atmospheric heating effects (cf. Chap. 7), the atmosphere may be modeled as a linear, nonabsorbing, isotropic, time-varying, inhomogeneous dielectric with constitutive laws

$$B(r, t) = \mu_0 H(r, t), \tag{6.1}$$

$$D(r, t) = \int d\tau \int df\, E(r, t - \tau)\varepsilon(r, t, f) \exp(i2\pi f\tau), \tag{6.2}$$

where $\varepsilon(r, t, f)$, the space-time-frequency permittivity of the atmosphere, is independent of E and H. Spatial variations in ε reflect the instantaneous spatial structure of the turbulent eddies. Temporal variations in ε result from the drift and evolution of the eddies. The frequency variations in ε represent dispersive effects. The spatial scales associated with turbulent eddies range from millimeters to hundreds of meters [6.14, 24, 25], which implies that from near-infrared to ultraviolet wavelengths the resulting scattering angles are small enough that at all depths in the half-space $z > 0$ the wave continues to propagate predominantly in the $+z$ direction [6.21]. Thus the appropriate boundary conditions for a rigorous solution of the atmospheric diffraction problem are to assume that $E(r, t)$ and $H(r, t)$ are prescribed on the $z = 0$ plane, and that E and H satisfy the Sommerfeld radiation condition at $|r| \to \infty$, $z \geqq 0$.

Although an explicit solution to the foregoing diffraction problem cannot be obtained for arbitrary $\varepsilon(r, t, f)$, it is nonetheless true that, because Maxwell's equations coupled with the given constitutive laws and boundary conditions comprise a linear spatio-temporal system, the diffraction problem admits to a dyadic Green's function solution [6.26, 27]. In particular, using the notation

$$E_i(\varrho, t) = E(r, t)|_{r = (\varrho, z = 0)}, \tag{6.3}$$

$$E_0(\varrho', t) = E(r, t)|_{r = (\varrho', z = L)}, \tag{6.4}$$

$$H_i(\varrho, t) = H(r, t)|_{r = (\varrho, z = 0)}, \tag{6.5}$$

$$H_0(\varrho', t) = H(r, t)|_{r = (\varrho', z = L)}, \tag{6.6}$$

to represent the field vectors in the $z = 0$ and $z = L$ planes, we have the general statement of the extended Huygens-Fresnel principle, viz, there exists a dyadic Green's function with components $\mathcal{G}_{EE}$, $\mathcal{G}_{EH}$, $\mathcal{G}_{HE}$, and $\mathcal{G}_{HH}$ such that for arbitrary E_i and H_i, E_0, and H_0 may be determined from the superposition integrals

$$E_0(\varrho', t)$$
$$= \int d\varrho \int d\tau [\mathcal{G}_{EE}(\varrho', \varrho, t, \tau) E_i(\varrho, t - \tau) + \mathcal{G}_{EH}(\varrho', \varrho, t, \tau) H_i(\varrho, t - \tau)], \tag{6.7}$$

$$H_0(\varrho', t)$$
$$= \int d\varrho \int d\tau [\mathcal{G}_{HE}(\varrho', \varrho, t, \tau) E_i(\varrho, t - \tau) + \mathcal{G}_{HH}(\varrho', \varrho, t, \tau) H_i(\varrho, t - \tau)]. \tag{6.8}$$

The physical import of (6.7) and (6.8) is that the response of the atmosphere to an arbitrary electromagnetic input can be decomposed into a weighted superposition of Green's function terms, which represent the response of the atmosphere to a spatio-temporal point source. Thus, the propagation problem for an arbitrary polychromatic extended source is reduced, in principle, to a point-source propagation problem. In practice, an explicit solution for the Green's function components in (6.7) and (6.8) is unattainable for arbitrary $\varepsilon(r, t, f)$. However, by applying known propagation characteristics of the turbulent atmosphere, the statement of the extended Huygens-Fresnel principle may be substantially simplified, and, using this reduced characterization, a statistical determination of the necessary Green's function may be made.

The earth's turbulent atmosphere has been found to exhibit negligible depolarization [6.28–30]. Thus, by restricting the source to be linearly polarized, or by considering two orthogonal polarization components independently, scalar diffraction theory may be employed. Because the magnetic field of a linearly polarized wave that propagates predominantly in the $+z$ direction may be calculated from the associated electric field, the extended Huygens-Fresnel principle reduces to a scalar E-field diffraction integral of the form

$$E_0(\varrho', t) = \int d\varrho \int d\tau E_i(\varrho, t - \tau) G(\varrho', \varrho, t, \tau), \qquad (6.9)$$

where E_i and E_0 are the scalar fields in the $z=0$ and $z=L$ planes, and the Green's function (spatio-temporal impulse response) $G(\varrho', \varrho, t, \tau)$ is the field received at time t location $r = (\varrho', z = L)$ from a spatio-temporal point source $E_i(\varrho_1, t_1) = \delta(\varrho_1 - \varrho)\delta(t_1 - t + \tau)$ in the $z=0$ plane. Inasmuch as we may require that $E_i(\varrho, t)$ vanish identically for $|\varrho| > d_1/2$, and limit any optical system in the $z=L$ plane to respond to $E_0(\varrho', t)$ only if $|\varrho'| \leq d_2/2$, (6.9) reduces the R_1 to R_2 propagation problem to determining the Greens's function G.

It is known [6.21, 25, 31–33] that propagation through the turbulent atmosphere does not appreciably increase the time duration of an optical pulse if, at the transmitter, its duration is in excess of a few picoseconds. Thus, if $E_i(\varrho, t)$ is limited in temporal frequency to a bandwidth $\Delta f \leq 10^{10}$–10^{11} Hz about an optical carrier frequency f_0, then the wave propagates through the atmosphere as though it were monochromatic. By introducing the complex envelope representations

$$E_i(\varrho, t) = \text{Re}[U_i(\varrho, t)\exp(-i2\pi f_0 t)], \qquad (6.10)$$

$$E_0(\varrho', t) = \text{Re}[U_0(\varrho', t)\exp(-i2\pi f_0 t)], \qquad (6.11)$$

the extended Huygens-Fresnel principle then takes the following quasimonochromatic form [6.15–21]:

$$U_0(\varrho', t) = \int_{R_1} d\varrho U_i(\varrho, t - L/c) h_{21}(\varrho', \varrho, t), \qquad (6.12)$$

where we have explicity included the restriction that U_i vanish identically outside of the R_1 pupil. In the remainder of the chapter we shall always presume that the quasimonochromatic condition is met; hence, according to (6.12), a complete characterization of the R_1 to R_2 propagation problem is afforded by a complete characterization of $h_{21}(\varrho', \varrho, t)$. The Green's function $h_{21}(\varrho', \varrho, t)$ is the response at time t, location $r = (\varrho', z = L)$ to a monochromatic (frequency f_0) spatial point source $U_i(\varrho_1, t) = \delta(\varrho_1 - \varrho)$ in the $z=0$ plane. It is related to the polychromatic Green's function $G(\varrho', \varrho, t, \tau)$ by a Fourier transform integral, i.e.,

$$h_{21}(\varrho', \varrho, t) = \int d\tau G(\varrho', \varrho, t, \tau) \exp(i2\pi f_0 \tau). \tag{6.13}$$

Once again, an explicit determination of the required Green's function is, for arbitrary deterministic $\varepsilon(r, t, f_0)$, impossible. However, by viewing $\varepsilon(r, t, f_0)$ as a random selection from an ensemble of possible permittivity distributions, a statistical characterization may be obtained. This approach will be pursued in Section 6.1.2.

The subscripts attached to the Green's function in (6.12) are to indicate that the wave is propagating from R_1 to R_2. In a bidirectional optical communication system in which, at each terminal, common optics are employed for both the transmitter and receiver, or in a monostatic imaging radar, we must of necessity be able to characterize wave propagation from R_2 to R_1 as well as from R_1 to R_2. Suppose that $V_i(\varrho', t)$ is the complex envelope of a quasimonochromatic scalar field that is transmitted from the R_2 pupil in the $-z$ direction. Following the same reasoning that led to (6.12), we deduce that $V_0(\varrho, t)$, the complex envelope of the field in the $z=0$ plane, is given by

$$V_0(\varrho, t) = \int\limits_{R_2} d\varrho' V_i(\varrho', t - L/c) h_{12}(\varrho, \varrho', t), \tag{6.14}$$

where h_{12} is the Green's function for R_2 to R_1 propagation. In the limiting case in which R_1 and R_2 are separated by free space, the Green's functions h_{21} and h_{12} are known; they satisfy the condition

$$h_{21}(\varrho', \varrho, t) = h_{12}(\varrho, \varrho', t) \quad \text{for all} \quad \varrho \in R_1, \varrho' \in R_2, t, \tag{6.15}$$

where we have retained the time parameter although the free-space Green's functions are time independent. Equation (6.15) is known as the reciprocity theorem of Helmholtz [6.34, 35]. It turns out that this theorem is also valid when the propagation medium is the turbulent atmosphere [6.17, 18, 36]. Specifically, if $h_{21}(\varrho', \varrho, t)$ and $h_{12}(\varrho, \varrho', t)$ are the Green's functions for R_1 to R_2 and R_2 to R_1 propagation, respectively, through the same instantaneous atmospheric state, i.e., the same frozen spatial array of eddies, then (6.15) is valid. Physically, the reciprocity theorem asserts the equivalence of the point field measurement at $r = (\varrho', z = L)$ of the wave received from a monochromatic point source at $r = (\varrho, z = 0)$ and the point field measurement at $r = (\varrho, z = 0)$ of the field from a monochromatic point source at $r = (\varrho', z = L)$ for each instantaneous atmospheric state. Because the coherence time associated with atmospheric turbulence is of the order of milliseconds or longer [6.21, 25], the reciprocity condition has become the basis for a number of turbulence compensation techniques [6.16, 36–38] (cf. Sect. 6.3.5).

6.1.2 Green's Function Statistics

In the preceding section, we formulated the extended Huygens-Fresnel principle atmospheric propagation model with the presumption that the space-time-frequency permittivity ε was a deterministic function. In order to make further headway, it is necessary and traditional [6.14, 15, 18–21, 24] to statistically characterize ε, and hence h_{21}. It is convenient at the outset of the statistical analysis to introduce the frozen-flow assumption (Taylor's hypothesis), wherein the time dependence in $\varepsilon(r, t, f_0)$, and hence $h_{21}(\varrho', \varrho, t)$, is attributed to a known transverse wind profile along the path blowing a fixed array of turbulent eddies by the R_1, R_2 pupils [6.14, 39, 40]. This assumption reduces the statistical propagation problem to the study of monochromatic spherical wave propagation in a stochastic time-independent inhomogeneous medium characterized by permittivity

$$\varepsilon(r) = \varepsilon_0 [1 + n_1(r)]^2 \, ,$$

where $n_1(r)$ is the random refractive index perturbation that is due to atmospheric turbulence. Because the remainder of this section deals only with a single atmospheric state (time-independent ε), we shall drop, for the moment, the t dependence from the Green's function h_{21}. We shall reinstate the t dependence in succeeding sections whenever it becomes important to relate time-averaged experimental quantities to ensemble-averaged theoretical quantities.

The stochastic Green's function $h_{21}(\varrho', \varrho)$ is the Rayleigh-Sommerfeld Green's function associated with the stochastic Helmholtz equation

$$\nabla^2 U(r) + k^2 [1 + n_1(r)]^2 U(r) = 0, \tag{6.16}$$

where $k = 2\pi f_0/c$ is the free-space wavenumber at the frequency of interest [6.17]. It is generally assumed that $n_1(r)$ is a zero-mean fluctuation with a locally homogeneous covariance function of the form [6.14, 24, 40]

$$\langle n_1(r_1) n_1(r_2) \rangle = C_n^2 [(z_1 + z_2)/2]$$
$$\cdot \int du\, S_n(|u|) \exp[iu \cdot (r_1 - r_2)] \, , \tag{6.17}$$

where $r_1 = (\varrho_1, z_1)$, $r_2 = (\varrho_2, z_2)$, angular brackets denote averaging over the turbulence ensemble, $C_n^2(z)$ is the refractive index structure constant (turbulence-strength profile) along the path from $R_1(z=0$ plane) to $R_2(z=L$ plane), and $S_n(u)$ is the normalized three-dimensional refractive index spatial frequency spectrum. We shall briefly summarize the statistical characterization of h_{21} that has been derived from (6.16) and (6.17).

Without loss of generality, let us express the atmospheric Green's function as a multiplicatively perturbed version of the free-space (paraxial approximation) Green's function, viz,

$$h_{21}(\varrho', \varrho) = \{\exp[ik(L + |\varrho' - \varrho|^2/2L)]/i\lambda L\}$$
$$\cdot \exp[\chi(\varrho', \varrho) + i\phi(\varrho', \varrho)], \tag{6.18}$$

where $\lambda = c/f_0$ is the operating wavelength, and χ and ϕ are, respectively, the stochastic turbulence-induced log-amplitude and phase fluctuations of the field received at ϱ', in R_2 from a monochromatic point source at ϱ in R_1. Within the region of validity of Rytov's method, χ and ϕ are jointly gaussian random fields and hence statistically determined by their mean functions and covariance functions [6.18–20]

$$m_\chi = \langle \chi(\varrho', \varrho) \rangle, \tag{6.19}$$

$$m_\phi = \langle \phi(\varrho', \varrho) \rangle, \tag{6.20}$$

$$C_\chi(\varrho', \varrho) = \langle [\chi(\varrho_1' + \varrho', \varrho_1 + \varrho) - m_\chi][\chi(\varrho_1', \varrho_1) - m_\chi] \rangle, \tag{6.21}$$

$$C_\phi(\varrho', \varrho) = \langle [\phi(\varrho_1' + \varrho', \varrho_1 + \varrho) - m_\phi][\phi(\varrho_1', \varrho_1) - m_\phi] \rangle, \tag{6.22}$$

$$C_{\chi, \phi}(\varrho', \varrho) = \langle [\chi(\varrho_1' + \varrho', \varrho_1 + \varrho) - m_\chi][\phi(\varrho_1', \varrho_1) - m_\phi] \rangle. \tag{6.23}$$

Energy conservation implies [6.41] that $m_\chi = -C_\chi(0, 0)$, and no loss in generality results from assuming $m_\phi = 0$; thus the key functions to determine are the covariances. It may be shown [6.15, 20] that for arbitrary $C_n^2(z)$ and $S_n(u)$ these functions are given by the following expressions in the weak perturbation regime:

$$C_\chi(\varrho', \varrho)$$
$$= 4\pi^2 k^2 \int_0^L dz \int_0^\infty du\, u C_n^2(z) S_n(u) J_0(du) \sin^2[u^2 z(L-z)/2kL], \tag{6.24}$$

$$C_\phi(\varrho', \varrho)$$
$$= 4\pi^2 k^2 \int_0^L dz \int_0^\infty du\, u C_n^2(z) S_n(u) J_0(du) \cos^2[u^2 z(L-z)/2kL], \tag{6.25}$$

$$C_{\chi, \phi}(\varrho', \varrho)$$
$$= 2\pi^2 k^2 \int_0^L dz \int_0^\infty du\, u C_n^2(z) S_n(u) J_0(du) \sin[u^2 z(L-z)/kL], \tag{6.26}$$

where $d = |\varrho'z + \varrho(L-z)|/L$. These auto- and cross-covariance functions are two-source spherical wave covariance functions, i.e., they measure the covariances between turbulence-induced logarithmic perturbations of the field received at the point $\varrho_1' + \varrho'$ in the R_2 pupil from a spherical wave source at $\varrho_1 + \varrho$ in R_1, and the field received at ϱ_1' in R_2 from another spherical wave source located at the point ϱ_1 in the R_1 pupil. The structure functions associated with the foregoing covariances, i.e.,

$$D_\chi(\varrho',\varrho)=2[C_\chi(0,0)-C_\chi(\varrho',\varrho)],\tag{6.27}$$

$$D_\phi(\varrho',\varrho)=2[C_\phi(0,0)-C_\phi(\varrho',\varrho)],\tag{6.28}$$

$$D_{\chi,\phi}(\varrho',\varrho)=2[C_{\chi,\phi}(0,0)-C_{\chi,\phi}(\varrho',\varrho)],\tag{6.29}$$

have been partially evaluated [6.42] when $S_n(u)$ is taken to be the Kolmogorov spectrum. More complete results are available when $\varrho=0$, so that (6.24)–(6.26) reduce to the log-amplitude and phase auto- and cross-covariance functions for a single spherical wave source [6.43, 44].

Thus far we have assumed that the region $z \leq L$ falls entirely within the weak perturbation regime, i.e., $z=L$ occurs before the onset of saturated scintillation [6.21, 40]. Subject to this limitation, we have, through (6.24)–(6.26), a complete statistical characterization of $h_{21}(\varrho',\varrho)$, which, via the extended Huygens-Fresnel principle, determines the statistics of $U_0(\varrho')$, for any particular $U_i(\varrho)$. It is almost always true that the phase variance $C_\phi(0,0)$ is large enough to imply [6.40, 45, 46]

$$\langle h_{21}(\varrho',\varrho)\rangle\approx 0,\tag{6.30}$$

$$\langle h_{21}(\varrho'_1+\varrho',\varrho_1+\varrho)h_{21}(\varrho'_1,\varrho_1)\rangle\approx 0,\tag{6.31}$$

which in turn means that for arbitrary input fields $U_i(\varrho)$ we have

$$\langle U_0(\varrho')\rangle\approx 0,\tag{6.32}$$

$$\langle U_0(\varrho'_1)U_0(\varrho'_2)\rangle\approx 0.\tag{6.33}$$

The simplest nontrivial output field moment is therefore the mutual coherence function $\langle U_0(\varrho'_1)U_0^*(\varrho'_2)\rangle$, which, for a given input field U_i, may be calculated from the mutual coherence function for h_{21} according to [6.47]

$$\langle U_0(\varrho'_1)U_0^*(\varrho'_2)\rangle$$
$$=\int_{R_1} d\varrho_1 \int_{R_1} d\varrho_2 U_i(\varrho_1)U_i^*(\varrho_2)\langle h_{21}(\varrho'_1,\varrho_1)h_{21}^*(\varrho'_2,\varrho_2)\rangle.\tag{6.34}$$

From the weak perturbation statistics we have presented, it is easily demonstrated that [6.19, 20]

$$\langle h_{21}(\varrho'_1,\varrho_1)h_{21}^*(\varrho'_2,\varrho_2)\rangle$$
$$=\{\exp[ik(|\varrho'_1-\varrho_1|^2-|\varrho'_2-\varrho_2|^2)/2L-D(\varrho'_1-\varrho'_2,\varrho_1-\varrho_2)/2]\}/(\lambda L)^2,\tag{6.35}$$

where $D(\varrho', \varrho) = D_\chi(\varrho', \varrho) + D_\phi(\varrho', \varrho)$ is the two-source spherical-wave wave structure function given by

$$D(\varrho', \varrho) = 2.91k^2 \int_0^L dz C_n^2(z)[|\varrho'z + \varrho(L-z)|/L]^{5/3}, \tag{6.36}$$

for the Kolmogorov spectrum. The calculation of the output field mutual coherence will be vital to our evaluation of atmospheric imaging systems; thus it is particularly important to note that use of (6.35) and (6.36) to obtain this mutual coherence from (6.34) has been shown [6.23] to be valid well into the strong perturbation regime wherein saturation of scintillation markedly affects the log-amplitude covariance function [6.21, 48–50]. Theoretical studies to date of the saturation regime, aside from the aforementioned mutual coherence result, have not addressed the problem of Green's function statistics per se. Because of deviations from log normal statistics that must occur in the saturation regime [6.51, 52], it is not obvious that (6.18) is well suited to obtaining the desired strong perturbation characterization of h_{21}. In particular, deep in the strong perturbation regime, $h_{21}(\varrho', \varrho)$ must itself become a gaussian random field [6.50].

6.1.3 Normal Mode Decomposition

The normal mode decomposition of the R_1 to R_2 propagation channel consists of regarding the frozen (instantaneous) propagation medium and the input and output pupils as a passive resonator, and deriving the input and output transverse modes (eigenfunctions) and the associated diffraction losses (eigenvalues) for a single pass from the transmitter to the receiver at the wavelength of interest. For free-space propagation, the normal mode approach has proved fruitful in the solution of maximum power transfer and apodization problems [6.53, 54], and in clarifying the role of noise in super-resolution imaging [6.55, 56]. It happens that for atmospheric propagation there are near-field and far-field propagation regimes for which the normal mode decompositions are quite similar to those associated with the corresponding free-space channel [6.57, 58], as discussed below. The implications of these results will be pursued in succeeding sections.

The instantaneous normal mode decomposition associated with pupils R_1, R_2 and atmospheric state $h_{21}(\varrho', \varrho)$ consists of a complete orthonormal (CON) set of input eigenfunctions $\{\Phi_m(\varrho):1 \leq m < \infty, \varrho \in R_1\}$, a CON set of output eigenfunctions $\{\phi_m(\varrho'):1 \leq m < \infty, \varrho' \in R_2\}$, and a set of eigenvalues $\{\eta_m:1 \leq m < \infty\}$ that satisfy the following equations:

$$\int_{R_1} d\varrho_2 K_{21}(\varrho_1, \varrho_2)\Phi_m(\varrho_2) = \eta_m \Phi_m(\varrho_1), \quad \varrho_1 \in R_1, \tag{6.37}$$

$$\eta_m^{1/2}\phi_m(\varrho') = \int_{R_1} d\varrho \Phi_m(\varrho)h_{21}(\varrho', \varrho), \quad \varrho' \in R_2, \tag{6.38}$$

where

$$K_{21}(\varrho_1,\varrho_2)= \int_{R_2} d\varrho' h_{21}^*(\varrho',\varrho_1)h_{21}(\varrho',\varrho_2), \tag{6.39}$$

and it is assumed that the eigenfunctions are arranged in order of decreasing eigenvalues. Because $\{\Phi_m\}$ and $\{\phi_m\}$ are CON sets on their respective domains, any pair of functions $U_i(\varrho)$ $\varrho \in R_1$, $U_0(\varrho')$ $\varrho' \in R_2$ may be expressed as generalized Fourier series (modal expansions)

$$U_i(\varrho)= \sum_{m=1}^{\infty} U_{im}\Phi_m(\varrho), \quad \varrho \in R_1, \tag{6.40}$$

$$U_0(\varrho')= \sum_{m=1}^{\infty} U_{0m}\phi_m(\varrho'), \quad \varrho' \in R_2, \tag{6.41}$$

where

$$U_{im}= \int_{R_1} d\varrho U_i(\varrho)\Phi_m^*(\varrho), \tag{6.42}$$

$$U_{0m}= \int_{R_2} d\varrho' U_0(\varrho')\phi_m^*(\varrho'). \tag{6.43}$$

If we suppose that $U_0(\varrho')$ is the field that results over R_2 when $U_i(\varrho)$ is transmitted from R_1 through the atmospheric state $h_{21}(\varrho' \varrho)$, then it follows from (6.12) and (6.38) that

$$U_{0m}=\eta_m^{1/2}U_{im}, \quad 1 \leq m < \infty. \tag{6.44}$$

In other words, the normal mode decomposition replaces the extended Huygens-Fresnel principle with an equivalent set of fictitious parallel channels, as shown in Fig. 6.2. It is worthwhile to note that, because of the reciprocity principle (6.15), $\{\phi_m^*(\varrho'): 1 \leq m < \infty, \varrho' \in R_2\}$, $\{\Phi_m^*(\varrho): 1 \leq m < \infty, \varrho \in R_1\}$, $\{\eta_m: 1 \leq m < \infty\}$ are the input eigenfunctions, output eigenfunctions, and eigenvalues, respectively, of the normal mode decomposition for R_2 to R_1 propagation through the same atmospheric state [6.16].

For free-space propagation, the Fredholm equation (6.37) has been solved [6.53, 59]. The free-space eigenvalues depend parametrically on the Fresnel number $D_{f0} =(\pi d_1 d_2/4\lambda L)^2$, and the input and output eigenfunctions are prolate spheroidal wave functions. In the far-field region, $D_{f0} \ll 1$, the maximum eigenvalue η_1 is approximately equal to D_{f0}, and all other eigenvalues are insignificant. Furthermore, the input and output eigenfunctions associated with η_1 may be approximated as follows:

$$\Phi_1(\varrho)\approx(2/\pi^{1/2}d_1)\exp(-ik|\varrho|^2/2L), \quad \varrho \in R_1 \tag{6.45}$$

$$\phi_1(\varrho')\approx(2/\pi^{1/2}d_2)\exp(ik|\varrho'|^2/2L), \quad \varrho' \in R_2, \tag{6.46}$$

where an absolute-phase factor has been suppressed in (6.46).

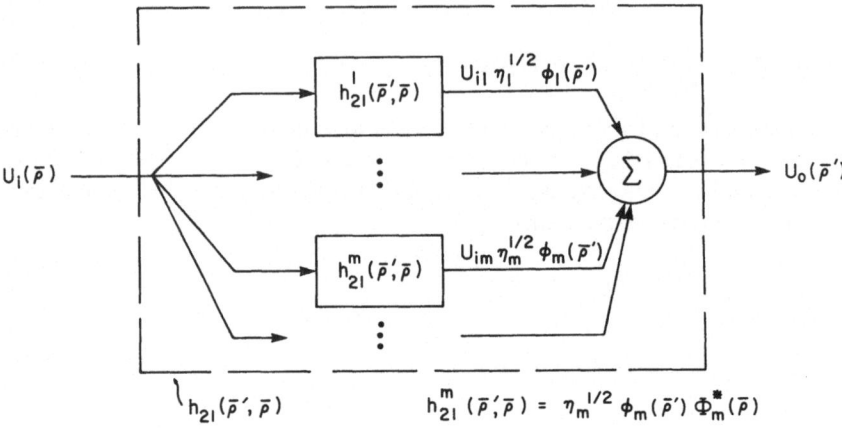

Fig. 6.2. Normal mode decomposition of the R_1-to-R_2 propagation channel

In the near-field region, $D_{f0} \gg 1$, there are D_{f0} near-unity eigenvalues; all others are approximately zero. The D_{f0} input eigenfunctions associated with the near-unity eigenvalues may be approximated by [6.60]

$$\Phi_m(\varrho) \approx (2/\pi^{1/2}d_1)\exp[-i(k|\varrho|^2/2L + 2\pi f_m \cdot \varrho)], \quad \varrho \in R_1, \quad 1 \le m < D_{f0}, \tag{6.47}$$

where $\{f_m\}$ is a set of spatial frequencies with magnitudes less than the Rayleigh resolution limit $d_2/2\lambda L$ that are chosen to approximately orthogonalize the functions.

For atmospheric propagation through instantaneous state $h_{21}(\varrho',\varrho)$ there are far-field and near-field propagation regimes that are distinguished by whether an effective instantaneous Fresnel number

$$D_f = \int_{R_1} d\varrho \int_{R_2} d\varrho' |h_{21}(\varrho',\varrho)|^2, \tag{6.48}$$

is less than or greater than unity [6.57]. In particular, when $D_f \ll 1$ we are in the far-field regime; the maximum instantaneous eigenvalue η_1 is then approximately D_f and all others are insignificant. Furthermore, in the far-field regime if R_1 lies within a single atmospheric coherence area, viz,

$$\chi(\varrho',\varrho) + i\phi(\varrho',\varrho) \approx \chi(\varrho',0) + i\phi(\varrho',0), \quad \varrho \in R_1, \quad \varrho' \in R_2, \tag{6.49}$$

then $\Phi_1(\varrho)$ is approximately given by (6.45); if R_2 lies within a single coherence area, then $\phi_1(\varrho')$ is approximately given by (6.46).

When $D_f \gg 1$, we are in the near-field regime; there are then D_f near-unity instantaneous eigenvalues and the rest are insignificant. Moreover, in the near-field regime $D_f \approx D_{f0}$ with high probability and when $d_1 \leqq d_2$, the atmospheric input eigenfunctions with significant eigenvalues are approximately given by the free-space result (6.47).

6.2 Imaging Applications

Atmospheric turbulence restricts the resolution attainable in long-exposure telescopic photography to the seeing limit of an atmospheric coherence length [6.10, 12]. Because the atmospheric coherence length is typically 10 cm for a vertical path, even quite modest telescopes are turbulence limited rather than diffraction limited. In recent years the use of predetection processing to improve the quality of images distorted by atmospheric turbulence has received considerable attention. In particular, both interferometric systems, in which the optical measurements that are made are fundamentally insensitive to atmospheric phase perturbations [6.61–66], and phase-compensation systems, in which real-time compensation of atmospheric effects is attempted [6.67–71], have been shown to yield diffraction-limited resolution under appropriate conditions.

In this section we shall survey the performance limitations of various atmospheric imaging systems. We shall assume ideal, noise-free signal processors and concentrate on the fundamental performance constraints that are due to the propagation medium itself.

The imaging problem that we shall consider is the following. The object to be imaged lies in the R_1 pupil and is self-luminous or transilluminated such that it radiates a quasimonochromatic scalar field of nominal wavelength λ. The field from the object propagates from the R_1 pupil to the $z = L$ plane and is collected over R_2, the entrance pupil of the imaging system. The task of the receiver, shown schematically in Fig. 6.3, is to use the field collected over R_2 to form an estimate (image) of the object distribution. We shall be interested in the average quality of the image formed as a function of the processor structure employed by the receiver. Before proceeding to evaluate any imaging system, we shall generalize our propagation model to include incoherent sources.

6.2.1 Propagation Model for Incoherent Sources

In Section 6.1 it was implicitly assumed that the source in the R_1 pupil was coherent in both space and time, i.e., $U_i(\varrho, t)$ was a deterministic function of its arguments. Such a situation may well prevail in a laser communication system, but it is not necessarily the case in an imaging system. For example, if, in Fig. 6.3, $U_i(\varrho, t)$ is generated by transillumination of an optically flat intensity transparency with the beam from a highly coherent laser, then the model of Section 6.1 is directly applicable. On the other hand, if $U_i(\varrho, t)$ is the result of diffuse reflection of a laser illuminator beam in an imaging radar, then it will

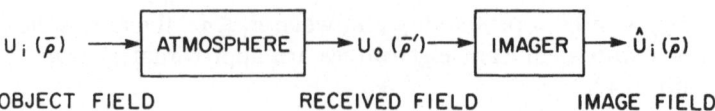

OBJECT FIELD   RECEIVED FIELD   IMAGE FIELD

Fig. 6.3. Block diagram of a general atmospheric imaging system

exhibit spatial incoherence (the well-known speckle phenomenon [6.72–75]), but it will retain temporal coherence. In this case the model of Section 6.1 does describe the instantaneous field distribution that may be detected, but it is appropriate and convenient to examine the average behavior of the field over a statistical ensemble of speckle patterns. Finally, if $U_i(\varrho, t)$ is the complex envelope leaving R_1 from a naturally illuminated or self-luminous object, it will be incoherent in both space and time. In this case, the coherence time of the source (reciprocal bandwidth) is usually far smaller than the intrinsic time constant or integration time of any photographic or photoelectric detector, so that the field distribution that is detected by such equipment is in fact describable as the average over a statistical ensemble of source functions.

Consider the propagation of a quasimonochromatic wave from R_1 to R_2 through a single atmospheric state, i.e.,

$$U_0(\varrho', t) = \int_{R_1} d\varrho\, U_i(\varrho, t - L/c) h_{21}(\varrho', \varrho, t), \tag{6.50}$$

and suppose that $U_i(\varrho, t)$ is the random complex envelope of a spatially incoherent source with the following statistical moments:

$$\langle U_i(\varrho, t) \rangle_s = 0, \tag{6.51}$$

$$\langle U_i(\varrho_1, t) U_i^*(\varrho_2, t) \rangle_s$$
$$= I_s(\varrho_1) A_s \delta(\varrho_1 - \varrho_2), \tag{6.52}$$

($\langle \rangle_s$ denotes averaging over the source ensemble). In (6.52) $I_s(\varrho_1)$ is the ensemble-averaged source radiant intensity $[I_s(\varrho_1) = \langle |U_i(\varrho_1, t)|^2 \rangle_s]$, and A_s is the source coherence area ($A_s \approx \lambda^2$). By standard stochastic process techniques we obtain from (6.50)–(6.52) the output field moments

$$\langle U_0(\varrho', t) \rangle_s = 0, \tag{6.53}$$

$$\langle U_0(\varrho_1', t) U_0^*(\varrho_2', t) \rangle_s$$
$$= \int_{R_1} d\varrho\, A_s I_s(\varrho) h_{21}(\varrho_1', \varrho, t) h_{21}^*(\varrho_2', \varrho, t). \tag{6.54}$$

Note that (6.53, 54) apply to the instantaneous atmospheric state $h_{21}(\varrho', \varrho, t)$, and that (6.54) is therefore an extended van Cittert-Zernike theorem [6.76].

In light of our earlier remarks, it is apparent that due care must be exercised in relating (6.54) to experimental results. Suppose that a detector in the R_2 pupil measures the t_d second time-averaged mutual coherence function

$$t_d^{-1} \int_{-t_d/2}^{t_d/2} d\tau U_0(\varrho_1', \tau) U_0^*(\varrho_2', \tau).$$

If t_d is less than the atmospheric coherence time t_c and much greater than the reciprocal source bandwidth, as is the case in short-exposure astronomical photography, then this time-averaged mutual coherence is correctly given by (6.54). On the other hand, in the laser radar problem we may easily achieve a t_d much less than both t_c and the reciprocal source bandwidth, so that (6.54) characterizes only the ensemble average of the measured mutual coherence.

6.2.2 Thin-Lens Imaging

In this section we consider the performance of conventional imaging systems. Inasmuch as results for incoherent illumination may be developed, via (6.54), from those for coherent illumination, we begin with the latter case. Suppose that the object in R_1 radiates a monochromatic field of complex envelope $U_i(\varrho)$, and, in order to form an image of the object, a well-corrected thin lens has been placed in the R_2 pupil, as shown in Fig. 6.4. Let us say that

$$U_i'(\alpha) = \exp(ik|\alpha|^2 L/2) U_i(\alpha L) \tag{6.55}$$

is the object distribution in angular coordinates $\alpha = \varrho/L$, where we have included a quadratic phase factor that simplifies the analysis, and let $\hat{U}_{il}'(\alpha)$ be the instantaneous angular coordinate image formed by the lens system. It can be shown [6.77] that

$$\hat{U}_{il}'(\alpha_2) = \int d\alpha_1 U_i'(\alpha_1) p_l(\alpha_2, \alpha_1), \tag{6.56}$$

where

$$p_l(\alpha_2, \alpha_1)$$
$$= \lambda^{-2} \int_{R_2} d\varrho' \exp\{\chi(\varrho', \alpha_1 L) + i[\phi(\varrho', \alpha_1 L) + k\varrho' \cdot (\alpha_2 - \alpha_1)]\}. \tag{6.57}$$

In other words, for the atmospheric state characterized by $\chi(\varrho', \varrho) + i\phi(\varrho', \varrho)$, the instantaneous coherent object thin-lens image is obtained from a superposition integral with shift-varying point spread function $p_l(\alpha_2, \alpha_1)$.

The analysis of atmospheric imaging systems simplifies considerably when it may be assumed that p_l is approximately shift invariant (isoplanatic). A sufficient condition for isoplanatism is that R_1 lies within a single atmospheric

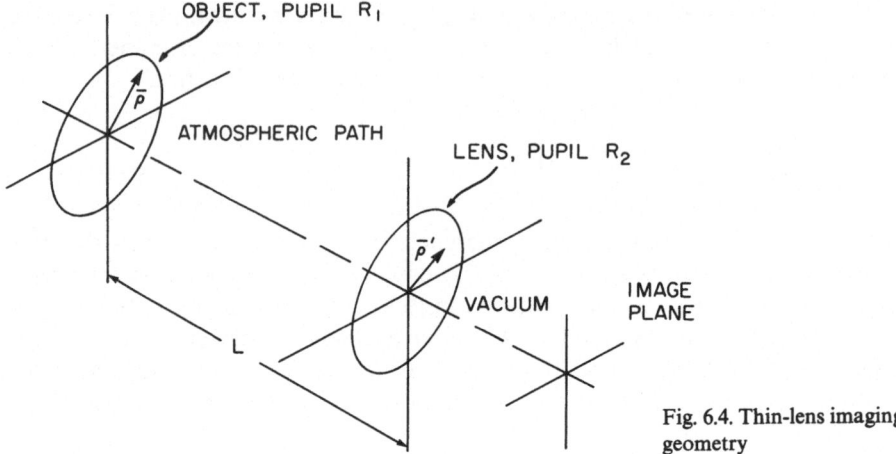

Fig. 6.4. Thin-lens imaging geometry

coherence area [cf. (6.49)], further discussion of isoplanatic conditions appears in [6.77–80]. Let us suppose that (6.49) applies so that the superposition integral (6.56) becomes a convolution, i.e., $\hat{U}'_{il}$ is related to U'_i as follows:

$$\hat{U}'_{il}(\alpha_2) = \int d\alpha_1\, U'_i(\alpha_1) p_l(\alpha_2 - \alpha_1),$$ (6.58)

where

$$p_l(\alpha) = \lambda^{-2} \int_{R_2} d\varrho'\, \exp\{\chi(\varrho',0) + i[\phi(\varrho',0) + k\varrho'\cdot\alpha]\}.$$ (6.59)

Because (6.58) is a convolution integral, introduction of angular Fourier transforms yields the equivalent transfer function equation

$$\tilde{U}'_{il}(f) = \tilde{U}'_i(f)T_l(f),$$ (6.60)

where $\tilde{U}'$ is the transform of U' and, by (6.59), the coherent object instantaneous transfer function $T_l(f)$ is

$$T_l(f) = \exp[\chi(\lambda f,0) + i\phi(\lambda f,0)]\,\mathrm{circ}(2\lambda|f|/d_2).$$ (6.61)

In the absence of turbulence, we have $\chi + i\phi \equiv 0$, and (6.59) and (6.61) reduce to the well-known diffraction-limited results (Ref. [6.81], pp. 110, 111)

$$p_0(\alpha) = d_2 J_1(\pi|\alpha| d_2/\lambda)/2\lambda|\alpha|,$$ (6.62)

$$T_0(f) = \mathrm{circ}(2\lambda|f|/d_2).$$ (6.63)

Thus, within an isoplanatic patch, the instantaneous effect of turbulence on coherent object image formation is to introduce a random multiplicative perturbation into the angular frequency domain. Using the results of Section 6.1 we find that

$$\langle T_i(f) \rangle \approx 0 , \tag{6.64}$$

$$\langle T_i(f_1) T_i^*(f_2) \rangle$$
$$= T_0(f_1) T_0^*(f_2) \exp[-(\lambda|f_1 - f_2|/\varrho_0')^{5/3}/2] , \tag{6.65}$$

where

$$\varrho_0' = \left[2.91 k^2 \int_0^L dz C_n^2(z)(z/L)^{5/3} \right]^{-3/5} , \tag{6.66}$$

is the spherical wave atmospheric coherence length in the $z = L$ plane. Note that holographically recording $\hat{U}_{il}'(\alpha)$ with an exposure interval whose duration is much greater than an atmospheric coherence time yields a measurement of $\langle \hat{U}_{il}'(\alpha) \rangle$; (6.64) implies $\langle \hat{U}_{il}'(\alpha) \rangle = 0$. Physically, the long-exposure holographic image is destroyed by atmospheric phase randomness, i.e., $C_\phi(0,0) \gg 1$. A long-exposure irradiance recording of $\hat{U}_{il}'$ yields the nonzero image

$$\langle |\hat{U}_{il}'(\alpha)|^2 \rangle$$
$$= \int df_1 \int df_2 \tilde{U}_i'(f_1) \tilde{U}_i'^*(f_2) \langle T_i(f_1) T_i^*(f_2) \rangle$$
$$\cdot \exp[i2\pi(f_1 - f_2) \cdot \alpha] . \tag{6.67}$$

If we have a sinusoidal object

$$U_i(\alpha) = A \cos(2\pi f_0 \cdot \alpha) , \tag{6.68}$$

where $|f_0| < d_2/2\lambda$, the long-exposure irradiance image is

$$\langle |\hat{U}_{il}'(\alpha)|^2 \rangle$$
$$= (A^2/2) \{1 + \cos(4\pi f_0 \cdot \alpha) \exp[-(2\lambda|f_0|/\varrho_0')^{5/3}/2]\} ; \tag{6.69}$$

hence atmospheric turbulence has reduced long-exposure fringe visibility by a factor of $\exp[-(2\lambda|f_0|/\varrho_0')^{5/3}/2]$. Clearly, if $|f_0| < \varrho_0'/\lambda$ or $d_2 < \varrho_0'$ then this fringe visibility reduction is at most $\exp(-2^{2/3}) = 0.20$. Thus, atmospheric turbulence does not appreciably affect the quality of a long-exposure coherent object irradiance image if either the object contains only angular frequencies that are resolved by a coherence length diameter pupil or the receiver pupil comprises a single atmospheric coherence area.

In the case of a spatially incoherent object with ensemble-averaged source radiant intensity (in angular coordinates) $I_s'(\alpha)$, application of the extended van Cittert-Zernike theorem to (6.58) yields the following result:

$$\hat{I}_{sl}'(\alpha_2) = \int d\alpha_1 L^{-2} A_s I_s'(\alpha_1) |p_l(\alpha_2 - \alpha_1)|^2 . \tag{6.70}$$

In (6.70), $\hat{I}'_{sl}$ is the image irradiance in angular coordinates for an instantaneous atmospheric state averaged over the source ensemble. Introduction of angular Fourier transforms gives the incoherent object transfer function equation

$$\tilde{I}'_{sl}(f) = \tilde{I}'_s(f)\tau_I(f),$$ (6.71)

where the instantaneous incoherent object transfer function $\tau_I(f)$ is obtained from $T_I(f)$ by an autocorrelation integral (Ref. [6.81], pp. 113–115)

$$\tau_I(f) = L^{-2}A_s \int df_1\, T_I(f_1 + f/2)\, T_I^*(f_1 - f/2).$$ (6.72)

In the absence of turbulence, substitution of (6.63) into (6.72) gives the well-known incoherent object diffraction limited result [6.81]

$$\tau_0(f) = (d_2^2 A_s/2\lambda^2 L^2)\,\mathrm{circ}(\lambda|f|/d_2)$$
$$\cdot \{\cos^{-1}(\lambda|f|/d_2) - (\lambda|f|/d_2)[1 - (\lambda|f|/d_2)^2]^{1/2}\}.$$ (6.73)

In the presence of turbulence, the results of Section 6.1 imply that [6.10, 12]

$$\langle\tau_I(f)\rangle = \tau_0(f)\exp[-(\lambda|f|/\varrho'_0)^{5/3}/2].$$ (6.74)

A time-averaged recording of image irradiance made with an integration time that greatly exceeds both the reciprocal source bandwidth and the atmospheric coherence time thus produces the following image:

$$\langle\hat{I}'_{sl}(\alpha)\rangle = \int df\,\tilde{I}'_s(f)\,\langle\tau_I(f)\rangle\exp(i2\pi f\cdot\alpha).$$ (6.75)

As in the case of long-exposure coherent object imaging, we see that long-exposure incoherent object imaging is severely degraded in the presence of turbulence unless $d_2 < \varrho'_0$ or the object contains only angular frequencies that satisfy $|f| < \varrho'_0/\lambda$.

6.2.3 Interferometric Imaging

The limitations on long-exposure atmospheric imaging with a thin lens are quite severe; at visible wavelengths ϱ'_0 is typically a few centimeters for horizontal paths, and ten centimeters for vertical paths. Physically, it is the turbulence-induced phase perturbations that are primarily responsible for this restricted resolution. The thin lens transforms the diverging spherical wavefront from a point in the object into a spherical wavefront converging towards the corresponding image point. When $d_2 > \varrho'_0$ in the atmosphere, the incoming wavefront at R_2 is phase distorted and, hence, it is brought to an imperfect focus (see Fig. 6.5). A natural approach for circumventing this problem is to use an imaging system which is insensitive to phase effects; this principle is the essence of interferometric imaging.

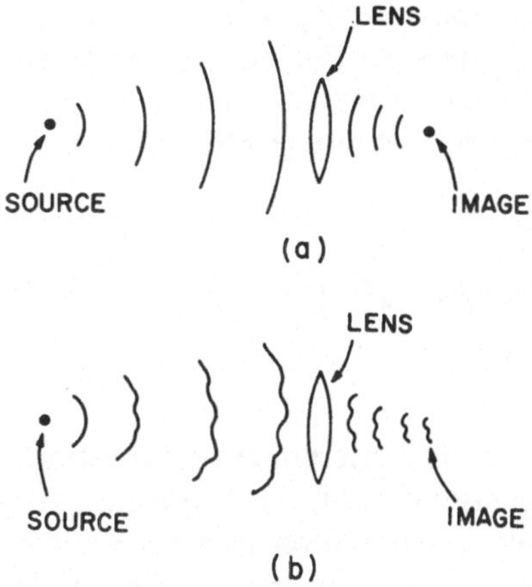

LENS

SOURCE   IMAGE

(a)

LENS

SOURCE   IMAGE

(b)

Fig. 6.5a and b. Point source image formation: (a) diffraction-limited vacuum lens system; (b) turbulence-limited atmosphere lens system

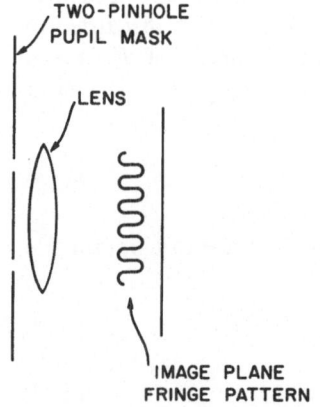

TWO-PINHOLE
PUPIL MASK

LENS

IMAGE PLANE
FRINGE PATTERN

Fig. 6.6. Michelson stellar interferometer

Interferometric imaging procedures include amplitude interferometry [6.65, 82], speckle interferometry [6.61, 62], and irradiance interferometry [6.63, 64, 83]. These techniques are, as we shall see, fundamentally insensitive to atmospheric phase perturbations and, in their simplest forms, produce images of the object autocorrelation function rather than the object itself. They all require that the object radiate a spatially incoherent wave, and, in addition, irradiance interferometry requires that the reciprocal source bandwidth be greater than the detector integration time.

Let us first consider amplitude interferometry. Suppose that the object in the R_1 pupil emits a spatially incoherent wave of average radiant intensity, in angular coordinates, $I'_s(\alpha)$. For convenience, let us say that R_1 comprises a single atmospheric coherence area to ensure isoplanatism [6.84], and let us assume the reciprocal source bandwidth is much smaller than the detector integration time, as it is in astronomical applications. Amplitude interferometry is easily explained in terms of the Michelson stellar interferometer shown in Fig. 6.6. In this system, light from the object is imaged by a thin-lens arrangement after passage through a pupil mask at R_2 which is opaque everywhere but at two pinholes located at ϱ'_1 and ϱ'_2. For an instantaneous atmospheric state, the image plane irradiance is a sinusoidal fringe pattern with angular frequency $(\varrho'_1 - \varrho'_2)/\lambda$, and complex fringe visibility [6.85]

$$\gamma = \exp[ik(|\varrho'_2|^2 - |\varrho'_1|^2)/2L]$$

$$\cdot \langle U_0(\varrho'_1, t) U_0^*(\varrho'_2, t) \rangle_s / [\langle |U_0(\varrho'_1, t)|^2 \rangle_s \langle |U_0(\varrho'_2, t)|^2 \rangle_s]^{1/2}$$

$$= \exp\{i[\phi(\varrho'_1, 0) - \phi(\varrho'_2, 0)]\} \tilde{I}'_s[(\varrho'_1 - \varrho'_2)/\lambda]/\tilde{I}'_s(0), \tag{6.76}$$

where the last equality follows from (6.54). Atmospheric phase perturbations prevent accurate direct measurement of the phase of $\tilde{I}'_s[(\varrho'_1 - \varrho'_2)/\lambda]$ from γ for $|\varrho'_1 - \varrho'_2| > \varrho'_0$; however, there is no atmospheric degradation in assessing the modulus of γ. By using a variety of pinhole spacings, or by means of multiple pinholes [6.65, 82], amplitude interferometry yields an unperturbed measurement of $|\tilde{I}'_s(f)|^2$ for all $|f| \leq d_2/\lambda$. Because the phase of $\tilde{I}'_s$ has been discarded in this procedure, reconstruction of a diffraction-limited image of I'_s is not possible. What we can obtain is a diffraction-limited image of the object autocorrelation function

$$\Gamma'_s(\alpha) = \int d\alpha_1 I'_s(\alpha_1 + \alpha/2) I'_s(\alpha_1 - \alpha/2). \tag{6.77}$$

In particular, because we can measure $|\tilde{I}'_s(f)|^2 \, \mathrm{circ}(\lambda|f|/d_2)$, we can obtain the autocorrelation image

$$\hat{\Gamma}'_s(\alpha_2) = \int df |\tilde{I}'_s(f)|^2 \, \mathrm{circ}(\lambda|f|/d_2) \exp(i2\pi f \cdot \alpha_2)$$

$$= \int d\alpha_1 \Gamma'_s(\alpha_1) p_A(\alpha_2 - \alpha_1), \tag{6.78}$$

where

$$p_A(\alpha) = d_2 J_1(2\pi|\alpha|d_2/\lambda)/\lambda|\alpha| \tag{6.79}$$

is the amplitude interferometer point spread function (for autocorrelation images). Although the object autocorrelation Γ'_s does not uniquely determine I'_s, it is of use in estimating stellar diameters and binary pair spacings [6.86, 87]. A variety of techniques have been suggested for retrieving the missing phase information [6.66, 88–91], but these methods will not be discussed herein.

The structure and performance of the speckle interferometer may be derived directly from our discussion of thin-lens atmospheric imaging. For a spatially incoherent object that comprises a single atmospheric coherence area, the short-exposure image irradiance $\hat{I}'_{si}(\alpha)$ is given by the inverse transform of the transfer function equation

$$\tilde{I}'_{si}(f) = \tilde{I}'_s(f) \tau_i(f), \tag{6.80}$$

where

$$\tau_l(f)$$
$$= L^{-2} A_s \int df_1 \, \text{circ}(2\lambda|f_1 - f/2|/d_2) \, \text{circ}(2\lambda|f_1 + f/2|/d_2)$$
$$\cdot \exp\{\chi[\lambda(f_1 - f/2), 0] + \chi[\lambda(f_1 + f/2), 0]$$
$$- i(\phi[\lambda(f_1 - f/2), 0] - \phi[\lambda(f_1 + f/2), 0]\} \tag{6.81}$$

and we have assumed that the detector integration time is long compared to the reciprocal source bandwidth. The severe resolution restriction that is incurred in long-exposure imaging when $d_2 > \varrho_0'$ is attributable to the averaging of random phase terms inside the integrand of (6.81). The speckle interferometer obtains reduced sensitivity (although not complete insensitivity) to these random phase terms by recording a series of short-exposure images and using optical Fourier transform and averaging techniques [6.61, 62] to measure

$$\langle |\tilde{I}_{sl}'(f)|^2 \rangle = |\tilde{I}_s'(f)|^2 \langle |\tau_l(f)|^2 \rangle, \tag{6.82}$$

where, within the region of validity of Rytov's method, the effective transfer function is

$$\langle |\tau_l(f)|^2 \rangle = L^{-4} A_s^2 \int df_1 \int df_2 \, \text{circ}(2\lambda|f_1 - f/2|/d_2)$$
$$\cdot \text{circ}(2\lambda|f_1 + f/2|/d_2) \, \text{circ}(2\lambda|f_2 - f/2|d_2)$$
$$\cdot \text{circ}(2\lambda|f_2 + f/2|/d_2)$$
$$\cdot \exp(-\{D(\lambda f, 0) + D[\lambda(f_1 - f_2), 0]$$
$$- D[\lambda(f_1 - f_2 + f), 0]/2 - D[\lambda(f_1 - f_2 - f), 0]/2\}$$
$$+ 2\{C_\chi[\lambda(f_1 - f_2 + f), 0] + iC_{\chi,\phi}[\lambda(f_1 - f_2 + f), 0]$$
$$+ C_\chi[\lambda(f_1 - f_2 - f), 0] - iC_{\chi,\phi}[\lambda(f_1 - f_2 - f), 0]\}). \tag{6.83}$$

The transfer function (6.83) has been evaluated [6.62], neglecting the log-amplitude covariance and log-amplitude and phase cross-covariance terms, with results as shown in Fig. 6.7. We can see that the speckle processor maintains a significant transfer function level all the way to the free-space diffraction limit even when $d_2 \gg \varrho_0'$. By compensating for the enhanced transfer function values at low angular frequencies, an autocorrelation image comparable to that of the amplitude interferometer can be realized. The speckle processor enjoys an implementation advantage over the simple two-pinhole amplitude interferometer owing to the fact that it measures all angular frequencies simultaneously; real-time speckle processing has been reported [6.91].

The term speckle interferometry derives from the nature of the short-exposure thin-lens image when $d_2 \gg \varrho_0'$; for a point source in R_1, the short-

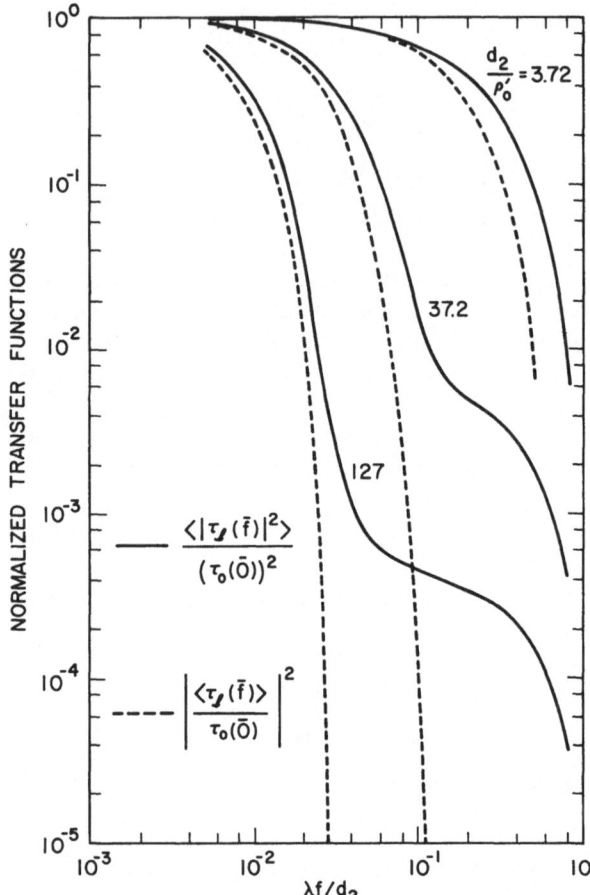

Fig. 6.7. Normalized transfer functions for speckle interferometry (solid lines), and long-exposure photography (broken lines) for various d_2/ϱ_0' values vs $\lambda f/d_2 \equiv \lambda|f|/d_2$; (after Ref. [6.62], Fig. 5)

exposure image is then a random pattern of λ/d_2 angular diameter speckles distributed over a nominal λ/ϱ_0' angular diameter region in image space. In a vacuum propagation laser radar system, a diffuse target in the R_1 pupil will radiate a spatially incoherent temporally coherent wave, so that an irradiance detector placed in the R_2 pupil (whose integration time is short compared to the reciprocal source bandwidth) will observe a speckle pattern [6.72–75] that is due to object roughness. Irradiance interferometry exploits this speckle pattern to form an autocorrelation image that, in the presence of turbulence, is completely independent of phase perturbations in the R_2 pupil. The technique employed, which is the spatial analog of the *Hanbury-Brown* and *Twiss* procedure [6.92], is as follows. It is assumed [6.72, 73] that the object in the R_1 pupil radiates a quasimonochromatic field whose complex envelope $U_i(\varrho, t)$ is gaussian distributed with statistical moments (6.51) and (6.52). An irradiance measurement is made in the R_2 pupil photographically, using an exposure time that is less than both the reciprocal source bandwidth and the atmospheric coherence time. For a source that comprises a single atmospheric coherence area, the measured irradiance is, from (6.12) and (6.49),

$$
\begin{aligned}
&|U_0(\varrho', t)|^2 \\
&= \exp[2\chi(\varrho', 0)] \left| \int\limits_{R_1} d\varrho\, U_i(\varrho, t - L/c) \exp(ik|\varrho' - \varrho|^2/2L)/i\lambda L \right|^2 .
\end{aligned}
\tag{6.84}
$$

Optical signal processing is then used to generate

$$
\left| \int\limits_{R_2} d\varrho' |U_0(\varrho', t)|^2 \exp(-i2\pi\varrho' \cdot f/L) \right|^2 ,
$$

and, via inverse Fourier transformation, we obtain the instantaneous R_2 pupil irradiance autocorrelation function

$$
\begin{aligned}
&\Gamma_0(\varrho', t) \\
&= (\pi d_2^2/4)^{-1} \int d\varrho_1' |U_0(\varrho_1' - \varrho'/2, t)|^2 |U_0(\varrho_1' + \varrho'/2, t)|^2 \\
&\quad \cdot \mathrm{circ}(2|\varrho_1' - \varrho'/2|/d_2)\mathrm{circ}(2|\varrho_1' + \varrho'/2|/d_2).
\end{aligned}
\tag{6.85}
$$

When d_2 greatly exceeds both the object speckle size and the log-amplitude coherence length in the $z = L$ plane, $\Gamma_0(\varrho', t)$ will approximately equal its average over both the source and turbulence ensembles; hence we find that

$$
\begin{aligned}
&\Gamma_0(\varrho', t) \\
&\approx (2A_s^2/\pi\lambda^4)[|\tilde{I}_s'(0)|^2 + |\tilde{I}_s'(\varrho'/\lambda)|^2] \exp[4C_\chi(\varrho', 0)] \\
&\quad \cdot [\cos^{-1}(|\varrho'|/d_2) - (|\varrho'|/d_2)(1 - |\varrho'|^2/d_2^2)^{1/2}] \mathrm{circ}(|\varrho'|/d_2),
\end{aligned}
\tag{6.86}
$$

where we have assumed that χ is gaussian distributed. To the extent that the log-amplitude covariance term in (6.86) may be neglected (as perhaps is the case in very weak turbulence) or may be corrected for [6.64, 93], irradiance interferometry yields a diffraction-limited measurement of the angular-frequency object spectrum $|\tilde{I}_s'(f)|^2$, which is equivalent to the autocorrelation image

$$
\hat{\Gamma}_s'(\alpha_2) = \int d\alpha_1 \Gamma_s'(\alpha_1)|p_0(\alpha_2 - \alpha_1)|^2 ,
\tag{6.87}
$$

where $p_0(\alpha)$ is the free-space coherent object point spread function (6.62).

6.2.4 Phase-Compensated Imaging

The rationale for interferometric imaging was to regain diffraction-limited (autocorrelation) resolution by use of phase-insensitive detection procedures. An alternative approach to diffraction-limited atmospheric imaging is to

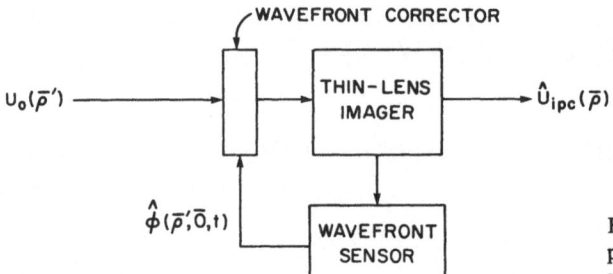

Fig. 6.8. Block diagram of a phase-compensated imaging system

actively measure and compensate for the turbulence-induced phase perturbations in real time. Real-time phase compensation is feasible because of the modest temporal bandwidth of the atmospheric phase process ($\lesssim 1\,\text{kHz}$), and recent advances in optical and electronic technology. A generic block diagram of a phase-compensated imaging system is shown in Fig. 6.8; it is composed of a wavefront corrector (or phase plate), a wavefront sensor (or phase-estimation array), and a conventional (thin-lens) imaging subsystem. The wavefront corrector is a deformable mirror [6.68, 94] or spatial-phase modulator [6.69, 95, 96] which attempts to subtract out the atmospheric phase perturbation $\phi(\varrho', 0, t)$ in an isoplanatic situation by means of a control signal (or phase estimate) $\hat{\phi}(\varrho', 0, t)$ supplied by the wavefront sensor. Recent studies have shown that atmospheric phase perturbations may be measured using image sharpness parameters [6.67], dithered phase modulation [6.69], or shearing interferometry [6.68], and have given some indication of phase-estimation performance in low light level conditions [6.97, 98]. In these studies it is generally assumed that the object to be imaged lies within a single isoplanatic patch, and that scintillation may be neglected. At present, the performance of real-time phase compensation systems is limited primarily by phase estimation errors that are due to finite wavefront sensor signal-to-noise ratio, and fitting errors that are due to finite number of spatial modes restorted by the wavefront corrector [6.99]. In keeping with our avowed desire to focus on propagation-medium effects, however, we shall assume that perfect phase measurement and correction are possible, and discuss the residual limitations imposed by the atmosphere. As the technology evolves, there will come a point at which propagation effects will limit phase-compensated imaging performance for bright sources; for the moment, the results of this section must be regarded as benchmarks against which operational systems are to be compared.

Let us first consider coherent object imaging. Suppose that the object in R_1 radiates a coherent wave of complex envelope [in angular coordinates, cf. (6.55)] $U_i'(\alpha)$, and in addition to $U_i'(\alpha)$ it also emits a point reference field $\delta(\alpha - \alpha_0)$. [Alternatively, we may consider an imaging radar in which there is an unresolvable specular reflection $\delta(\alpha - \alpha_0)$, in addition to the diffuse object field $U_i'(\alpha)$, emanating from the target in the $z = 0$ plane]. We assume that in R_2 we can separate the field component that is due to the point reference from $U_0(\varrho')$, the field component that is due to $U_i'(\alpha)$, and because the source is bright we can thereby make a perfect measurement of $\phi(\varrho', \alpha_0 L)$ vs ϱ', for the current instantaneous atmospheric state. This instantaneous phase measurement is used to drive a wavefront corrector located in the entrance pupil of a thin-lens imaging system, which, for convenience, we take to be a 1:1 image of R_2. This system produces the instantaneous angular coordinate image [cf. (6.56) and (6.57)]

$$\hat{U}_{ipc}'(\alpha_2) = \int d\alpha_1 \, U_i'(\alpha_1) p_{pc}(\alpha_2, \alpha_1), \tag{6.88}$$

where

$$p_{pc}(\alpha_2, \alpha_1)$$
$$= \lambda^{-2} \int_{R_2} d\varrho' \exp\{\chi(\varrho', \alpha_1 L) + i[\phi(\varrho', \alpha_1 L) - \phi(\varrho', \alpha_0 L) + k\varrho' \cdot (\alpha_2 - \alpha_1)]\}. \tag{6.89}$$

From the results of Section 6.1, it follows that the long-exposure holographic phase-compensated image is given by

$$\langle \hat{U}_{ipc}'(\alpha_2) \rangle = \exp[-C_\chi(0,0)/2 + iC_{\chi,\phi}(0,0)]$$
$$\cdot \int d\alpha_1 \, U_i'(\alpha_1) p_0(\alpha_2 - \alpha_1) \exp\{-D_\phi[0, (\alpha_1 - \alpha_0)L]$$
$$- iC_{\chi,\phi}[0, (\alpha_1 - \alpha_0)L]\}, \tag{6.90}$$

where we have assumed χ and ϕ are jointly gaussian. Because within the weak perturbation regime we have [6.40]

$$D_\phi(0, \varrho) \approx D(0, \varrho) = (|\varrho|/\varrho_0)^{5/3}, \tag{6.91}$$

where

$$\varrho_0 = \left[2.91 k^2 \int_0^L dz \, C_n^2(z) (1 - z/L)^{5/3} \right]^{-3/5}, \tag{6.92}$$

is the atmospheric coherence length in the $z = 0$ plane, (6.90) implies that $\langle \hat{U}_{ipc}' \rangle$ provides a diffraction-limited image of that portion of the object within the coherence area in the $z = 0$ plane containing the reference point $\alpha_0 L$. Note that outside of the region of high resolution we have $\langle \hat{U}_{ipc}' \rangle \approx 0$. This behavior is not surprising inasmuch as the transmitted reference compensated imaging system we have under consideration is in essence recording a lensless Fourier transform (LFT) hologram [6.100, 101], and such systems are known to have a field-of-view limitation (not a resolution limit) that is set by modulation transfer function rolloff [6.102]. In the vacuum LFT case, this rolloff is due to the resolution limit of the recording medium (film); in our phase-compensated atmospheric imager, it is due to the finite coherence area in the $z=0$ plane. It can be shown [6.77, 100] that the long-exposure irradiance image $\langle |\hat{U}'_{ipc}(\alpha)|^2 \rangle$ consists of a diffraction-limited image of the coherence area that contains the reference point, surrounded by a low-resolution image of the rest of the object.

The foregoing transmitted reference phase-compensated imaging system may also be employed when the object to be imaged radiates a spatially incoherent field. In particular, the irradiance image obtained using an exposure time that is much greater than both the reciprocal source bandwidth and the atmospheric coherence time is [cf. (6.70)]

$$\langle \hat{I}'_{spc}(\alpha_2) \rangle = L^{-2} A_s \int d\alpha_1 I'_s(\alpha_1) \langle |p_{pc}(\alpha_2, \alpha_1)|^2 \rangle, \tag{6.93}$$

Because for any complex-valued random variable v, we have

$$\langle |v|^2 \rangle = |\langle v \rangle|^2 + \text{Var}(v), \tag{6.94}$$

and $\text{Var}(v) \geq 0$, it follows that $\langle \hat{I}'_{spc}(\alpha_2) \rangle$ contains a diffraction-limited image of that portion of the object that lies within the $z=0$ plane coherence area containing the reference point [6.77]. It is interesting to note that for spatially incoherent objects which comprise a single atmospheric coherence area and have reciprocal bandwidths much smaller than the atmospheric coherence time, the point reference field is not necessary. For such objects we can obtain the required atmospheric phase information by observing, interferometrically [6.68], the short-exposure mutual coherence in R_2, viz,

$$\begin{aligned}
\langle U_0(\varrho'_1, t) U_0^*(\varrho'_2, t) \rangle_s \\
= \exp\{\chi(\varrho'_1, 0) + \chi(\varrho'_2, 0) + i[\phi(\varrho'_1, 0) - \phi(\varrho'_2, 0)]\} \\
\cdot L^{-2} A_s \exp[ik(|\varrho'_1|^2 - |\varrho'_2|^2)/2L] \tilde{I}'_s[(\varrho'_1 - \varrho'_2)/\lambda],
\end{aligned} \tag{6.95}$$

for a series of $|\varrho'_1 - \varrho'_2|$ values. This technique works because the R_2 pupil phase contribution that is due to atmospheric turbulence enters into the source-averaged mutual coherence in a way distinctly different from the R_2 pupil phase contribution that is due to the object. Self-referenced operation is impossible with coherent light unless outside information is available to distinguish object phase terms from atmospheric phase terms.

6.2.5 Modal Theory of Optimum Imaging

The preceding analysis shows that, whenever the performance of a phase-compensated imaging system is not limited by finite wavefront sensor signal-to-noise ratio or wavefront corrector fitting errors, the principal restriction on such a system is due to finite isoplanatic diameter, i.e., the object to be imaged must lie within a single atmospheric coherence area. There is some theoretical indication [6.80] that the field contributions from the different isoplanatic elements of a large object may be isolated from one another and processed in parallel to yield a diffraction-limited image. In this section, we shall use the normal mode description of atmospheric propagation to reinforce the preceding view that phase-compensated imaging achieves optimum resolution, within the limits set by equipment performance and isoplanatism.

Consider the highly idealized problem of estimating a coherent source distribution, $U_i(\varrho)$ for $\varrho \in R_1$, based on a noiseless measurement of the field $U_0(\varrho')$, received over R_2 after propagation through instantaneous atmospheric state $h_{21}(\varrho', \varrho)$. We shall assume that the instantaneous Green's function and the associated mode decomposition are known, *a priori*, to the receiver. Although no real receiver can in fact have this *a priori* channel knowledge, the performance of the optimum known channel imaging system will exceed that of any real system. Expanding $U_i(\varrho)$ and $U_0(\varrho')$ in the generalized Fourier series (6.40) and (6.41) and using the propagation relation (6.44), it is apparent that whenever a perfect measurement of $U_0(\varrho')$ is made for ϱ' in R_2, $U_i(\varrho)$ may be reconstructed with arbitrary precision.

For free-space propagation, the foregoing statement comprises one of the mathematical formulations for super-resolution imaging [6.55]. However, because the free-space eigenvalues are, in the near-field regime, clustered about one and zero with few falling between, the division algorithm

$$U_{im} = \eta_m^{-1/2} U_{0m}, \qquad 1 \leq m < \infty, \tag{6.96}$$

which retrieves the object mode amplitudes $\{U_{im}\}$ from the received mode amplitudes $\{U_{0m}\}$ greatly magnifies any inaccuracies that occur in measuring a U_{0m} associated with a small eigenvalue. Thus, the realistic inclusion of spatially white noise in modeling the measurement process implies that reliable object mode amplitude estimates can be obtained, in the usual case, only for the D_{f0} modes whose eigenvalues are close to one [6.55, 56]. Moreover, from (6.47) it follows that these D_{f0} resolvable modes span the spatial-frequency region resolved by a thin lens in the R_2 pupil [6.60]. In fact, the mode amplitude representation of free-space thin-lens imaging is known to be [6.58]

$$\hat{U}_{iml} = \int\limits_{R_1} d\varrho \, \hat{U}'_{il}(\varrho/L) \Phi_m^*(\varrho)$$
$$= \eta_m^{1/2} U_{0m} = \eta_m U_{im}, \qquad 1 \leq m < \infty, \tag{6.97}$$

in the absence of measurement noise, which confirms our previous statement that free-space thin-lens imaging accurately estimates the D_{f0} object mode amplitudes associated with near-unity eigenvalues.

In the presence of turbulence, the instantaneous eigenvalues still exhibit the aforementioned clustering about one and zero in the near-field regime. Thus, reliable object mode amplitude estimation is again limited to the modes with near-unity eigenvalues. Because of the turbulence-induced phase and log-amplitude perturbations in R_2, the thin-lens system no longer has the mode amplitude representation (6.97). However, the channel-matched filter system [6.58]

$$\hat{U}'_{\text{icf}}(\alpha_2) = \int_{R_2} d\varrho'\, U_0(\varrho')h^*_{21}(\varrho', \alpha_2 L)$$
$$= \int d\alpha_1\, U'_{\text{i}}(\alpha_1)p_{\text{cf}}(\alpha_2, \alpha_1), \tag{6.98}$$

where

$$p_{\text{cf}}(\alpha_2, \alpha_1) = L^2 K_{21}(\alpha_2 L, \alpha_1 L)\exp[\text{i}kL(|\alpha_2|^2 - |\alpha_1|^2)/2], \tag{6.99}$$

does realize the mode amplitude relation

$$\int_{R_1} d\varrho\, \hat{U}'_{\text{icf}}(\varrho/L)\Phi^*_m(\varrho) = \eta_m U_{\text{im}}, \quad 1 \leq m < \infty, \tag{6.100}$$

and is therefore essentially the optimum known channel imaging system. It is easily shown that [6.58]

$$\langle p_{\text{cf}}(\alpha_2, \alpha_1)\rangle = p_0(\alpha_2 - \alpha_1)\exp\{-D[0, (\alpha_2 - \alpha_1)L]/2\}, \tag{6.101}$$

and thus in the limit as d_2 approaches infinity the average channel-matched filter point spread function is exactly equal to the diffraction-limited point spread function p_0. Finally, if we suppose that scintillation may be neglected, and that the object comprises a single atmospheric coherence area, then (6.99) is achieved by a phase-compensated imaging system [6.77].

6.3 Communication Applications

As an optical communication channel, the turbulent atmosphere may be characterized, superficially, as a log normal fading channel having an enormous coherence bandwidth ($\geq 10^{10}$ Hz), a long coherence time ($\geq 10^{-3}$ s), and significant dispersion in space and spatial frequency [6.8, 25]. A more detailed analysis reveals that turbulence-induced beam spreading limits the minimum average far-field beamwidth of a collimated laser transmitter [6.18, 20, 77], and that angle-of-arrival (phase) fluctuations limit the power collection capability of a diffraction-limited receiver [6.5, 77]. Moreover, scintillation, which was found to play a minor role in imaging systems, is especially significant in the communication problem, because deep signal fades have a disastrous effect on communication performance [6.7, 8, 13]. In this section, we shall investigate the interplay between these propagation effects and the performance of various optical communication systems for the specific case of bidirectional earth-space transmission. Similar results are obtained for terrestrial links in which far-field propagation conditions prevail.

Consider a bidirectional earth-space optical communication system using common optics at each terminal for both transmission and reception. In particular, let us say that the R_1 pupil (in Fig. 6.1) is aboard a spacecraft, and serves both as the exit pupil for the space-to-earth (downlink) transmitter and the entrance pupil for the earth-to-space (uplink) receiver. The pupil R_2 is assumed to be at a ground terminal, and serves the dual functions of uplink transmitter exit pupil and downlink receiver entrance pupil. We assume that the spacecraft lies far above the atmosphere, so that all the turbulence in the propagation path is concentrated near the R_2 pupil (the ground terminal). Because of this concentration of turbulence near R_2, the earth-space channel behaves as though vacuum conditions prevail for $0 \leq z < L$, with turbulence creating a random phase and amplitude screen in the R_2 pupil. As a result, uplink transmission is plagued by atmospheric beam spreading, whereas downlink reception suffers from angle-of-arrival fluctuations. Both links are subject to scintillation fading, but, because of the disparate atmospheric coherence lengths in the $z=0$ and $z=L$ planes, spatial diversity reception to combat fading is possible only on the downlink. We shall begin the communication analysis by abstracting a quantitative formulation for the foregoing earth-space propagation effects from the extended Huygens-Fresnel principle.

6.3.1 Earth-Space Propagation Channel

For typical pupil diameters on the ground and the spacecraft, and altitudes at or beyond synchronous orbit we may assume $(d_1^2 + d_2^2)/\lambda L \ll 1$. Moreover, for earth-space transmission, the atmospheric coherence lengths in the $z=0$ and $z=L$ planes are, respectively,

$$\varrho_0 = \left[2.91 k^2 (\sec\beta)^{8/3} L^{-5/3} \int_0^\infty dh c_n^2(h) h^{5/3} \right]^{-3/5}, \tag{6.102}$$

and

$$\varrho_0' = \left[2.91 k^2 \sec\beta \int_0^\infty dh c_n^2(h) \right]^{-3/5} = \varrho_0 h_0 \sec\beta/L, \tag{6.103}$$

where β is the zenith angle of the R_1 to R_2 path, $c_n^2(h) = C_n^2(L - h \sec\beta)$ is the turbulence strength distribution vs height above the ground terminal, and

$$h_0 = \left[\int_0^\infty dh c_n^2(h) h^{5/3} \Big/ \int_0^\infty dh c_n^2(h) \right]^{3/5}, \tag{6.104}$$

is an atmospheric scale height [6.103]. Thus, because ϱ_0' is typically 10 cm for near-zenith paths at visible wavelengths and h_0 is on the order of 10 km [6.104], the spacecraft (R_1) pupil will always lie within a single atmospheric coherence area, although the ground (R_2) pupil may contain many atmospheric coherence areas. Under these conditions, the downlink and uplink versions of the extended Huygens-Fresnel principle for a single atmospheric state, (6.12) and (6.14), reduce to the following far-field expressions [6.16, 103]:

$$U_0(\varrho', t)$$
$$= \exp[\chi(\varrho', 0) + i\phi(\varrho', 0)] \int_{R_1} d\varrho \, U_i(\varrho, t - L/c)/i\lambda L, \quad \varrho' \in R_2, \tag{6.105}$$

and

$$V_0(\varrho, t)$$
$$= \int_{R_2} d\varrho' \exp[\chi(\varrho', 0) + i\phi(\varrho', 0)] V_i(\varrho', t - L/c)/i\lambda L, \quad \varrho \in R_1, \tag{6.106}$$

where an absolute phase factor has been suppressed. As previously surmised, turbulence presents a random multiplicative perturbation in the R_2 pupil for both downlink and uplink transmission.

According to (6.105), the instantaneous downlink power gain (fractional power transfer) is given by

$$G_d = \int_{R_2} d\varrho' |U_0(\varrho', t)|^2 \Big/ \int_{R_1} d\varrho |U_i(\varrho, t - L/c)|^2$$
$$= \left\{ \left[\int_{R_2} d\varrho' \exp[2\chi(\varrho', 0)] \right] \left| \int_{R_1} d\varrho \, U_i(\varrho, t - L/c) \right|^2 \right\}$$
$$\div \left[\int_{R_1} d\varrho |U_i(\varrho, t - L/c)|^2 (\lambda L)^2 \right]. \tag{6.107}$$

For any atmospheric state, the collimated transmitter,

$$U_i(\varrho, t) = s(t) \, \text{circ}(2|\varrho|/d_1), \tag{6.108}$$

where $s(t)$ is an information-bearing modulation, achieves the maximum instantaneous downlink power gain [6.16]

$$G_d^0 = [\pi d_1^2 / 4(\lambda L)^2] \int_{R_2} d\varrho' \exp[2\chi(\varrho', 0)], \tag{6.109}$$

whose average value

$$\langle G_d^0 \rangle = (\pi d_1 d_2 / 4\lambda L)^2 \equiv D_{f0}, \tag{6.110}$$

equals the diffraction-limited free-space performance. Equation (6.110) confirms that downlink beam spreading is negligible. Note, however, that scintillation does affect instantaneous downlink power transfer, and spatial diversity is available at R_2 if d_2 exceeds the log-amplitude coherence length.

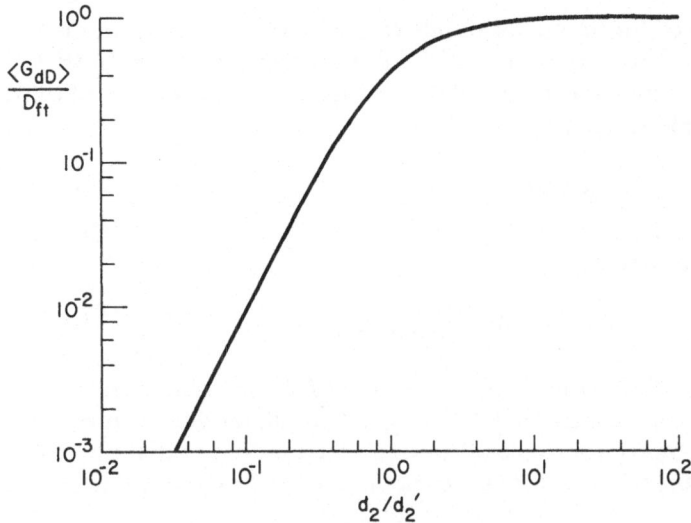

Fig. 6.9. Normalized average diffraction-limited downlink power gain vs normalized receiver diameter

For free-space propagation, the optimum downlink power gain $G_d^0 = D_{f0}$ may be realized with a receiver that responds only to the normally incident plane wave component of $U_0(\varrho', t)$. In the presence of turbulence, the power gain of this diffraction-limited receiver G_{dD} satisfies [6.5]

$$G_{dD} = (d_1/d_2\lambda L)^2 \left| \int_{R_2} d\varrho' \exp[\chi(\varrho', 0) + i\phi(\varrho', 0)] \right|^2, \tag{6.111}$$

and has average value

$$\begin{aligned} \langle G_{dD} \rangle &= [d_1^2/2(\lambda L)^2] \int d\varrho' \{\cos^{-1}(|\varrho'|/d_2) - (|\varrho'|/d_2)[1 - (|\varrho'|/d_2)^2]^{1/2}\} \\ &\cdot \mathrm{circ}(|\varrho'|/d_2) \exp[-(|\varrho'|/\varrho_0')^{5/3}/2], \end{aligned} \tag{6.112}$$

when U_i is the collimated transmitter, (6.108). Comparison of (6.111) and (6.109) reveals that diffraction-limited downlink reception does realize power gain G_d^0 whenever R_2 lies within a single atmospheric coherence area. Physically, this condition guarantees that angle-of-arrival fluctuations may be neglected. When $d_2 \gg \varrho_0'$, R_2 comprises many atmospheric coherence areas, and (6.112) reduces to the turbulence-limited result [6.5]

$$\langle G_{dD} \rangle \approx (\pi d_1 d_2'/4\lambda L)^2 \equiv D_{ft}, \tag{6.113}$$

where $d_2' = 3.18\varrho_0'$ is the effective receiver diameter (see Fig. 6.9).

The power gain characteristics of the uplink may be developed directly from (6.106) by using arguments analogous to those employed in the downlink analysis. However, complete uplink results can also be obtained from the far-field downlink mode decomposition and the reciprocity theorem [6.16, 77], simply by interchanging the roles of transmitter and receiver. Specifically, instantaneous uplink power gain,

$$G_u = \int_{R_1} d\varrho |V_0(\varrho, t)|^2 \Big/ \int_{R_2} d\varrho' |V_i(\varrho', t - L/c)|^2, \qquad (6.114)$$

is maximized by transmitting

$$V_i(\varrho', t) = s(t) \exp[\chi(\varrho', 0) - i\phi(\varrho', 0)] \operatorname{circ}(2|\varrho'|/d_2), \qquad (6.115)$$

from R_2, and using a diffraction-limited receiver at R_1. This system realizes the optimum instantaneous uplink power gain $G_u^0 \equiv G_d^0$, but, for $d_2 > \varrho_0'$, it requires that an adaptive transmitter be used to track the random spatial wavefront $\exp[\chi(\varrho', 0) - i\phi(\varrho', 0)]$ (cf. Sect. 6.3.5). The collimated uplink transmitter,

$$V_i(\varrho', t) = s(t) \operatorname{circ}(2|\varrho'|/d_2), \qquad (6.116)$$

has power gain $G_{uc} \equiv G_{dD}$ [6.36], and thus it achieves $G_{uc} = G_u^0$ whenever R_2 lies within a single atmospheric coherence area. When $d_2 \gg \varrho_0'$, atmospheric beam spreading limits average collimated transmitter uplink power gain to

$$\langle G_{uc} \rangle \approx (\pi d_1 d_2'/4\lambda L)^2 \equiv D_{ft}, \qquad (6.117)$$

where $d_2' = 3.18\varrho_0'$ is now the effective transmitter diameter. Note that diffraction-limited reception is always optimum at R_1, i.e., no spatial diversity is available at the uplink receiver.

6.3.2 Statistical Models for Optical Detection

We have heretofore assumed noise-free signal field detection. Continued use of this assumption in the communication problem will lead us inexorably to conclude that error-free information transmission through the atmosphere is possible at arbitrarily high data rates. We must therefore include in our analysis realistic models for the noise encountered in optical detection. In this section, we shall present statistical descriptions for direct detection and heterodyne detection optical receivers.

The heart of any optical communication receiver is a photodetector. The principal photodetectors for use in optical receivers are photomultipliers, semiconductor photodiodes, avalanche photodiodes, and bulk photoconductors [6.105–107]. In all of these devices, charge carriers are created by photoabsorption at a rate that is proportional to the instantaneous optical power incident on the detector's active region. The intrinsically discrete nature of the current thus generated gives rise to the shot noise observed in low light

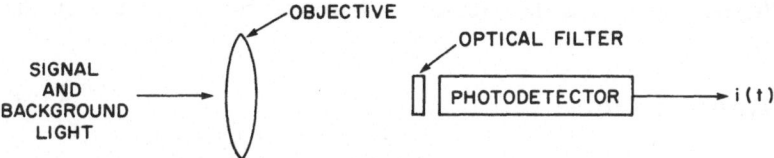

Fig. 6.10. Direct detection optical configuration

level detection [6.105, 108]. Additional shot noise results from any extraneous (background) light which falls on the detector, as well as from spontaneously generated charge carriers (dark current). In photomultipliers and avalanche photodiodes, the charge carriers generated by photoabsorption undergo a current multiplication process within the device that is intended to bring the output signal above the level of post-detection thermal noise. Any random gain fluctuations that occur in this current multiplication process add so-called excess noise to the output [6.105, 108].

Consider the idealized direct detection optical receiver shown in Fig. 6.10. It consists of a circular objective lens (focal length l, diameter d_r) to collect and focus signal light, an optical frequency filter (bandwidth $\Delta\lambda$ in wavelength units) to discriminate against background light, and a photodetector (circular active region of diameter d_D) to produce an electrical signal proportional to the incident optical power. We shall neglect dark current, thermal noise, and excess noise so that, conditioned on knowledge of the instantaneous optical power impinging on the photodetector, the photocurrent generated, $i(t)$, is a Poisson point process [6.108, 109]. In particular, assuming the receiver objective is illuminated by a known quasimonochromatic signal field of complex envelope $U_s(\varrho, t)$, center frequency f_0, plus a background field from a broadband spatially distributed source (e.g., scattered sunlight), then

$$N(t) = \int_{t_0}^{t} d\tau q^{-1} i(\tau), \quad t \geq t_0, \tag{6.118}$$

which is the number of photons counted between t_0 and t, is a Poisson random variable with mean value (over the shot-noise ensemble)

$$\langle N(t) \rangle_n = \int_{t_0}^{t} d\tau \mu(\tau), \quad t \geq t_0, \tag{6.119}$$

where

$$\mu(t) = (\alpha/hf_0)$$
$$\cdot [(c\varepsilon_0/2) \int d\varrho_2 \, \mathrm{circ}(2|\varrho_2|/d_r) | \int d\varrho_1 \, \mathrm{circ}(2|\varrho_1|/d_r) U_s(\varrho_1, t)$$
$$\cdot d_D J_1(\pi d_D |\varrho_1 - \varrho_2|/\lambda l)/2\lambda l |\varrho_1 - \varrho_2||^2 + P_b] \tag{6.120}$$

is the effective average photon arrival rate at time t [6.108]. In (6.118) and (6.120), q is the charge released per photoabsorption, α is the detector's quantum efficiency, and P_b is the average background power incident on the detector, i.e.,

$$P_b = \lambda^2 N_\lambda \Delta\lambda (\pi d_r d_D / 4\lambda l)^2 , \tag{6.121}$$

where N_λ is the background light spectral radiance [6.110, 111].

When $d_D = 4\lambda l/\pi d_r$, the system of Fig. 6.10 has a diffraction-limited field of view; in this case (6.120) and (6.121) reduce to

$$\mu(t) \approx (\alpha/hf_0)$$
$$\cdot [(2c\varepsilon_0/\pi d_r^2)| \int d\varrho \ \mathrm{circ}(2|\varrho|/d_r) U_s(\varrho, t)|^2 + P_{bD}] , \tag{6.122}$$

and

$$P_{bD} = \lambda^2 N_\lambda \Delta\lambda , \tag{6.123}$$

respectively. When d_D is just large enough to encompass the entire angular spectrum of the signal field, (6.120) and (6.121) reduce to

$$\mu(t) \approx (\alpha/hf_0)$$
$$\cdot [(c\varepsilon_0/2) \int d\varrho \ \mathrm{circ}(2|\varrho|/d_r)|U_s(\varrho, t)|^2 + P_{bs}] , \tag{6.124}$$

and

$$P_{bs} = \lambda^2 N_\lambda \Delta\lambda (d_r/d_r')^2 , \tag{6.125}$$

where $d_r' \leq d_r$ is the effective coherence diameter of $U_s(\varrho, t)$. For a signal field U_s that is a normally incident plane wave we have $d_r' = d_r$, hence a diffraction-limited receiver collects all available signal power with a minimum amount of background light. If, on the other hand, $d_r' \ll d_r$, then a diffraction-limited receiver collects a miniscule fraction of the available signal light; a wide field-of-view receiver will collect all the signal light, but with this increased signal power comes an enormously increased background level.

In a heterodyne detection receiver, background light is seldom a significant contributor to the post-detection signal-to-noise ratio. Consider the idealized heterodyne receiver shown in Fig. 6.11. This system differs in structure from that of Fig. 6.10 only in that a local oscillator signal $U_l(\varrho) \exp(\mathrm{i}2\pi f_1 t)$ is injected at the objective lens, and an intermediate frequency (IF) filter is employed to isolate photocurrent beat frequencies in the neighborhood of f_1 Hz [5.108, 109]. The structural similarity between direct detection and heterodyne detection, however, belies significant differences in their post-detection noise characteristics. In heterodyne reception, the local oscillator power is made large enough to ensure that its shot noise dominates all other noise components, and, as a result, gaussian rather than Poisson statistics apply [5.108, 112]. Specifically, assuming that the detector collects all of the local oscillator power

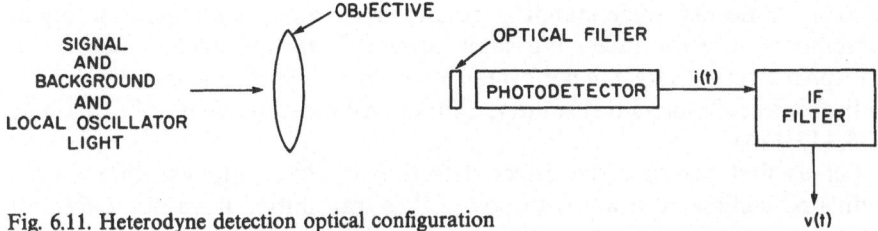

Fig. 6.11. Heterodyne detection optical configuration

and that the IF filter does not distort the beat frequency components that are due to signal light, the normalized heterodyne receiver output $v(t)$ is given by

$$v(t) = (c\varepsilon_0/2)^{1/2} \operatorname{Re}[\int d\varrho\, U_s(\varrho, t) u_l^*(\varrho) \exp(-i2\pi f_1 t)]$$
$$+ n(t), \tag{6.126}$$

where

$$u_l(\varrho) = \operatorname{circ}(2|\varrho|/d_r) U_l(\varrho)/[\int d\varrho \operatorname{circ}(2|\varrho|/d_r)|U_l(\varrho)|^2]^{1/2}, \tag{6.127}$$

and $n(t)$ is a zero-mean white gaussian noise process of spectral height $hf_0/4\alpha$. When a plane wave local oscillator,

$$u_l(\varrho) = (4/\pi d_r^2)^{1/2} \operatorname{circ}(2|\varrho|/d_r), \tag{6.128}$$

is employed and $d_D \geq 4\lambda l/\pi d_r$, the foregoing heterodyne receiver has a diffraction-limited field of view, viz, it responds only to the normally incident plane wave component of $U_s(\varrho, t)$, and (6.126) becomes

$$v(t) \approx (c\varepsilon_0/2)^{1/2} \operatorname{Re}[\int d\varrho \operatorname{circ}(2|\varrho|/d_r)(4/\pi d_r^2)^{1/2} U_s(\varrho, t)$$
$$\cdot \exp(-i2\pi f_1 t)] + n(t). \tag{6.129}$$

This diffraction-limited system will collect all the available signal power if and only if the signal field is also a normally incident plane wave.

6.3.3 Diffraction-Limited Reception

We shall begin the synthesis of Sections 6.3.1 and 6.3.2 by discussing downlink communication systems which employ collimated beam transmitters and diffraction-limited (direct detection or heterodyne detection) receivers. In the absence of turbulence, these systems achieve maximum power gain with, in the case of direct detection, a minimum of received background light. In the presence of turbulence, system performance is degraded by angle-of-arrival fluctuations and scintillation. Some quantitative appreciation of this performance degradation may be gleaned from simple carrier-to-noise ratio evaluations. A deeper understanding results from an investigation of digital communication error rates. We shall develop both approaches. Because of atmospheric reciprocity, the performance results to be presented herein are also applicable to collimated transmitter, diffraction-limited receiver uplink systems [6.36, 113, 114].

Let us first consider the direct detection receiver. Suppose that an unmodulated collimated beam with power P_t is transmitted from the spacecraft, and the photocurrent $i(t)$ resulting from ground-based diffraction-limited direct detection of this carrier signal is low-pass filtered to limit the shot-noise bandwidth. In terms of this filtered photocurrent $i'(t)$, a standard carrier-to-noise ratio definition is the following [6.115]:

$$\text{CNR}_d = \frac{[\langle\langle i'(t)\rangle_n\rangle - q\alpha P_{bD}/hf_0]^2}{\langle\langle[i'(t)-\langle\langle i'(t)\rangle_n\rangle]^2\rangle_n\rangle}, \tag{6.130}$$

where, as previously, $\langle\rangle_n$ and $\langle\rangle$ denote averaging with respect to the shot-noise and turbulence ensembles. With the aid of conditional Poisson process analysis [6.116], the CNR_d follows from (6.111) and (6.122), and is given by

$$\text{CNR}_d = \frac{P_t\langle G_{dD}\rangle\alpha/2hf_0\Delta f}{1+P_{bD}/P_t\langle G_{dD}\rangle + \text{Var}(G_{dD})P_t\alpha/2\langle G_{dD}\rangle hf_0\Delta f}, \tag{6.131}$$

where $2\Delta f$ is the bilateral filter bandwidth, $\langle G_{dD}\rangle$ satisfies (6.112), and, within the weak perturbation regime, $\text{Var}(G_{dD})$ satisfies [6.44, 117]

$$\begin{aligned}
\text{Var}(G_{dD}) &= (d_1/d_2\lambda L)^4 \\
&\quad \cdot \int_{R_2} d\varrho_1' \int_{R_2} d\varrho_2' \int_{R_2} d\varrho_3' \int_{R_2} d\varrho_4' \\
&\quad \cdot \exp\{-[D(\varrho_1'-\varrho_2',0)+D(\varrho_3'-\varrho_4',0)]/2\} \\
&\quad \cdot (\exp\{-[D(\varrho_1'-\varrho_4',0)+D(\varrho_2'-\varrho_3',0)-D(\varrho_1'-\varrho_3',0) \\
&\quad -D(\varrho_2'-\varrho_4',0)]/2+2[C_\chi(\varrho_1'-\varrho_3',0)-iC_{\chi,\phi}(\varrho_1'-\varrho_3',0) \\
&\quad +C_\chi(\varrho_2'-\varrho_4',0)+iC_{\chi,\phi}(\varrho_2'-\varrho_4',0)]\}-1).
\end{aligned} \tag{6.132}$$

In the absence of turbulence, (6.131) collapses to the well-known free-space carrier-to-noise ratio [6.109, 115]

$$\text{CNR}_d^0 = \frac{P_t D_{f0}\alpha/2hf_0\Delta f}{1+P_{bD}/P_t D_{f0}}. \tag{6.133}$$

Equations (6.131) and (6.133) are both of the form

$$\text{CNR}_d = \frac{\langle P_r\rangle\alpha/2hf_0\Delta f}{1+\xi^{-1}+\text{Var}(P_r)\alpha/2\langle P_r\rangle hf_0\Delta f}, \tag{6.134}$$

where $\langle P_r \rangle$ and $\mathrm{Var}(P_r)$ are the mean and variance of the received power (over channel statistics), and ξ is the average signal-to-background power ratio. The numerator in (6.134) is the so-called quantum-limited CNR_d, which prevails whenever signal light shot noise is the predominant contributor to the noise in $i'(t)$. The denominator in (6.134) quantifies the decrement from quantum-limited operation that is due to background light shot noise, and the excess noise that accompanies signal power fluctuations. Thus, according to (6.131) and (6.133), atmospheric propagation reduces diffraction-limited CNR_d from its free-space value in three ways: average signal power is reduced by a factor of $\langle G_{dD} \rangle / D_{f0}$ [cF. (6.111)–(6.113)]; average signal-to-background ratio is reduced by a factor of $\langle G_{dD} \rangle / D_{f0}$; and excess noise is introduced by the random variations in G_{dD}. When $d_2 < \varrho'_0$, we have

$$G_{dD} \approx D_{f0} \exp[2\chi(\mathbf{0},\mathbf{0})] ; \tag{6.135}$$

hence

$$CNR_d \approx \frac{P_t D_{f0} \alpha / 2 h f_0 \varDelta f}{1 + P_{bD}/P_t D_{f0} + [\exp(4\sigma_\chi^2) - 1] P_t D_{f0} \alpha / 2 h f_0 \varDelta f} , \tag{6.136}$$

where $\sigma_\chi^2 = C_\chi(\mathbf{0},\mathbf{0})$. In physical terms, (6.135) and (6.136) imply that scintillation is the sole cause of atmospheric CNR_d degradation for a coherence area receiver pupil. At the opposite extreme, we have the following results for the case $d_2 \gg \varrho'_0$ [6.44, 117]:

$$\langle G_{dD} \rangle \approx D_{ft} \ll D_{f0}, \tag{6.137}$$

$$\mathrm{Var}(G_{dD}) \approx \langle G_{dD} \rangle^2 , \tag{6.138}$$

and

$$CNR_d \approx \frac{P_t D_{ft} \alpha / 2 h f_0 \varDelta f}{1 + P_{bD}/P_t D_{ft} + P_t D_{ft} \alpha / 2 h f_0 \varDelta f} . \tag{6.139}$$

In this situation, angle-of-arrival fluctuations have severely reduced both the average signal level and the signal-to-background ratio from their free-space values, and the nature of the excess noise term has changed because of the transition from log normal to exponential statistics for G_{dD} [6.44, 118, 119].

The foregoing CNR_d calculations apply to the reception of an unmodulated carrier, and as such they need not be truly indicative of the performance that may be obtained with a modulated source. For example, (6.139) predicts $CNR_d \lesssim 1$ for arbitrary P_t, and (6.136) predicts $CNR_d \lesssim 1$ whenever $\sigma_\chi^2 \gtrsim 0.15$. It might seem, therefore, that reliable diffraction-limited reception is thwarted by the atmosphere, but this is not so. Because atmospheric fluctuations have kHz bandwidths, *information* transmitted via subcarrier modulation at MHz to GHz bandwidths may be retrieved, neglecting the effects of noise, from the turbulence-induced fading by employing a receiver with an automatic gain control (AGC). In particular, for $d_2 < \varrho'_0$, the idealized AGC which generates

$$i''(t) = \langle G_{dD} \rangle [i'(t) - q\alpha P_{bD}/hf_0] G_{dD}^{-1}, \tag{6.140}$$

has carrier-to-noise ratio

$$\text{CNR}_{dAGC} \approx \frac{\exp(-4\sigma_\chi^2)P_t D_{f0}\alpha/2hf_0\varDelta f}{1 + \exp(8\sigma_\chi^2)P_{bD}/P_t D_{f0}}, \tag{6.141}$$

which compares favorably with the free-space CNR_d^0, (6.133). However, when $d_2 \gg \varrho'_0$, the AGC (6.140) does not lead to a meaningful CNR_d because $\langle G_{dD}^{-1} \rangle = \infty$ when G_{dD} is exponentially distributed. Moreover, even with an appropriate AGC selection, a CNR_d analysis still suppresses an important distinction between free-space and atmospheric receiver operation. In the former instance, the post-AGC noise level is time independent, whereas in the latter instance, the instantaneous post-AGC noise level varies inversely with atmospheric fading. Hence, for the atmospheric receiver there will be occasional periods during which the post-AGC signal is buried in a greatly enhanced noise level, even though the performance of a free-space system with the same CNR_d is at all times satisfactory.

A comprehensive assessment of direct detection analog communication performance through atmospheric turbulence can be developed from filtering theory [6.120–122], but this approach will not be pursued herein because of the mathematical sophistication it entails. Instead, we shall develop performance results for a simple direct detection binary communication system. We shall forego exact performance calculations in favor of approximation techniques which permit a parametric comparison between free-space and atmospheric diffraction-limited links, and generalize in an obvious manner to the case of diversity reception [6.123–125]. Other treatments of direct detection digital communication through turbulence [6.7, 120, 126–129], which rely on numerical evaluation, afford greater accuracy but less insight.

Consider a binary pulse position modulation (PPM) system in which the spacecraft transmits an information bit every T seconds using the field envelope (6.108), where $s(t)$ is as shown in Fig. 6.12. We shall assume that T is less than the atmospheric coherence time, that each bit is equally likely to be 0 or 1, and that the receiver has a diffraction-limited field of view and it uses bit-by-bit decoding without channel tracking to make bit decisions based on observation of $i(t)$. In this case, the minimum error probability decision rule is to assume a zero was sent during the interval $(m-1)T < t \leq mT$, if and only if

$$\int_{(m-1)T}^{(m-1/2)T} d\tau q^{-1} i(\tau - L/c) \geq \int_{(m-1/2)T}^{mT} d\tau q^{-1} i(\tau - L/c), \tag{6.142}$$

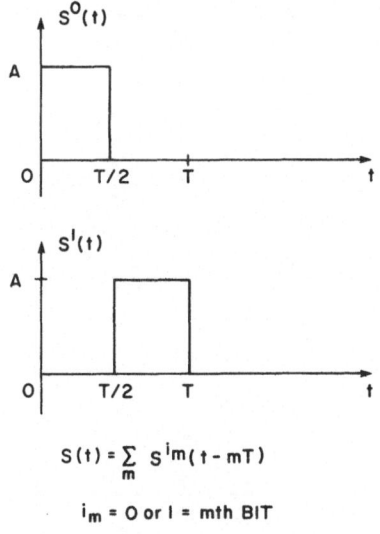

$$S(t) = \sum_m s^{i_m}(t - mT)$$

$i_m = 0$ or $1 = $ mth BIT

$A = (8P_t/\pi d_1^2 c\epsilon_0)^{1/2}$

Fig. 6.12. Pulse position modulation signal set

i.e., the receiver counts photons in each half of the bit interval, and chooses the message that corresponds to the larger count [6.120, 129]. Note that the optimality of this receiver is unaffected by the amount of turbulence in the propagation path; its error probability is not, however, similarly insensitive to turbulence strength.

For free-space propagation, the error probability $Pr(e)^0$ of the diffraction-limited photon-counting receiver is bounded by the following expression [6.130–132]

$$Pr(e)^0 \leqq (\exp\{2n_{bD}[(1+\xi^0)^{1/2} - 1] - n_s^0\})/2, \tag{6.143}$$

where $n_{bD} = \alpha P_{bD} T/2hf_0$ and $n_s^0 = \alpha P_t D_{f0} T/2hf_0$ are the average number of background and signal photons detected in the half bit interval which contains the signal, and $\xi^0 = n_s^0/n_{bD}$ (see Fig. 6.13). When $\xi^0 \gg 1$, the error rate is quantum limited and (6.143) becomes

$$Pr(e)^0 \leqq \exp(-n_s^0)/2, \tag{6.144}$$

which equals the exact result in the limit as $\xi^0 \to \infty$ [6.132]. When $\xi^0 \ll 1$, the error rate is background limited and (6.143) reduces to

$$Pr(e)^0 \leqq \exp(-n_s^0\xi^0/4)/2, \tag{6.145}$$

which is known to be an exponentially tight bound in the limit as $\xi^0 \to 0$ with $n_s^0\xi^0$ fixed [6.132]. For a given mean signal count n_s^0, it is obviously quite

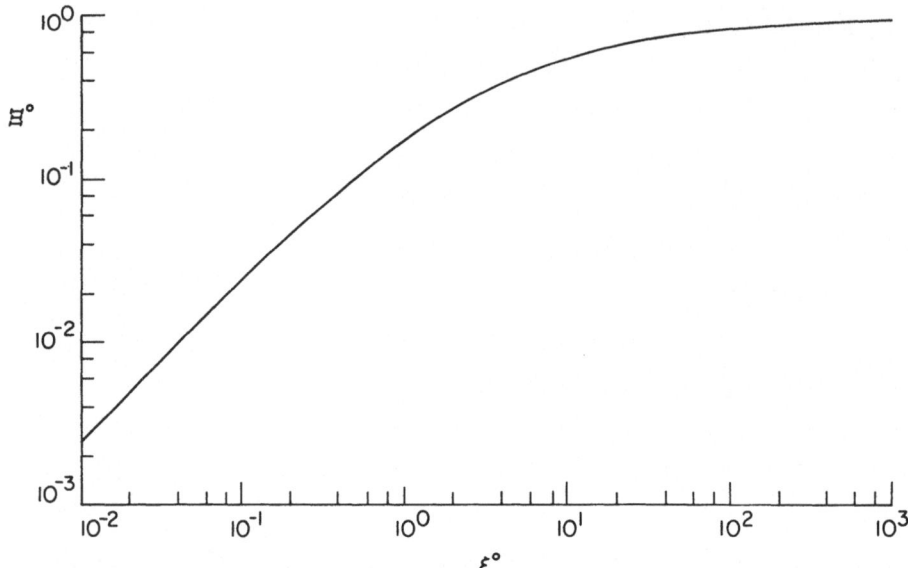

Fig. 6.13. Error exponent for free-space PPM direct detection, (6.143), $\Pr(e)^0 \leq \exp(-n_s^0 \Xi^0)/2$ where $\Xi^0 \equiv -2[(1+\xi^0)^{1/2}-1]/\xi^0 +1$

disadvantageous to operate in the background-limited regime. This precipitous loss of performance that transpires in the vicinity of $\xi^0 = 1$ is referred to as the background noise threshold.

In the presence of turbulence, (6.143) can be used as an upper bound for the atmospheric error rate conditioned on knowledge of G_{dD}, by replacing n_s^0 with $n_{sD} = \alpha P_t G_{dD} T/2hf_0$ and ξ^0 with n_{sD}/n_{bD}. An upper bound for the unconditional error rate follows from averaging over the statistics of G_{dD} [6.123]. Simpler expressions are obtained, however, from the asymptotic forms (6.144) and (6.145) by sacrificing detailed knowledge of the conditional error rate in the neighborhood of the background noise threshold [6.124, 125]. Specifically, for $d_2 < \varrho_0'$ we have that the atmospheric error rate $\Pr(e)$ satisfies

$$\Pr(e) \leq \mathrm{Fr}(\langle n_{sD} \rangle, 0; \sigma_\chi)/2, \qquad (6.146)$$

when $\xi_D = \langle n_{sD} \rangle/n_{bD} \gg 1$, and

$$\Pr(e) \leq \mathrm{Fr}[\exp(4\sigma_\chi^2)\langle n_{sD} \rangle \xi_D/4, 0; 2\sigma_\chi]/2, \qquad (6.147)$$

when $\xi_D \ll 1$, where $\langle n_{sD} \rangle = \alpha P_t D_{r0} T/2hf_0$ is the average signal count, and

$$\mathrm{Fr}(a,b;c) = \int_0^\infty dy I_0(2ba^{1/2}y)\exp(-ay^2)$$
$$\cdot \exp[-(\ln y + c^2)^2/2c^2]/(2\pi c^2)^{1/2}y, \qquad (6.148)$$

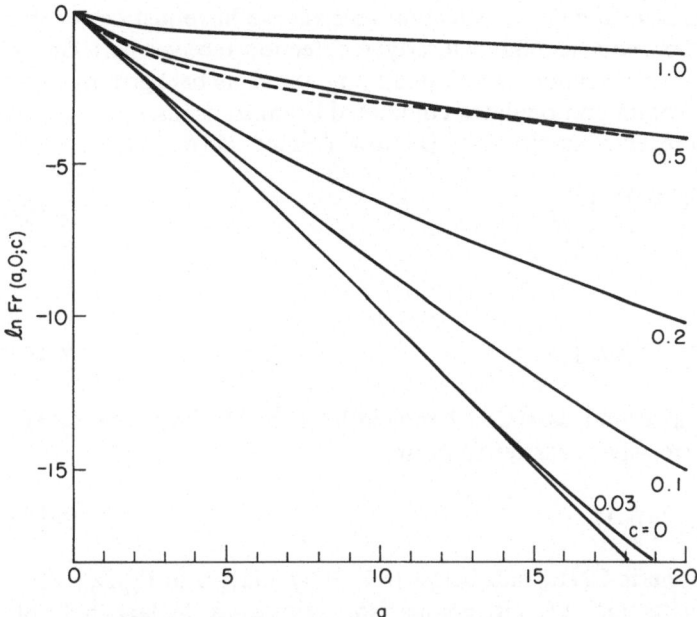

Fig. 6.14. Error bounds for atmospheric PPM direct detection: the log normal density frustration function, $\ln \mathrm{Fr}(a, 0; c)$, for various c values vs a (solid lines); the exponential density performance, $-\ln(1+a)$, vs a (broken line); (after Ref. [6.118], Fig. 26)

is the log normal density frustration function [6.120, 123]. From the behavior of $\mathrm{Fr}(a, 0; c)$, which we have plotted in Fig. 6.14, it is apparent that even modest amounts of turbulence cause binary PPM error rates for diffraction-limited direct detection to be well in excess of the corresponding free-space values for a coherence area receiver pupil. On the other hand, for $d_2 \gg \varrho_0'$, G_{dD} becomes exponentially distributed and we find that

$$\Pr(e) \leqq [2(1 + \langle n_{\mathrm{sD}} \rangle)]^{-1}, \tag{6.149}$$

when $\xi_{\mathrm{D}} \gg 1$, and

$$\Pr(e) \leqq (\pi/4 \langle n_{\mathrm{sD}} \rangle \xi_{\mathrm{D}})^{1/2}, \tag{6.150}$$

when $\xi_{\mathrm{D}} \ll 1$, where $\langle n_{\mathrm{sD}} \rangle$, the average signal count, now satisfies $\langle n_{\mathrm{sD}} \rangle = \alpha P_{\mathrm{t}} D_{\mathrm{ft}} T / 2 h f_0$. For future reference, (6.149) has been included in Fig. 6.14. For the moment we need only note that performance is independent of d_2 once $d_2 \gg \varrho_0'$ is achieved, and that in this case $\langle n_{\mathrm{sD}} \rangle \ll n_{\mathrm{s}}^0$, so the atmospheric error rates are substantially inferior to the corresponding free-space expressions, (6.144) and (6.145).

Let us now examine the heterodyne detection receiver. We shall assume a plane wave local oscillator so that the receiver has a diffraction-limited field of view identical to that of the direct detection receiver we have just considered [cf. (6.129)]. As a consequence, the heterodyne detection receiver performs as though it were a direct detection system operating above its background noise threshold ($\xi_D \gg 1$). For an unmodulated collimated beam transmission of power P_t, the heterodyne receiver carrier-to-noise ratio defined as [6.134]

$$\mathrm{CNR_h} = \frac{\langle [\langle v(t) \rangle_n]^2 \rangle}{\langle \langle [v(t) - \langle v(t) \rangle_n]^2 \rangle_n \rangle}, \tag{6.151}$$

satisfies [6.5, 6]

$$\mathrm{CNR_h} = \alpha P_t \langle G_{dD} \rangle / 2hf_0 \Delta f, \tag{6.152}$$

where $2\Delta f$ is the unilateral passband bandwidth of the IF filter. For $d_2 < \varrho_0'$, (6.152) equals the free-space carrier to noise

$$\mathrm{CNR_h^0} = \alpha P_t D_{f0} / 2hf_0 \Delta f. \tag{6.153}$$

For $d_2 \gg \varrho_0'$, atmospheric $\mathrm{CNR_h}$ falls below (6.153) by a factor of D_{ft}/D_{f0}. As in the case of direct detection, a $\mathrm{CNR_h}$ comparison suppresses the fact that only the atmospheric system is subject to signal fading. Thus, let us investigate binary PPM error rates for diffraction-limited heterodyne reception.

Consider the binary PPM transmitter we have previously described. Assuming T less than the atmospheric coherence time, each bit equally likely to be 0 or 1, and bit-by-bit decoding without channel tracking, the minimum error probability diffraction-limited heterodyne receiver decides a zero was sent during the interval $(m-1)T < t \leq mT$ if and only if [6.118, 120]

$$\left[\int_{(m-1)T}^{(m-1/2)T} dt v(t-L/c) \cos 2\pi f_1 t \right]^2 + \left[\int_{(m-1)T}^{(m-1/2)T} dt v(t-L/c) \sin 2\pi f_1 t \right]^2$$

$$\geq \left[\int_{(m-1/2)T}^{mT} dt v(t-L/c) \cos 2\pi f_1 t \right]^2 + \left[\int_{(m-1/2)T}^{mT} dt v(t-L/c) \sin 2\pi f_1 t \right]^2. \tag{6.154}$$

This decision rule may be implemented by means of passband matched filters and envelope detection [6.135].

For free-space propagation, the foregoing receiver has error probability [6.136]

$$\mathrm{Pr}(e)^0 = \exp(-n_s^0/2)/2, \tag{6.155}$$

which equals the error rate of a quantum-limited direct detection free-space system with transmitter power $P_t/2$. In the presence of turbulence, (6.155) gives the conditional error rate when n_s^0 is replaced with $n_{sD} = \alpha P_t G_{dD} T/2hf_0$; the unconditional error rate can then be computed by averaging over the statistics of G_{dD}. We find that the atmospheric error probability is

$$\Pr(e) = \mathrm{Fr}(\langle n_{sD} \rangle/2, 0; \sigma_\chi)/2, \tag{6.156}$$

for $d_2 < \varrho_0'$, where $\langle n_{sD} \rangle = \alpha P_t D_{f0} T/2hf_0$, and

$$\Pr(e) = [2(1 + \langle n_{sD} \rangle/2)]^{-1}, \tag{6.157}$$

for $d_2 \gg \varrho_0'$, where $\langle n_{sD} \rangle = \alpha P_t D_{ft} T/2hf_0$. Both of these expressions equal the corresponding bounds for a direct detection system with transmitter power $P_t/2$ that is operating above its background noise threshold.

6.3.4 Diversity Reception

A cursory examination of the preceding error rates for diffraction-limited downlink reception reveals the following significant features. For $d_2 < \varrho_0'$, even a modest level of scintillation ($\sigma_\chi^2 = 0.04$) implies atmospheric error probabilities which are markedly higher than those of an otherwise identical free-space system. This loss of performance may be alleviated, in principle, through increased transmitter power; for $\sigma_\chi^2 = 0.25$, the necessary power increase may exceed 10 dB. An alternative to increased transmitter power is increased receiver pupil area. To the extent that d_2 remains smaller than ϱ_0', a ten fold increase in pupil area is functionally equivalent, insofar as CNR and $\Pr(e)$ are concerned, to a ten fold increase in transmitter power. However, inasmuch as ϱ_0' is typically 10 cm for near-zenith paths at visible wavelengths, increasing d_2 leads rapidly to the situation $d_2 \gg \varrho_0'$ wherein diffraction-limited performance is independent of receiver area. Physically, the saturation of diffraction-limited receiver performance that occurs in the neighborhood of $d_2 = \varrho_0'$ is due to the suboptimal nature of downlink diffraction-limited gain G_{dD}, when angle-of-arrival fluctuations exceed d_2^{-1}. It is reasonable, therefore, to suppose that effective utilization of the signal power incident on a receiver pupil which comprises many atmospheric coherence areas requires that we open the receiver's field of view to encompass the angular spectrum of the signal field (cf. Sect. 6.3.1). Wide field-of-view direct detection is readily accomplished by matching the detector diameter in Fig. 6.10 to the focal spot formed by the signal. Such a receiver is generally referred to as an aperture integrator or photon bucket [6.120, 129, 137]. Wide field-of-view heterodyne detection cannot be accomplished using a plane wave local oscillator and a single photodetector.

It has long been recognized by communication theorists that to effect reliable information transmission over a fading channel with minimum transmitter power, diversity techniques are required [6.138–140]. For the particular case of the atmospheric downlink, the typical coherence lengths in space, time, and frequency dictate that spatial diversity reception be employed to combat turbulence-induced fading [6.8, 13, 120]. In communication parlance, the aperture integration receiver, which we have developed from a physical viewpoint, is a direct detection spatial diversity receiver; it is equivalent to linearly combining the photocurrents that would be generated by an array of $4\lambda l/d_2$ diameter photodetectors distributed over the $4\lambda l/d_2'$ diameter focal spot formed by the signal. There are, of course, a host of other possible direct detection diversity receivers which employ nonlinear weightings to combine the photocurrents generated by a detector array. Moreover, there are also heterodyne detection diversity receivers which use one or more local oscillators, and a spatial array of photodetectors. In this section we shall present performance analyses for a number of downlink diversity receivers [6.123–125]: the aperture integrator, the channel-matched filter receiver, and the phase-compensated receiver. Other discussions of atmospheric diversity combining include [6.13, 109, 114, 118, 120, 129, 141, 142].

The aperture integrator may be realized with the structure shown in Fig. 6.10 by choosing $d_D = 4\lambda l/\pi d_2'$, when $d_2 > \varrho_0'$. For an unmodulated collimated beam transmitter with power P_t and $d_2 > \varrho_0'$, this system has carrier-to-noise ratio

$$\text{CNR}_d = \frac{P_t D_{f0}\alpha/2hf_0\Delta f}{1 + P_{bs}/P_t D_{f0} + \text{Var}(G_d^0)P_t\alpha/2D_{f0}hf_0\Delta f}, \tag{6.158}$$

where we have again assumed a $2\Delta f$ Hz bilateral post-detection bandwidth, and, within the weak perturbation regime

$$\text{Var}(G_d^0) = D_{f0}^2 \int d\varrho' \{\exp[4C_\chi(\varrho', 0)] - 1\} (8/\pi^2 d_2^2)$$
$$\cdot [\cos^{-1}(|\varrho'|/d_2) - (|\varrho'|/d_2)(1 - |\varrho'|^2/d_2^2)^{1/2}] \text{circ}(|\varrho'|/d_2). \tag{6.159}$$

We note the following characteristics of (6.158) and (6.159): average received power equals free-space received power; average signal-to-background ratio, $\xi_s = P_t D_{f0}/P_{bs}$, equals diffraction-limited signal-to-background ratio, $\xi_D = P_t D_{ft}/P_{bD}$, [cf. (6.123) and (6.125)]; and excess noise is driven towards zero as d_2 increases by aperture averaging of scintillation [6.11, 143].

For binary PPM reception under the assumptions that led to (6.142), the optimum decision rule for the aperture integrator is still to count photons in each half of the bit interval and choose the message that corresponds to the larger count [6.120]. The resulting error probability satisfies the following bounds [6.124, 125, 144, 145], where $d_2 > \varrho_0'$ is assumed:

$$\Pr(e) \leqq \text{Fr}(\langle n_s\rangle, 0; \sigma)/2, \tag{6.160}$$

when $\xi_s \gg 1$, and

$$\Pr(e) \leqq \text{Fr}[\exp(4\sigma^2)\langle n_s\rangle\xi_s/4, 0; 2\sigma]/2, \tag{6.161}$$

when $\xi_s \ll 1$. In (6.160) and (6.161), $\langle n_s \rangle = \alpha P_t D_{f0} T / 2 h f_0$ is the average signal count,

$$\sigma^2 = \{\ln[1 + \text{Var}(G_d^0)/D_{f0}^2]\}/4, \tag{6.162}$$

is the aperture-averaged log-amplitude variance, and implicit use has been made of the convergence properties of sums of log normal variates [6.114, 118, 146, 147].

Whereas the performance of the diffraction-limited receiver improves with increasing receiver pupil area only in the regime $d_2 < \varrho_0'$, the performance of the aperture integrator continues to improve with increasing d_2 (well beyond $d_2 = \varrho_0'$) until sufficient aperture averaging has occurred to ensure $\sigma^2 \approx 0$ in (6.160)–(6.162). Because of practical difficulties, $\sigma^2 \approx 0$ may not be achievable, e.g., in the regime of saturated scintillation, aperture-averaged log-amplitude variance decreases quite slowly with increasing pupil area [6.142]. It is nonetheless instructive to consider the limiting performance of the aperture integrator when $\sigma^2 \approx 0$; in this case (6.160) and (6.161) reduce to

$$\Pr(e) \leqq \exp(-\langle n_s \rangle)/2, \tag{6.163}$$

when $\xi_s = D_{ft} \xi^0 / D_{f0} \gg 1$, and

$$\Pr(e) \leqq \exp(-\langle n_s \rangle \xi_s/4)/2, \tag{6.164}$$

when $\xi_s \ll 1$. Note that because $\langle n_s \rangle = n_s^0$, (6.163) is exactly equal to the corresponding quantum-limited free-space performance (6.144). Because $\xi_s \ll \xi^0$, the background light threshold is encountered at a higher signal light level by the atmospheric receiver, and (6.164) is significantly inferior to the background-limited free-space performance (6.145). From a physical point of view, these results are easily justified. When there is sufficient aperture averaging to ignore downlink scintillation, we have $G_d^0 \approx D_{f0}$ for all atmospheric states; thus the equivalence of atmospheric and free-space quantum-limited performance. However, in order to achieve atmospheric gain G_d^0, the aperture integrator opens its field of view substantially beyond the diffraction limit; hence the atmospheric system encounters background light threshold at a higher signal power than does the diffraction-limited free-space receiver.

The aperture integrator is easily implemented and makes effective use of all the available signal power. It turns out to be the optimum downlink direct detection receiver when it operates above the background light threshold [6.120, 129]. When the aperture integrator is background limited, improved downlink performance may be obtained with (direct detection or heterodyne detection) array receivers [6.120, 129, 142]. A simple bound on the maximum downlink performance improvement obtainable with array reception is afforded by a performance analysis of the channel-matched filter receiver [6.124], which is assumed to have perfect a priori knowledge of the atmospheric perturbation $\exp[\chi(\varrho', 0) + i\phi(\varrho', 0)]$ for ϱ' in R_2. This receiver, shown schemati-

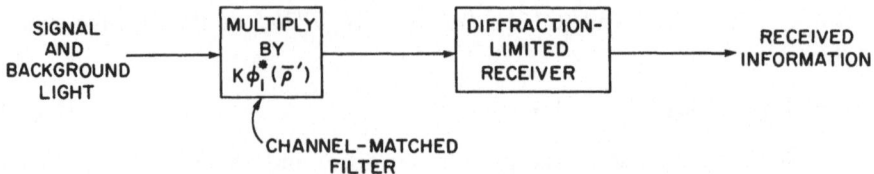

$$K = (\pi d_2^2/4)^{1/2}$$

Fig. 6.15. Block diagram of a channel-matched filter communication receiver

cally in Fig. 6.15, multiplies the field received over the R_2 pupil by $(\pi d_2^2/4)^{1/2}\phi_1^*(\varrho')$, where

$$\phi_1^*(\varrho') = \exp[\chi(\varrho',0) - i\phi(\varrho',0)]$$

$$\cdot\left\{\int_{R_2} d\varrho' \exp[2\chi(\varrho',0)]\right\}^{-1/2}, \tag{6.165}$$

and follows this transformation by diffraction-limited (direct detection or heterodyne detection) reception. Because ϕ_1^* is the conjugate of the only downlink output eigenfunction with nonzero eigenvalue, multiplication by ϕ_1^* has the effect of concentrating, as nearly as possible, all of the signal energy into the normally incident plane wave component of the field entering the diffraction-limited receiver. As a result, the binary PPM error rates for channel-matched filter reception satisfy [6.124, 125]

$$\Pr(e) \leq \mathrm{Fr}(\langle n_\mathrm{s}\rangle, 0; \sigma)/2, \tag{6.166}$$

for direct detection when $\xi_{\mathrm{cf}} = \langle n_\mathrm{s}\rangle/n_{\mathrm{bD}} \gg 1$,

$$\Pr(e) \leq \mathrm{Fr}[\exp(4\sigma^2)\langle n_\mathrm{s}\rangle\xi_{\mathrm{cf}}/4, 0; 2\sigma]/2, \tag{6.167}$$

for direct detection when $\xi_{\mathrm{cf}} \ll 1$, and

$$\Pr(e) = \mathrm{Fr}(\langle n_\mathrm{s}\rangle/2, 0; \sigma)/2, \tag{6.168}$$

for heterodyne detection, where $\langle n_\mathrm{s}\rangle$ and σ^2 are as given for the aperture integrator.

We may draw the following conclusions from (6.166)–(6.168): because $\langle n_\mathrm{s}\rangle = n_\mathrm{s}^0$, and $\xi_{\mathrm{cf}} = \xi^0$, (6.166)–(6.168) reduce exactly to the corresponding free-space results, (6.144), (6.145), and (6.155), whenever there is sufficient aperture averaging to set $\sigma^2 \approx 0$; the use of channel-matched filter direct detection, instead of aperture integration, drives the signal power for background light threshold down by a factor of $D_{\mathrm{ft}}/D_{\mathrm{f0}}$. In principle, the channel-matched filter heterodyne receiver may be implemented by using channel measurements and an adaptive local oscillator to achieve $u_l(\varrho') \approx \phi_1(\varrho')$, for ϱ' in R_2. There appears to be no convenient implementation for the channel-matched filter direct detection receiver, unless the need for both amplitude and phase control in the receiver pupil is dropped. Specifically, by using a phase plate (cf. Sect. 6.2.4) to multiply the field received over R_2 by $\phi_1^*(\varrho')/|\phi_1(\varrho')| = \exp[-i\phi(\varrho', 0)]$ prior to diffraction-limited reception, we obtain a phase-compensated communication receiver.

The technology of phase-compensated communication is virtually identical to that of phase-compensated imaging although the purposes are somewhat different; in imaging the aim of phase compensation is to restore diffraction-limited resolution, in communication the aim of phase compensation is to discriminate against background light. Assuming perfect phase compensation (i.e., no wavefront sensing or fitting errors), the binary PPM error rates for phase-compensated direct detection satisfy [6.124]

$$\Pr(e) \leq \mathrm{Fr}[\exp(\sigma'^2 - \sigma_\chi^2)\langle n_s\rangle, 0; \sigma']/2, \tag{6.169}$$

when $\xi_{pc} = \exp(\sigma'^2 - \sigma_\chi^2)\langle n_s\rangle/n_{bD} \gg 1$, and

$$\Pr(e) \leq \mathrm{Fr}[\exp(5\sigma'^2 - \sigma_\chi^2)\langle n_s\rangle\xi_{pc}/4, 0; 2\sigma']/2, \tag{6.170}$$

when $\xi_{pc} \ll 1$. In (6.169) and (6.170), σ'^2 is the aperture-averaged log-amplitude variance for a phase-compensated pupil which is given by [6.148]

$$\sigma'^2 = \ln\{\int d\varrho'(8/\pi^2 d_2^2)\exp[C_\chi(\varrho', 0)]\,\mathrm{circ}(|\varrho'|/d_2)$$
$$\cdot[\cos^{-1}(|\varrho'|/d_2) - (|\varrho'|/d_2)(1 - |\varrho'|^2/d_2^2)^{1/2}]\}, \tag{6.171}$$

where use has been made of the convergence properties of sums of lognormal variates. For d_2 sufficiently large we have $\sigma'^2 \approx 0$, so that, except for a signal power attenuation factor $\exp(-\sigma_\chi^2)$, the phase-compensated receiver achieves free-space performance. For $d_2 < \varrho_0'$, we find $\sigma'^2 = \sigma_\chi^2$ so that phase-compensated performance equals diffraction-limited performance as it must whenever the receiver pupil comprises a single coherence area.

6.3.5 Reciprocity Pointing

We shall conclude our discussion of atmospheric optical communications with a brief treatment of uplink transmission. Because of atmospheric reciprocity, the downlink results of Section 6.3.3 also characterize the performance of a collimated transmitter, diffraction-limited receiver uplink system. In particular, because of scintillation and beam spreading, the foregoing uplink system performs substantially worse in the presence of turbulence than it does for free-space propagation. Improved downlink performance can be obtained (without increased transmitter power) by means of spatial diversity reception. However, as we have previously noted, there is no spatial diversity available at the uplink receiver. The key to significantly improved uplink performance (at constant transmitter power) is an adaptive spatial diversity transmitter, i.e., a reciprocity pointing system [6.16, 37], as described below.

Recall from Section 6.3.1 that instantaneous uplink power gain is maximized by transmitting

$$V_i(\varrho', t) = s(t)\phi_1^*(\varrho') \tag{6.172}$$

from R_2, where $s(t)$ is an information-bearing modulation, and ϕ_1^* satisfies (6.165). Because the round trip propagation delay down from the top of the atmosphere to the ground terminal and back up to the top of the atmosphere is far smaller than an atmospheric coherence time, the atmospheric state information needed to implement (6.172) when $d_2 > \varrho_0'$ can be obtained from measurements of the downlink received signal field, which is proportional to $\phi_1(\varrho')$. Complex conjugation of the received signal field corresponds, physically, to reversing the local direction of wave propagation [6.16, 36]; hence the term reciprocity pointing. Note that the same equipment required to realize a channel-matched filter downlink receiver can be used to realize a reciprocity pointing transmitter. Furthermore, in the absence of wavefront sensing and fitting errors the binary PPM error rates achievable with a reciprocity pointing uplink satisfy (6.166)–(6.168), and so in the limit of $\sigma^2 \approx 0$ there is no turbulence-induced performance loss.

Our description of reciprocity pointing as a field conjugation operation implies that both amplitude and phase control are employed by the transmitter. A simpler implementation results if we use the phase conjugation transmitter,

$$V_i(\varrho', t) = s(t)(4/\pi d_2^2)^{1/2} \exp[-i\phi(\varrho', 0)], \tag{6.173}$$

at the ground terminal. Neglecting wavefront sensing and fitting errors, the binary PPM error rates achievable with the phase conjugation uplink are given by (6.169)–(6.171). Thus, as in the case of the phase-compensated downlink, a minimal performance loss is incurred when amplitude control is deleted.

In closing, we should note that the foregoing performance analyses have neglected spacecraft motion. When the free-space far-field beamwidth of the R_2 pupil is significantly smaller than the point-ahead angle needed to compensate for spacecraft motion, the beam transmitted from the ground will encounter a somewhat different atmospheric state than that probed by the downlink. To a first approximation [6.103], when $d_2 > \varrho_0'$ the point-ahead requirement reduces average reciprocity-pointing uplink power gain by a factor of $\exp[-(2|v_r|L/c\varrho_0)^{5/3}/2]$, where v_r is the transverse relative spacecraft velocity, and ϱ_0 satisfies (6.102).

Acknowledgement. The preparation of this chapter was supported in part by the National Science Foundation (NSF Grant ENG 74-03996-A01). This support is gratefully acknowledged.

References

6.1   T.H.Maiman: Nature **187**, 493 (1960)
6.2   P.Beckmann: Radio Sci. **69**D, 629 (1965)
6.3   I.Goldstein, P.A.Miles, A.Chabot: Proc. IEEE **53**, 1172 (1965)
6.4   J.I.Davis: Appl. Opt. **5**, 139 (1966)
6.5   D.L.Fried: Proc. IEEE **55**, 57 (1967)
6.6   D.L.Fried: IEEE J. QE-3, 213 (1967)
6.7   D.L.Fried, R.A.Schmeltzer: Appl. Opt. **6**, 1729 (1967)
6.8   E.V.Hoversten: IEEE Intl. Conv. Record **15** (11), 137 (1967)
6.9   D.L.Fried: J. Opt. Soc. Am. **55**, 1427 (1965)
6.10 D.L.Fried: J. Opt. Soc. Am. **56**, 1380 (1966)
6.11 D.L.Fried: J. Opt. Soc. Am. **57**, 169 (1967)
6.12 R.E.Hufnagel, N.R.Stanley: J. Opt. Soc. Am. **54**, 52 (1964)
6.13 R.S.Kennedy, E.V.Hoversten: IEEE Trans. IT-14, 716 (1968)
6.14 V.I.Tatarskii: *Wave Propagation in a Turbulent Medium* (McGraw-Hill, New York 1961)
6.15 A.Kon, V.Feizulin: Radiophys. Quant. Electron. **13**, 51 (1970)
6.16 J.H.Shapiro: IEEE Trans. COM-19, 410 (1971)
6.17 J.H.Shapiro: J. Opt. Soc. Am. **61**, 492 (1971)
6.18 R.F.Lutomirski, H.T.Yura: Appl. Opt. **10**, 1652 (1971)
6.19 H.T.Yura: Appl. Opt. **11**, 1399 (1972)
6.20 H.S.Lin: "Communication Model for the Turbulent Atmosphere"; Ph.D. Thesis, Case Western Reserve Univ. (1973)
6.21 R.L.Fante: Proc. IEEE **63**, 1669 (1975)
6.22 H.T.Yura: J. Opt. Soc. Am. **62**, 889 (1972)
6.23 R.L.Fante, J.L.Poirier: Appl. Opt. **12**, 2247 (1973)
6.24 J.W.Strohbehn: Proc. IEEE **56**, 1301 (1968)
6.25 E.Brookner: IEEE Trans. COM-18, 396 (1970)
6.26 C.T.Tai: *Dyadic Green's Functions in Electromagnetic Theory* (Intext, Scranton 1971) Chap. 4
6.27 J.A.Kong: *Theory of Electromagnetic Waves* (Wiley-Interscience, New York 1975) Chap. 6
6.28 A.A.M.Saleh: IEEE J. QE-3, 540 (1967)
6.29 D.L.Fried, G.E.Mevers: J. Opt. Soc. Am. **55**, 740 (1965)
6.30 J.W.Strohbehn, S.F.Clifford: IEEE Trans. AP-15, 416 (1967)
6.31 D.L.Fried: Appl. Opt. **10**, 721 (1971)
6.32 H.H.Su, M.A.Plonus: J. Opt. Soc. Am. **61**, 256 (1971)
6.33 C.S.Gardner, M.A.Plonus: J. Opt. Soc. Am. **64**, 68 (1974)
6.34 M.Born, E.Wolf: *Principles of Optics*, 5th ed. (Pergamon Press, Oxford 1975) p. 381
6.35 J.W.Goodman: *Introduction to Fourier Optics* (McGraw-Hill, New York 1968) p. 41
6.36 J.H.Shapiro: "Optimal Spatial Modulation for Reciprocal Channels"; Res. Lab. Electron. Tech. Rept. 476, M.I.T. (1970)
6.37 D.L.Fried, H.T.Yura: J. Opt. Soc. Am. **62**, 600 (1972)
6.38 W.T.Cathey, C.L.Hayes, W.C.Davis, V.F.Pizzuro: Appl. Opt. **9**, 701 (1970)
6.39 R.W.Lee, J.C.Harp: Proc. IEEE **57**, 375 (1969)
6.40 R.S.Lawrence, J.W.Strohbehn: Proc. IEEE **58**, 1523 (1970)
6.41 D.L.Fried, J.D.Cloud: J. Opt. Soc. Am. **56**, 1667 (1966)
6.42 R.L.Fante: J. Opt. Soc. Am. **66**, 74 (1976)
6.43 D.L.Fried: J. Opt. Soc. Am. **57**, 176 (1967)
6.44 G.P.Massa: "Fourth-Order Moments of an Optical Field that has Propagated through the Clear Turbulent Atmosphere"; S.M. Thesis, M.I.T. (1975)
6.45 R.F.Lutomirski, H.T.Yura: J. Opt. Soc. Am. **61**, 482 (1971)
6.46 J.B.Breckinridge: J. Opt. Soc. Am. **66**, 143 (1976)
6.47 A.Papoulis: *Systems and Transforms with Applications in Optics* (McGraw-Hill, New York 1968) Chap. 10

6.48   H.T.Yura: J. Opt. Soc. Am. **64**, 59 (1974)
6.49   S.F.Clifford, G.R.Ochs, R.S.Lawrence: J. Opt. Soc. Am. **64**, 148 (1974)
6.50   R.L.Fante: Radio Sci. **10**, 77 (1975)
6.51   T.I.Wang, J.W.Strohbehn: J. Opt. Soc. Am. **64**, 583 (1974)
6.52   T.I.Wang, J.W.Strohbehn: J. Opt. Soc. Am. **64**, 994 (1974)
6.53   D.Slepian: J. Opt. Soc. Am. **55**, 1110 (1965)
6.54   G.V.Borgiotti: IEEE Trans. **AP-14**, 158 (1966)
6.55   C.K.Rushforth, R.W.Harris: J. Opt. Soc. Am. **58**, 539 (1968)
6.56   G.Toraldo di Francia: J. Opt. Soc. Am. **59**, 799 (1969)
6.57   J.H.Shapiro: Appl. Opt. **13**, 2614 (1974)
6.58   J.H.Shapiro: Appl. Opt. **13**, 2609 (1974)
6.59   D.Slepian: Bell Syst. Tech. J. **43**, 3009 (1964)
6.60   C.Pask: J. Opt. Soc. Am. **66**, 68 (1976)
6.61   A.Labeyrie: Astron. Astrophys. **6**, 85 (1970)
6.62   D.Korff: J. Opt. Soc. Am. **63**, 971 (1973)
6.63   P.H.Deitz: J. Opt. Soc. Am. **65**, 279 (1975)
6.64   M.Greenebaum: "The Residual Effects of Atmospheric Turbulence on a Class of
       Holographic Imaging and Correlography Systems"; presented at the Opt. Soc. Am. Topical
       Meeting on Optical Propagation through Turbulence (1974)
6.65   D.G.Currie: "Amplitude Interferometry and High Resolution Image Formation"; presented
       at the Opt. Soc. Am. Topical Meeting on Optical Propagation through Turbulence (1974)
6.66   W.T.Rhodes, J.W.Goodman: J. Opt. Soc. Am. **63**, 647 (1973)
6.67   R.A.Muller, A.Buffington: J. Opt. Soc. Am. **64**, 1200 (1974)
6.68   J.W.Hardy, J.Feinleib, J.C.Wyant: "Real-Time Phase Correction of Optical Imaging
       Systems"; presented at the Opt. Soc. Am. Topical Meeting on Optical Propagation through
       Turbulence (1974)
6.69   W.B.Bridges, P.T.Brunner, S.P.Lazzara, T.A.Nussmeier, T.R.O'Meara, J.A.Sanguinet,
       W.P.Brown,Jr.: Appl. Opt. **13**, 291 (1974)
6.70   J.E.Pearson: Appl. Opt. **15**, 622 (1976)
6.71   R.P.Urtz,Jr., J.W.Justice: "Compensated Imaging"; presented at the Opt. Soc. Am. Topical
       Meeting on Imaging in Astronomy (1975)
6.72   L.I.Goldfischer: J. Opt. Soc. Am. **55**, 247 (1965)
6.73   J.W.Goodman: Proc. IEEE **53**, 1688 (1965)
6.74   J.W.Goodman: *Introduction to Fourier Optics* (McGraw-Hill, New York 1968) p. 132
6.75   J.C.Dainty (ed.): *Laser Speckle and Related Phenomena*, Topics in Applied Physics, Vol. 9
       (Springer, Berlin, Heidelberg, New York 1975)
6.76   M.Born, E.Wolf: *Principles of Optics*, 5th ed. (Pergamon Press, Oxford 1975) p. 510
6.77   J.H.Shapiro: J. Opt. Soc. Am. **66**, 460 (1976)
6.78   D.Korff, G.Dryden, R.P.Leavitt: J. Opt. Soc. Am. **65**, 1321 (1975)
6.79   D.L.Fried: "Isoplanatism in Predetection Compensation Imaging"; presented at the Opt.
       Soc. Am. Topical Meeting on Optical Propagation through Turbulence (1974)
6.80   J.H.Shapiro: J. Opt. Soc. Am. **66**, 469 (1976)
6.81   J.W.Goodman: *Introduction to Fourier Optics* (McGraw-Hill, New York 1968)
6.82   J.C.Moldon: Pattern Recog. J. **2**, 79 (1970)
6.83   P.H.Deitz, F.P.Carlson: J. Opt. Soc. Am. **63**, 274 (1973)
6.84   R.L.Fante: J. Opt. Soc. Am. **66**, 574 (1976)
6.85   M.Born, E.Wolf: *Principles of Optics*, 5th ed. (Pergamon Press, Oxford 1975) pp. 501–503
6.86   D.Y.Gezari, A.Labeyrie, R.V.Stachnik: Astrophys. J. **173**, L1 (1972)
6.87   A.M.Schneiderman, D.P.Karo: "The Uses and Limitations of Speckle Interferometry";
       presented at the Opt. Soc. Am. Topical Meeting on Imaging in Astronomy (1975)
6.88   K.T.Knox, B.J.Thompson: Astrophys. J. **193**, L45 (1974)
6.89   P.H.Deitz, F.P.Carlson: J. Opt. Soc. Am. **64**, 11 (1974)
6.90   J.C.Moldon: "Imaging of Objects Viewed through a Turbulent Atmosphere"; Res. Lab.
       Electron. Tech. Rept. 469, M.I.T. (1969)

6.91   D.C.Ehn, P.Nisenson: J. Opt. Soc. Am. **65**, 1196A (1975)

6.92   R.Hanbury-Brown, R.Q.Twiss: Nature **161**, 777 (1948)

6.93   M.King, M.Greenebaum, M.Elbaum: "Influence of the Atmosphere on Laser Produced Speckle Patterns"; presented at the Intern. Opt. Comp. Conf., Zurich (1974)

6.94   J.Feinleib, S.G.Lipson, P.F.Cone: Appl. Phys. Lett. **25**, 311 (1974)

6.95   G.Q.McDowell: "Pre-Distortion of Local-Oscillator Wavefront for Improved Optical Heterodyne Detection through a Turbulent Atmosphere"; Sc.D. Thesis, M.I.T. (1971)

6.96   V.N.Mahajan: J. Opt. Soc. Am. **65**, 271 (1975)

6.97   F.J.Dyson: J. Opt. Soc. Am. **65**, 551 (1975)

6.98   S.R.Robinson: "Spatial Phase Compensation Receivers for Optical Communication"; Ph.D. Thesis, M.I.T. (1975)

6.99   J.C.Wyant: "Active Wavefront Compensation for Astronomical Applications"; presented at the Opt. Soc. Am. Topical Meeting on Imaging in Astronomy (1975)

6.100 J.D.Gaskill: J. Opt. Soc. Am. **58**, 600 (1968)

6.102 H.T.Yura: Appl. Opt. **12**, 1188 (1973)

6.102 J.W.Goodman: *Introduction to Fourier Optics* (McGraw-Hill, New York 1968) pp. 227–230

6.103 J.H.Shapiro: J. Opt. Soc. Am. **65**, 65 (1975)

6.104 J.L.Bufton: Appl. Opt. **12**, 1785 (1973)

6.105 W.K.Pratt: *Laser Communication Systems* (Wiley and Sons, New York 1969) Chap. 5

6.106 H.Melchior, M.B.Fisher, F.R.Arams: Proc. IEEE **58**, 1466 (1970)

6.107 A.Yariv: *Introduction to Optical Electronics* (Holt, Rinehart, Winston, New York 1971) Chap. 11

6.108 R.M.Gagliardi, S.Karp: *Optical Communications* (Wiley-Interscience, New York 1976) Chap. 3

6.109 E.V.Hoversten: "Optical Communication Theory", in *Laser Handbook*, ed. by F.T.Arecchi, E.O.Schulz-DuBois (North-Holland, Amsterdam 1972)

6.110 W.K.Pratt: *Laser Communication Systems* (Wiley and Sons, New York 1969) Chap. 6

6.111 N.S.Kopeika, J.Bordogna: Proc. IEEE **58**, 1571 (1970)

6.112 C.W.Helstrom: J. Opt. Soc. Am. **57**, 353 (1967)

6.113 H.V.Hance, D.L.Fried: J. Opt. Soc. Am. **63**, 1015 (1973)

6.114 B.K.Levitt: "Variable-Rate Optical Communication through the Turbulent Atmosphere"; Res. Lab. Electron. Tech. Rept. 483, M.I.T. (1971)

6.115 R.M.Gagliardi, S.Karp: *Optical Communications* (Wiley-Interscience, New York 1976) Chap. 5

6.116 D.L.Snyder: *Random Point Processes* (Wiley-Interscience, New York 1975) Chap. 4

6.117 J.H.Shapiro, G.P.Massa: J. Opt. Soc. Am. **65**, 1218A (1975)

6.118 S.J.Halme: "Efficient Optical Communication in a Turbulent Atmosphere"; Res. Lab. Electron. Tech. Rept. 474, M.I.T. (1970)

6.119 B.K.Levitt: "Detector Statistics for Optical Communication through the Turbulent Atmosphere"; Res. Lab. Electron. Quart. Prog. Rept. 99, pp. 114–123, M.I.T. (1970)

6.120 E.V.Hoversten, R.O.Harger, S.J.Halme: Proc. IEEE **58**, 1626 (1970)

6.121 D.L.Snyder: *Random Point Processes* (Wiley-Interscience, New York 1975) Chap. 6

6.122 R.M.Gagliardi, S.Karp: *Optical Communications* (Wiley-Interscience, New York 1976) pp. 156–159

6.123 E.V.Hoversten: "More on the Performance of Direct Detection Optical Communication Systems for the Turbulent Atmosphere"; presented at the IEEE Intern. Sympos. on Inform. Theory, Notre Dame (1974)

6.124 J.H.Shapiro, M.Tebyani: "A Unified Analysis of PPM Error Rates for Optical Communication through the Turbulent Atmosphere"; presented at the IEEE Intern. Sympos. on Inform. Theory, Ronneby (1976)

6.125 M.Tebyani: "Robust Receivers for Optical Communication through Atmospheric Turbulence"; S.M. Thesis, M.I.T. (1976)

6.126 W.N.Peters, R.J.Arguello: IEEE J. **QE-3**, 532 (1967)

6.127 S.Solimeno, E.Corti, B.Nicoletti: J. Opt. Soc. Am. **60**, 1245 (1970)

6.128 W. E. Webb, J. T. Marino, Jr.: Appl. Opt. **14**, 1413 (1975)
6.129a. M. C. Teich, S. Rosenberg: Appl. Opt. **12**, 2616 (1973)
     b. S. Rosenberg, M. C. Teich: Appl. Opt. **12**, 2625 (1973)
6.130 E. A. Bucher: Appl. Opt. **11**, 884 (1972)
6.131 R. S. Kennedy: Proc. IEEE **58**, 1651 (1970)
6.132 D. L. Snyder: *Random Point Processes* (Wiley-Interscience, New York 1975) Chap. 2
6.133 S. J. Halme, B. K. Levitt, R. S. Orr: "Bounds and Approximations for some Integral Expressions Involving Lognormal Statistics"; Res. Lab. Electron. Quart. Prog. Rept. **93**, pp. 163–175, M.I.T. (1969)
6.134 R. M. Gagliardi, S. Karp: *Optical Communications* (Wiley-Interscience, New York 1976) Chap. 6
6.135 J. M. Wozencraft, I. M. Jacobs: *Principles of Communication Engineering* (Wiley and Sons, New York 1965) pp. 517–519
6.136 W. K. Pratt: *Laser Communication Systems* (Wiley and Sons, New York 1969) Chap. 12
6.137 R. S. Kennedy, S. Karp (eds.): "Optical Space Communication"; NASA Rept. SP-217 (1969)
6.138 J. M. Wozencraft, I. M. Jacobs: *Principles of Communication Engineering* (Wiley and Sons, New York 1965) pp. 533–550
6.139 R. S. Kennedy: *Fading Dispersive Communication Channels* (Wiley-Interscience, New York 1969)
6.140 H. L. Van Trees: *Detection, Estimation, and Modulation Theory*, Part III (Wiley and Sons, New York 1971) Sect. 11.3
6.141 E. V. Hoversten, D. L. Snyder, R. O. Harger, K. Kurimoto: IEEE Trans. COM-**22**, 17 (1974)
6.142 C. M. McIntyre, J. R. Kerr, J. R. Dunphy: "Practical Diversity Receivers in Turbulence: Experimental Performance and Relationship to Rigorous Likelihood Systems"; presented at the IEEE Nation. Telecommun. Conf., New Orleans (1975)
6.143 G. E. Homstad, J. W. Strohbehn, R. H. Berger, J. M. Heneghan: J. Opt. Soc. Am. **64**, 162 (1974)
6.144 J. N. Bucknam: "Direct Detection of Optical Signals in the Presence of Atmospheric Turbulence"; S.M. Thesis, M.I.T. (1969)
6.145 P. A. Humblet: "Two-Way Adaptive Optical Communication through the Clear Turbulent Atmosphere"; S.M. Thesis, M.I.T. (1975)
6.146 R. L. Mitchell: J. Opt. Soc. Am. **58**, 1267 (1968)
6.147 R. Barakat: J. Opt. Soc. Am. **66**, 211 (1976)
6.148 R. L. Fante: J. Opt. Soc. Am. **66**, 730 (1976)

7. Thermal Blooming in the Atmosphere

J. L. Walsh and P. B. Ulrich

With 15 Figures

The essential features of the phenomenon of thermal blooming can best be understood by considering a simple example. Let us imagine a laser beam propagating through a medium which is an ideal gas having some absorption at the laser wavelength. If the medium is initially in thermal equilibrium and the laser beam is turned on at $t=0$, the absorption will give rise to a local heating which in turn will produce a small pressure increase. The medium then expands at the speed of sound so as to restore pressure balance leaving behind a small decrease in the density. The local refractive index will decrease in proportion to the local density change. If the cumulative effect of these negative refractive index changes is large enough, we may expect them to change the shape of the laser beam profile as it propagates away from the source.

To understand how these modifications might change the laser propagation, let us consider a beam intensity profile which is peaked on the axis of the beam and falls to zero at the edges. The heating profile will show a maximum on the axis of the laser beam while the corresponding refractive index profile will exhibit a minimum. Thus the phase fronts which may originally have been normal to the beam axis move with slightly greater speed on the beam axis and take on a tilt toward the outside of the beam. The rays of the beam, which constitute the normals to the phase fronts, can thus be thought of as diverging outward from their original direction parallel to the beam axis. The entire beam "blooms" or grows in size as the heating of the medium continues. It is this phenomenon of thermal blooming which is the subject of this chapter.

The earliest published work on this topic is the 1964 paper by *Leite* et al. [7.1]. These authors used the effect for the measurement of small absorption coefficients in a number of liquids and solids. In their experiments, a sample was placed inside the cavity of a helium-neon laser, and the change in spot size on the resonator mirrors produced by these relatively thin thermal lenses was measured.

The subject of thermal blooming is an interdisciplinary one. It draws heavily on certain aspects of fluid mechanics, on wave propagation in fluids, and on both geometrical and physical optics. In addition, because many of the problems of interest are analytically intractable, we find that the applications often require extensive computer analyses of the coupled fluid mechanical and optical propagation equations. The material contained herein is presented at the advanced undergraduate or early graduate level. The reader is not assumed to have an extensive background in all of these topics. We attempt to develop relevant background material in the beginning of each section so as to make the entire chapter reasonably self-contained. We present simple analytical solutions to idealized problems in order to gain physical insight into the nature of the basic processes. In general we have not attempted to extract all possible analytical solutions.

The question of credit for original ideas deserves mention. In general we have attempted to present a tutorial treatment rather than a historical one. As a result we have not given a complete bibliography but instead we present only those references which are useful to our discussion. Those readers interested in a more complete bibliography may consult *Ulrich* [7.2].

7.1 An Overview

7.1.1 An Order of Magnitude Estimate

We begin our discussion by presenting a simple order-of-magnitude estimate for thermal blooming in an ideal gas. Consider an unfocused cw laser beam of total power P being emitted from an aperture of radius a into air at 1 atm pressure. The beam is assumed to have an intensity distribution which is maximum on the axis and falls to zero at radius a. We begin by writing an expression for N, the number of wavelengths of the radiation contained in a path of length L on the beam axis. N is given by the relation

$$N = (nL)/\lambda, \tag{7.1}$$

where n is the refractive index and λ is the vacuum wavelength of the radiation. After the beam is turned on, the refractive index will vary with time because of heating produced by the beam. The variation in N is given by

$$\delta N = (L/\lambda)\delta n. \tag{7.2}$$

We shall show later in this chapter that $\delta N \approx 1$ wavelength is an appropriate criterion for the onset of thermal blooming. This should not be surprising since a phase change of one wavelength is the criterion for onset of many phenomena in optics.

For air, over a wide range of values for the temperature, pressure and electric field, we find that the refractive index n can be expressed by the relation

$$(n-1) = k\varrho, \tag{7.3}$$

where ϱ is the density and k is the Gladstone-Dale constant. Under standard conditions $(n-1) \simeq 3 \times 10^{-4}$.

If the density varies by a small amount $\delta\varrho$, the corresponding change in refractive index can be obtained from (7.3). We find

$$\frac{\delta(n-1)}{(n_0-1)} = \frac{\delta\varrho}{\varrho_0} = s, \tag{7.4}$$

where ϱ_0 and n_0 are the initial quiescent values and s is the condensation. This equation can be written more compactly as

$$\delta n = (n_0 - 1)s. \tag{7.5}$$

We must now develop an expression for the change in condensation with heating produced by laser beam absorption. We make use of several of the properties of ideal gases [7.3]. The equation of state is simply

$$pV = \mu RT, \tag{7.6}$$

where p is the pressure, V is the volume occupied by μ moles of the gas, T is the absolute temperature, and R is the universal gas constant. From this expression we easily obtain the thermal expansion coefficient β defined by the relation

$$\beta = \frac{1}{V}\left(\frac{\partial V}{\partial T}\right)_p = \frac{1}{T}. \tag{7.7}$$

Finally we must consider the temperature rise when heating at constant pressure. If γ is the ratio of the molar specific heats c_p and c_v at constant pressure and volume, respectively, then

$$c_p = \frac{\gamma}{\gamma-1}R. \tag{7.8}$$

If we slowly add a small amount of heat dQ per mole to an ideal gas and assume that it is free to expand at constant pressure, then from (7.7) and (7.8) we find that the density will vary according to the relation

$$\frac{\delta\varrho}{\varrho} = -\frac{dV}{V} = -\beta\frac{dQ}{c_p} = -\frac{\gamma-1}{\gamma RT}dQ. \tag{7.9}$$

Let us now rewrite the differentials $d\varrho$ and dQ in terms of the small quantities s and q_v, i.e.,

$$\frac{d\varrho}{\varrho} = s \quad \text{and} \quad dQ = \frac{RT}{p_0}q_v, \tag{7.10}$$

where p_0 is the ambient pressure and q_v is the heat energy deposited per unit volume. Finally let us write

$$q_v = \alpha I t_0, \tag{7.11}$$

where α is the laser beam absorption coefficient, I is the beam intensity at the point in question, and t_0 is the heating time. Combining (7.7–11) we obtain an expression for the condensation at an arbitrary point r in the beam after heating for a time t_0

$$s(r) = -\frac{(\gamma - 1)}{\gamma p_0} \alpha I(r) t_0. \tag{7.12}$$

As it stands, (7.12) assumes that there is no air movement through the beam and that the intensity at the point in question remains constant during the heating time t_0. Let us consider the case in which there is a flow of air at velocity v normal to the axis of the beam. The heating rate for a parcel of air moving through the beam varies with the intensity distribution across the beam. Nevertheless we may obtain an order-of-magnitude estimate of the condensation from (7.12) by setting $I(r)$ equal to the peak intensity and replacing t_0 by the expression

$$t_0 = a/v, \tag{7.13}$$

where a is the characteristic transverse dimension of the beam.

If we now combine (7.2, 5, 12, 13) we obtain

$$\delta N = -(n_0 - 1)\frac{L}{\lambda}\frac{\gamma - 1}{\gamma p_0}\alpha I_0 \frac{a}{v}, \tag{7.14}$$

where I_0 is the peak beam intensity. To complete our estimate we must make a suitable choice for the path length L. In order to estimate the lowest threshold at which blooming could occur we would like to choose L as large as possible. Let us set

$$L \simeq (2a)^2/\lambda, \tag{7.15}$$

a distance which is called the "Rayleigh range". This is the distance beyond which an initially collimated beam begins to spread due to diffraction. Finally we write the relation between the total beam power P and the on-axis intensity I_0,

$$P \simeq \pi a^2 I_0. \tag{7.16}$$

If we now combine (7.14–16) and set $\delta N = -1$ we may obtain an expression for P_t the total beam power at threshold for the onset of thermal blooming in a transverse flow

$$P_t \simeq \frac{\pi}{4}\frac{\gamma p_0}{\gamma - 1}\frac{1}{\alpha(n_0 - 1)}\frac{\lambda^2 v}{a}. \tag{7.17}$$

As an example let us consider a 10 μm beam in air at 1 atm. Thus the constants in (7.17) have the values

$$p_0 = 10^5 \text{ newtons m}^{-2} \qquad n_0 - 1 = 3 \times 10^{-4}$$
$$\gamma = 1.4 \qquad\qquad\qquad \lambda = 10^{-5} \text{ m}.$$

Inserting these values into (7.17) we obtain

$$P_t = 0.1 \frac{v}{\alpha a} \tag{7.18}$$

in SI units. Let us choose

$$\alpha = 3 \times 10^{-3} \text{ m}^{-1}$$
$$a = 10^{-2} \text{ m} \tag{7.19}$$
$$v = 0.3 \text{ m s}^{-1}.$$

Substituting these values into (7.18), we obtain for the conditions chosen, $P_t = 10^3$ W as the threshold level for thermal blooming in air. The choices for α and v are representative for a 10.6 μm laser in quiet air. Thus thermal blooming can set in at fairly modest power levels. The Rayleigh range in this example is of the order of 40 m and at the range the beam has been exponentially attenuated only about 13 %.

The temperature change can be estimated from the expression

$$\frac{\delta T}{T} = -\frac{\delta \varrho}{\varrho} = -s \tag{7.20}$$

for the heating of an ideal gas at constant pressure. For this example we find $s = -9.1 \times 10^{-4}$ and $\delta T = 0.27$ K. The temperature change we have calculated is comparable with what one could find associated with atmospheric turbulence. But a given value of δT associated with turbulence will be much less effective in disrupting the beam than the same value of δT for blooming because for turbulence the value of δT and its corresponding δn take on both positive and negative values along the beam path, thus producing a significant amount of net cancellation. The δT associated with thermal blooming, on the other hand, produces a δn which is of the same sign all along the beam path and hence is capable of producing a steady cumulative effect.

Finally, it should be pointed out that thermal blooming is a phenomenon of importance primarily because of lasers and their ability to produce relatively intense collimated light beams.

7.1.2 Overview of Our Treatment of Thermal Blooming

In Section 7.2 we review the propagation of optical beams in the scalar paraxial approximation but with a general refractive index. The treatment includes both wave optics and the geometric optics approximation. The properties of gaussian beams with arbitrary choice of the origin of coordinates are then summarized. The gaussian beam profile serves as a prototype for our analysis of blooming phenomena.

In Section 7.3 we develop the basic partial differential equation for the density variation in a fluid medium with time- and space-dependent heat sources. These heat sources are intended to represent the heating by laser beam absorption. A number of solutions for the time- and space-dependent density disturbance are presented for various heat source distributions. There is no attempt made here to obtain a self-consistent optical-fluid mechanical treatment because no such solutions have been presented in the literature. Nevertheless by solving a number of special cases we are able to gain considerable insight into the fluid mechanical effects.

In Section 7.4 we present simple analytic solutions to model problems which combine linearized hydrodynamics and first-order perturbation theory for the propagation using either wave optics or geometric optics. The models are based on the results of the density disturbances calculated in Section 7.3 for various heating profiles.

Section 7.5 is devoted to a discussion of the computer methods and algorithms which have been applied to the thermal blooming problem. A number of results from the computer analyses are also presented.

7.1.3 Architecture of Thermal Blooming

Before we proceed with our detailed discussion of the various topics outlined above, we shall consider in more detail how each of these topics contributes to our overall understanding of thermal blooming. Our discussion will focus on the two major elements: 1) initial parameters and method of approach and 2) nature of the resulting solutions.

Initial Parameters and Method of Approach

The important ingredients of the problem which must be considered include the following:

Time Dependence of the Beam. The important cases are single pulse, multiple pulse, and cw. The method of analysis depends on the relationship between the parameters characterizing the time dependence of the beam and the characteristic times associated with either the medium or the geometry.

The Beam Spatial Dependence. Often the gaussian or partially truncated gaussian profile at the source is chosen as representative. The beam may be focused at infinity or at some closer distance. In some cases variable focus may be considered. The question of whether there is one or several wavelengths present is also important. And finally the beam quality is an important element. This parameter is usually expressed simply by stating that the beam is a certain number of times diffraction limited.

The Type of Cooling. We use the term "cooling" to indicate that there are generally one or several factors which limit the exposure time of a given parcel of the medium to the laser beam. This is a separate consideration from the fact that the beam itself may be only of limited duration. The various types of cooling include a steady wind, a slewed beam, convective cooling, and, of course, no cooling as in the case of a short pulse.

The Degree of Self-Consistency Away from the Aperture. There are two limiting cases which may be considered. In the first case the spatial dependence of the heating profile at the instant the laser beam is turned on is retained for all later times. The propagation through this disturbed medium is allowed to unfold with time as the amount of heating increases. This approach is the basis for all analytic calculations on thermal blooming currently in the literature.

The second possibility allows the spatial dependence of beam heating to follow the actual propagation of the beam as it evolves with time. At this writing, this procedure has been carried out successfully only by computer calculation.

Electromagnetic Theory. The two possibilities are the pure wave optics treatment and the geometric optics approximation. The latter is often useful in that a physical picture is often easier to construct and the calculations are more readily carried out. The wave optics treatment of course is the most accurate but it may require a computer calculation.

Nature of the Resulting Solutions

There are several levels at which thermal blooming problems can be attacked and the solutions presented.

Most Rigorous. The most rigorous solutions require the use of wave optics and the full fluid mechanical equations in a self-consistent way. This is the least tractable approach analytically, and for that reason much computer work has been done in this area. In principle, this methodology permits the inclusion of totally realistic models for both laser beam shape and the absorbing media. Initially however, only fully coherent diffraction-limited gaussian beams at a single wavelength were considered. More realistic problems have been considered recently. An example is the case of two or more simultaneous beams from the same aperture, each having a different wavelength, beam shape, and absorption coefficient. The treatment of non-diffraction-limited beams has also been carried out. This problem presents some unique difficulties because the simple statement that a beam is, say, m times diffraction limited is not sufficient to characterize the initial amplitude and phase distribution for computer calculation. Instead, one must take an average of the results of many calculations, each of which has its initial conditions chosen randomly subject to the constraint that it produce an m times diffraction-limited beam if propagated in vacuum.

Analytic Results. There are a number of limiting cases in which analytic results can be obtained with varying degrees of difficulty. They often have the advantage that the dependence of the results on variations in the key parameters is more visible than in the computer results. Taken together they provide a great deal of insight into the nature of the solutions to be expected in the general case.

Empirical Computer Results. In this approach, repetitive computer runs are made with all parameters fixed except total beam power, which varies between successive runs. From the results of these calculations, empirical formulae can be derived for important beam parameters after blooming. The parameters which may be considered include maximum on-axis intensity, beam cross section, and beam deflection. In some cases similar results can be derived by the direct analytical approach described above.

Characterization of Blooming Results. When a computer calculation is completed, one has available an enormous amount of very detailed information about intensity as a function of transverse position within the beam profile, the range, and the time. Generally the information presented is much more detailed than is needed. It is important therefore to choose appropriately one or several summary parameters which extract the important features. Examples include the deflection of some feature of the beam such as the peak intensity, the centroid or the centroid of that portion exceeding a previously chosen threshold. Other features of interest may include various parameters describing the beam spread around either the peak or the centroid.

7.2 Electromagnetic Theory

In this section we review those aspects of electromagnetic theory which are important to the problem of thermal blooming. Thus we are concerned primarily with paraxial beams in a medium which may have small variations in the local refractive index. These beams will be considered from both the wave optics and ray optics points of view.

7.2.1 The Electromagnetic Wave Equation

We begin by writing the basic Maxwell equations in the absence of free charges or external currents. In rationalized mks units we may write [7.4]

$$\boldsymbol{V} \times \boldsymbol{E} = -\frac{\partial \boldsymbol{B}}{\partial t}$$

$$\boldsymbol{V} \times \boldsymbol{H} = \frac{\partial \boldsymbol{D}}{\partial t}. \tag{7.21}$$

In addition we have the constitutive relations

$$D = \varepsilon E \quad \text{and} \quad B = \mu H,$$ (7.22)

where ε is the dielectric constant and μ is the permeability. In the following discussion, μ will be taken as a constant but ε will be allowed to vary throughout the medium.

We begin by taking the curl of the first equation in (7.21) and $\partial/\partial t$ of the second and substituting the second result into the first. If we make use of the vector identity

$$V \times V \times E = V(V \cdot E) - V^2 E,$$ (7.23)

we obtain

$$V^2 E - \mu \frac{\partial^2}{\partial t^2}(\varepsilon E) = V(V \cdot E).$$ (7.24)

We may obtain an expression for $V \cdot E$ by noting that in the absence of free charge, $V \cdot D = 0$. From the first equation in (7.22) we can write

$$E \cdot (V\varepsilon) + \varepsilon(V \cdot E) = 0,$$ (7.25)

so that (7.24) can be written

$$V^2 E - \mu \frac{\partial^2}{\partial t^2}(\varepsilon E) = -V[E \cdot V(\log \varepsilon)].$$ (7.26)

A slightly different equation is obtained for H, but this will not concern us here. Finally we may put

$$\varepsilon(r, t) = \varepsilon_0 n^2(r, t),$$ (7.27)

where ε_0 is the vacuum value and $n(r, t)$ is the time- and space-dependent refractive index. If c is the speed of light in vacuum with

$$c^2 = \frac{1}{\mu \varepsilon_0},$$ (7.28)

we can write (7.26) in the form

$$V^2 E - \frac{1}{c^2} \frac{\partial^2}{\partial t^2}(n^2 E) = -2V[E \cdot V(\log n)].$$ (7.29)

Let us now consider the relative magnitudes of the terms in (7.29). First of all, we want to remove n^2 from inside the partial derivative on the left side. This procedure is valid if the temporal variation of n takes place on a time scale long compared to the period of the electromagnetic field. To estimate the magnitude of these effects, we note first that significant variations in n cannot occur faster than an acoustic transit time for the beam profile. For a 1 cm diameter beam in air, the acoustic transit time is about 3×10^{-5} s. This is quite long compared to the periods of oscillation for laser beams of interest which might vary typically between 10^{-14} and 10^{-16} s.

To evaluate the term on the right side of (7.29), consider a paraxial beam traveling in the positive z direction and initially having a polarization E_x only. As a term to modify E_x, this term is quite small. But as a source term for a polarization E_z for example, the term is still small but may not be inconsequential.

For our present discussion we shall drop the term on the right-hand side of (7.29) and write

$$\nabla^2 E - (n^2/c^2)\partial^2 E/\partial t^2 = 0. \tag{7.30}$$

This is the basic wave equation from which the following discussion will proceed.

7.2.2 Scalar Wave Equation for Paraxial Beams

Let us consider the case of a monochromatic collimated beam $E(x, y, z, t)$ propagating in the z direction and having a single polarization in, say, the x direction. By a beam we mean a field configuration which is significantly different from zero only in a confined region in the xy plane. The rapid oscillation with wavelength in the z direction can be written as a separate factor. Thus the complete field can be taken as the real part of the expression

$$E_x(x, y, z, t) = \psi(x, y, z) \exp(\mathrm{i}kz - \mathrm{i}\omega t). \tag{7.31}$$

The wave number k corresponds to that associated with the quiescent refractive index n_0. Modifications to the propagation because of departures from n_0 are contained within ψ. If we substitute (7.31) into (7.30) we obtain

$$\nabla_T^2 \psi + 2\mathrm{i}k \frac{\partial \psi}{\partial z} - k^2 \psi + \frac{n^2 \omega^2}{c^2} \psi = 0, \tag{7.32}$$

where ∇_T^2 is the transverse Laplacian operator and $\partial^2 \psi/\partial z^2$ has been dropped because for a beam field configuration, the variation of ψ with z is slow enough so that it can be neglected in comparison with the other terms. Now by definition

$$\omega = \frac{kc}{n_0} \quad \text{and} \quad k = \frac{2\pi n_0}{\lambda_{\text{vac}}}, \tag{7.33}$$

and therefore (7.32) can be written

$$V_T^2\psi + 2ik\frac{\partial\psi}{\partial z} + k^2\left[\frac{n^2(x,y,z,t)}{n_0^2} - 1\right]\psi = 0.$$ (7.34)

This is the basic differential equation from which wave optics calculations for paraxial beams may be carried out, given a refractive index profile $n(x,y,z,t)$.

7.2.3 Gaussian Beam Modes in a Uniform Medium

The gaussian beam modes [7.5] constitute the natural field configuration for the output of many lasers. Because of this and also because of their relative analytical simplicity, they will be used frequently in this chapter for representative analytical calculations involving the laser beam profile before blooming effects have set in. We wish now to consider briefly the properties of these modes. We shall develop the results in a somewhat different form than that which is usually presented. The form chosen here is one which is better suited to our discussion of thermal blooming.

If we set $n(x,y,z) = n_0$ in (7.34) and ask for solutions $\psi(r,z)$ possessing cylindrical symmetry around the z axis, (7.34) becomes

$$\frac{\partial^2\psi}{\partial r^2} + \frac{1}{r}\frac{\partial\psi}{\partial r} + 2ik\frac{\partial\psi}{\partial z} = 0.$$ (7.35)

We shall solve (7.35) for the gaussian beam modes having cylindrical symmetry by using the method of integral transforms. The properties of integral transforms which are important for the topics of this chapter are summarized in the Appendix.

If we take the Hankel transform (see Appendix) of the entire equation and let $\bar\psi(\lambda, z)$ equal the Hankel transform with respect to r of $\psi(r,z)$, we obtain an ordinary differential equation for $\bar\psi$

$$-\lambda^2\bar\psi(\lambda, z) + 2ik\frac{d\bar\psi}{dz} = 0.$$ (7.36)

This equation has the solution

$$\bar\psi(\lambda, z) = A(\lambda)\exp[-i\lambda^2 z/(2k)],$$ (7.37)

where $A(\lambda)$ is a function of λ to be determined from the initial conditions.

For $\psi(r,z)$ at the plane $z = 0$, let us choose a gaussian amplitude profile and a spherical wave front converging toward the point on the axis located at $z = f$. For our chosen $\exp(-i\omega t)$ time dependence, the required distribution will be given by the expression

$$\psi(r,0) = \psi_0\exp\left(\frac{-r^2}{r_0^2} - \frac{ikr^2}{2f}\right),$$ (7.38)

where ψ_0 is the field amplitude at the origin and r_0 is the $1/e$ point of the electric field amplitude. If we take the Hankel transform of (7.38) we find

$$\bar{\psi}(\lambda, 0) = A(\lambda) = \psi_0 \frac{r_0^2 f}{2f + ikr_0^2} \exp\left[\frac{-\lambda^2 f r_0^2}{2(2f + ikr_0^2)}\right]. \qquad (7.39)$$

We may now put the results of (7.39) into (7.37) and calculate the inverse transform. The result can be written in the form

$$\psi(r, z) = \frac{\psi_0}{(1 - z/f) + i2z/(kr_0^2)} \exp\left[\frac{-r^2}{w^2(z)} - \frac{ikr^2}{2R(z)}\right], \qquad (7.40)$$

where $w(z)$ is the $1/e$ point of the beam amplitude profile and $R(z)$ is the radius of curvature of the wave front. Wave fronts concave toward the positive z axis are taken as having positive curvature. The function $w(z)$ is given by the expression

$$w^2(z) = r_0^2(1 - z/f)^2 + 4z^2/(k^2 r_0^2). \qquad (7.41)$$

In general this expression has a minimum value at some point on the z axis. This point is called the beam waist and is designated by $z = z_w$. If we take the derivative of (7.41) with respect to z and set the result equal to zero, we find that the beam waist is located at

$$z_w = \frac{k^2 r_0^4 f}{k^2 r_0^4 + 4f^2}. \qquad (7.42)$$

The radius of curvature of the phase front $R(z)$ can be written as

$$R(z) = \frac{f - 2z + z^2/z_w}{1 - z/z_w}. \qquad (7.43)$$

We see from (7.43) that $R(0) = f$ in accordance with our initial conditions. At the beam waist $R(z_w)$ becomes infinite, corresponding to the fact that the constant phase surface has become a plane. For $z > z_w$, $R(z) < 0$ and the phase surfaces have become convex toward $z > 0$. In the limit as z goes to infinity, $R(z)$ goes to $-(z - z_w)$ so that the outgoing waves are spherical with centers of curvature on the axis at the position of the beam waist. These results are of course consistent with those obtained by *Kogelnik* and *Li* [7.5].

The intensity distribution associated with this field configuration is rather easily obtained. If we write for the intensity

$$I(r, z) = |\psi(r, z)|^2, \qquad (7.44)$$

we find from (7.40) and (7.41)

$$I(r, z) = \frac{r_0^2}{w^2(z)} |\psi_0|^2 \exp[-2r^2/w^2(z)]. \qquad (7.45)$$

Later in this chapter there are occasions where we may wish to simplify this expression even further. For example we will often write

$$I(r, z) = I_0 \exp(-r^2/a^2), \qquad (7.46)$$

where "a" is taken as a constant. This is equivalent to assuming that the variations in beam size given by (7.41) take place over a distance large compared to that required to observe significant blooming effects.

7.2.4 Geometric Optics

The geometric optics approximation is used in thermal blooming calculations because it permits the development of both basic insights and simple analytic results which are often not readily accessible from the more vigorous wave optics calculations. In this subsection we summarize only those elements which will be required in our later developments. Our discussion is based in large part on the treatment of *Born* and *Wolf* [7.6].

The geometric optics approximation is based on the idea that in certain regions of space and for sufficiently short wavelengths, one may find surfaces called phase fronts such that the normals to these phase fronts describe the direction of local energy flow and also represent the direction in which the electromagnetic wave (including its phase) varies most rapidly. Thus we take as a trial form for the electric field, the real part of the expression

$$E(r, t) = e(r) \exp[i k_0 S(r) - i \omega t], \qquad (7.47)$$

where S is intended to be a scalar function of position independent of $e(r)$, k_0 is the vacuum wave number, and $e(r)$ is a slowly varying function of position which carries polarization information. A similar expression is written for the magnetic field.

When (7.47) and a corresponding expression for the magnetic field are inserted into Maxwell's equations, we find that in the short wavelength limit

$$[\nabla S(r)]^2 = n^2(r), \qquad (7.48)$$

where $n(r)$ is the position-dependent refractive index and $S(r) = \text{const}$ are the surfaces of constant phase. If we let t denote a unit vector along the direction normal to the surfaces $S(r)$, we can write

$$t = dr/ds, \qquad (7.49)$$

where r is the position vector of a point along a given ray and s is the scalar measure of distance along the ray. Using (7.48) we obtain

$$nt = \nabla S(r). \qquad (7.50)$$

By differentiating this expression and writing $S(r)$ in terms of n, we can obtain a differential equation for the light rays themselves. The result obtained by *Born* and *Wolf* is simply

$$\frac{d}{ds}\left(n\frac{dr}{ds}\right) = \nabla n(r),$$ (7.51)

which can be written using (7.49)

$$n\frac{dt}{ds} + \frac{dn}{ds}t = \nabla n.$$ (7.52)

Let us now define K as the curvature of the ray by the expression

$$K = dt/ds.$$ (7.53)

The magnitude of K is the reciprocal of the radius of curvature and it lies in the osculating plane of the ray. Combining (7.52) and (7.53) we can write

$$nK = \nabla n - \frac{dn}{ds}t.$$ (7.54)

The expression on the right side of (7.54) is the difference between the gradient of n and its component along the ray. Thus the direction of the vector indicated by the right side of (7.54) is always normal to the ray and the rays bend toward regions of higher refractive index.

Let us consider now the intensity distribution associated with the rays we have just described. The intensity (power/area) at a point can be thought of as a vector quantity of magnitude $I(r)$ directed along the ray trajectory. If there are absorption losses αI per unit volume, then for an arbitrary volume V bounded by a surface A and free of sources we could write

$$\int_A It \cdot d\sigma = -\int_V \alpha I dv,$$ (7.55)

where $d\sigma$ and dv are the elements of surface area and volume, respectively. Applying Gauss' theorem we have

$$\nabla \cdot (It) + \alpha I = 0,$$ (7.56)

which is the basic equation for the intensity distribution in geometric optics.

7.2.5 Scalar Diffraction Theory in a Uniform Medium

In order to develop analytical estimates for thermal blooming phenomena, it will be useful to have available some results which can be developed from Kirchoff diffraction theory. *Jackson* [Ref. 7.4, p. 282] gives an expression for

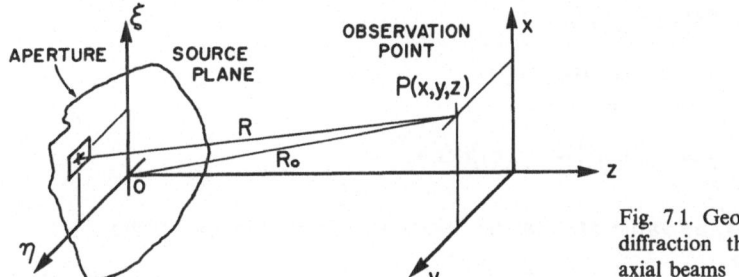

Fig. 7.1. Geometry for scalar diffraction theory with paraxial beams

the scalar field diffracted into a region by an illuminated aperture S_1. He obtains

$$\psi(R) = -\frac{1}{4\pi} \int_{S_1} \frac{\exp(ikR)}{R} \, n \cdot \left[\nabla\psi + ik\left(1 + \frac{i}{kR}\right) \frac{R}{R} \psi \right] ds' \,, \tag{7.57}$$

where da' is the differential element of area in the aperture, n is a unit normal directed into the observation region, R is the vector distance from the aperture element da' to the observation point, and $\nabla'\psi$ is evaluated in the aperture. An $\exp(-i\omega t)$ time dependence is assumed. In the Kirchoff approximation one assumes that ψ and its normal derivative have the same value in the aperture region that they would have if the aperture were not present.

For our present purpose we are interested primarily in paraxial beams of the type described by (7.31). Let us choose $z = 0$ as the plane of the aperture as illustrated in Fig. 7.1. Then, since the dominant contribution to $[\partial\psi/\partial z]_{z=0}$ comes from differentiating the $\exp(ikz)$ term, we set $n \cdot (\nabla'\psi)$ equal to $ik\psi$. For beam fields well removed from the aperture, we have $kR \gg 1$ and $n \cdot (R/R) = 1$. Under these conditions, (7.57) can be written

$$\psi(R) = -\frac{ik}{2\pi} \int_{S_1} \psi(S_1) \frac{\exp(ikR)}{R} da' \,. \tag{7.58}$$

Let us now introduce the cartesian coordinates (ξ, η) to locate points within the aperture. The coordinates of the observation point P will be designated (y, y, z). Then we can write

$$R^2 = (x - \xi)^2 + (y - \eta)^2 + z^2 \,. \tag{7.59}$$

If R_0 is the distance from the origin to the observation point P, then (7.59) can be rewritten

$$R^2 = R_0^2 \left[1 - 2\frac{(\xi x + \eta y)}{R_0^2} + \frac{\xi^2 + \eta^2}{R_0^2} \right]. \tag{7.60}$$

In cases of interest to us here, R_0 is always much greater than any of the variables x, y, ξ, or η. With this assumption and after some manipulation we obtain from (7.58) and (7.60)

$$\psi(x, y, z) = -\frac{ik}{2\pi}\frac{\exp(ikR_0)}{R_0}$$
$$\cdot \int_{S_1} u(\xi, \eta)\exp[ikf(\xi, \eta)]d\xi d\eta,$$

(7.61)

where $u(\xi, \eta)$ is the value of ψ in the aperture, $f(\xi, \eta)$ is given by the relation

$$f(\xi, \eta) = -(l\xi + m\eta) + (\xi^2 + \eta^2)/(2R_0),$$

(7.62)

and l and m are the direction cosines

$$l = x/R_0 \quad m = y/R_0.$$

(7.63)

Finally let us consider the case of a one-dimensional source which is independent of η and extends from $-\infty$ to $+\infty$ along the η axis. To obtain the one-dimensional diffraction pattern we may integrate (7.61) with respect to η. Because of the nature of the problem the result should be independent of y. Choosing $y = 0$, the required integral is

$$\int_{-\infty}^{\infty} \exp[ik\eta^2/(2R_0)]d\eta = (1 + i)\sqrt{\pi R_0/k}.$$

(7.64)

When this result is substituted into (7.61) we obtain

$$\psi(x, z) = \frac{(1 - i)}{2}\sqrt{\frac{k}{\pi R_0}}\exp(ikR_0)$$
$$\cdot \int_{S_1} u(\xi)\exp\left[ik\left(-l\xi + \frac{\xi^2}{2R_0}\right)\right]d\xi.$$

(7.65)

Let us apply this result to the calculation of the diffraction pattern form a slit having an amplitude profile

$$u(\xi) = u_0(1 - \xi^2/a^2)^{1/2}\exp\left(-\frac{ik\xi^2}{2f}\right), \quad |\xi| < a.$$

(7.66)

This distribution is even in ξ, having the peak value u_0 at $\xi = 0$ and is focused at $z = f$. In Section 7.4 we shall see how the result we obtain here is modified by the presence of thermal blooming. If we set $R_0 = f$ in (7.65) and substitute (7.66) we obtain the amplitude in the focal plane

$$\psi(x, f) = \frac{(1 - i)}{2}\sqrt{\frac{k}{\pi f}}\exp(ikf)$$
$$\cdot u_0 \int_{-a}^{a}\left(1 - \frac{\xi^2}{a^2}\right)^{1/2}\cos(kl\xi)d\xi.$$

(7.67)

The term $\sin(kl\xi)$ has been dropped from the factor $\exp(-ikl\xi)$ because it is odd in ξ and will not contribute to the integral. Using *Erdelyi* et al. [7.7], we obtain

$$\psi(x,f) = \frac{(1-i)}{2}\sqrt{\frac{k}{\pi f}}\exp(ikf)u_0\pi\frac{J_1(kla)}{kl},\tag{7.68}$$

where J_1 (kla) is the Bessel function of the first kind of order one.

The intensity distribution in the focal plane

$$I(x,f) = |\psi(x,f)|^2\tag{7.69}$$

can be written

$$I(x,f) = \frac{u_0^2}{2}\frac{\pi f}{kx^2}J_1^2(kax/f).\tag{7.70}$$

The intensity distribution peaks on the z axis at $x=0$ where it has the value

$$I(0,f) = \frac{u_0^2}{8}\frac{\pi ka^2}{f}.\tag{7.71}$$

The peak value falls off as $1/f$, which we expect from a one-dimensional distribution. Finally we see that the complete distribution contains a series of nulls given by the consecutive zeros of the Bessel function.

7.2.6 Pasted Phase Approximation

In our summary of diffraction theory thus far we have assumed that the refractive index has a constant value in the region between the aperture and the observation point. In the presence of thermal blooming, however, this assumption is no longer valid and we must make appropriate modifications to the theory. We shall show that (7.58) is still provided that we replace the product ikr in the term $\exp(ikr)$ by an integral over the optical path between the source plane and the observation point.

We begin by considering the scalar field radiated by a point source in a medium with small disturbances in the refractive index. By analogy with (7.30) we have

$$\nabla^2\psi - \frac{n^2}{c^2}\frac{\partial^2\psi}{\partial t^2} = 0,\tag{7.72}$$

where $\psi(r,\theta,\phi)$ is a scalar disturbance and $n(r,\theta,\phi)$ has small fluctuations around its quiescent value. The coordinates $r,\theta,$ and ϕ are spherical polar coordinates with respect to what will eventually be our observation point. If we assume an $\exp(-i\omega t)$ time dependence and write (7.72) in polar coordinates, we obtain

$$\frac{1}{r}\frac{\partial^2}{\partial r^2}(r\psi)+\frac{1}{r^2\sin^2\theta}\frac{\partial}{\partial\theta}\left(\sin\theta\frac{\partial\psi}{\partial\theta}\right)+\frac{1}{r^2\sin^2\theta}\frac{\partial^2\psi}{\partial\theta^2}+\frac{n^2\omega^2}{c^2}\psi=0. \tag{7.73}$$

In order to simplify matters, we shall assume that by far the largest contribution to the partial derivatives of ψ comes from its variation with wavelength in the r direction. Thus we drop all $\partial/\partial\theta$ and $\partial/\partial\phi$ terms in (7.73). This of course does not mean that ψ is independent of θ and ϕ but only that the partial derivatives with respect to these variables in (7.73) can be neglected. In addition we set

$$F(r,\theta,\phi)=r\psi(r,\theta,\phi), \tag{7.74}$$

where θ and ϕ are now considered as parameters. We also write

$$\omega=k_\text{v}c, \tag{7.75}$$

where k_v is the vacuum wave number. With these assumptions, (7.73) becomes

$$\frac{d^2F}{dr^2}+n^2k_\text{v}^2F=0. \tag{7.76}$$

When n is constant, (7.76) has the familiar solutions $F=\exp(\pm inkr)$. When n is not constant we may obtain an approximate solution by use of the Wentzel-Kramers-Brillouin (WKB) approximation [7.8]. We begin with the substitution

$$F(r)=\exp\left[g(r)\right]. \tag{7.77}$$

Inserting this expression into (7.76) we obtain

$$g''(r)+[g'(r)]^2+k_\text{v}^2n^2(r)=0. \tag{7.78}$$

For a first-order solution we may neglect $g''(r)$ to obtain the solution

$$g_1(r)=\pm ik_\text{v}\int n(r)dr. \tag{7.79}$$

The second-order solution may be obtained by differentiating (7.79) twice with respect to r, inserting the result in (7.78) in place of $g''(r)$, and solving for the new $g'(r)$. Thus we obtain

$$[g_2'(r)]^2=-n^2(r)k_\text{v}\left(1\pm\frac{i}{n^2k_\text{v}}\frac{dn}{dr}\right). \tag{7.80}$$

Taking the square root of this expression and remembering that the second term in the brackets is small because n is slowly varying on the scale of a wavelength, we find

$$g_2'(r) = \pm ink_v\left(1 \pm \frac{i}{2n^2k_v}\frac{dn}{dr}\right). \tag{7.81}$$

Integrating this expression and substituting the result into (7.74) and (7.77) we find

$$\psi(r, \theta, \phi) = \frac{\exp\left[\pm ik_v \int n(r, \theta, \phi)dr\right]}{\sqrt{nr}}. \tag{7.82}$$

Equation (7.82) tells us that the phase of the disturbance at any point is given simply by the integral of the rate of change of phase over the path. The amplitude is also modified in second order by the factor $\sqrt{n}$. It turns out that with this modification, we obtain the same power flow through successive spherical surfaces independent of $n(r)$. Since the variation of n is generally small however, we will neglect this factor in the following discussion.

Having obtained an expression for the nearly spherical disturbance from a point source in a slightly non uniform medium, we proceed now to develop an expression for scalar diffraction in a slightly non uniform medium. Our treatment parallels those given for Kirchoff diffraction in a uniform medium by *Born* and *Wolf* [Ref. 7.6, pp. 375–382] and by *Baker* and *Copson* [7.9].

We begin with Green's identity

$$\int_v (U\nabla^2 V - V\nabla^2 U)dv = -\int_s\left(U\frac{\partial V}{\partial t} - V\frac{\partial U}{\partial t}\right)ds, \tag{7.83}$$

where s is any closed surface bounding the volume v, U, and V are arbitrary functions with continuous first and second partial derivatives within and on the surface s, and t is a normal to the surface directed into the volume v. Less stringent conditions on one of these functions can be obtained [Ref. 7.9, p. 23] and are generally assumed in the applications. If both U and V satisfy the expression

$$\nabla^2\psi + n^2k_v^2\psi = 0, \tag{7.84}$$

then the integrand on the left side of (7.83) vanishes.

Let us now identify the scalar function U with one polarization of an electric field incident on the surface and whose disturbance at an interior observation point P we wish to calculate. For V, we insert the function ψ defined by (7.82). Because this function is singular at P, we must exclude the point P from the volume. We do this by adding a small spherical inner surface of radius ε surrounding P. The right side of (7.83) then goes over into two surface integrals as follows:

$$\int_s + \int_p\left(U\frac{\partial V}{\partial t} - V\frac{\partial U}{\partial t}\right)ds = 0. \tag{7.85}$$

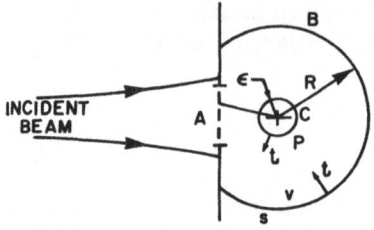

Fig. 7.2. Calculation of disturbance at P due to illuminated aperture A

Let us consider the case where the surface s contains a planar aperture illuminated normally by the incident radiation and a spherical segment with radius R centered on P. These elements are designated by A and B in Fig. 7.2. Finally we designate by C the small sphere surrounding our observation point. We wish to evaluate the surface integrals in (7.85).

Let U_A denote the amplitude in the aperture. For a normally incident beam, the major contribution to $\partial U/\partial t$ will come from the rapid phase variation in the direction normal to the aperture. Thus we may write

$$\frac{\partial U_A}{\partial t} = ikU_A. \tag{7.86}$$

Similarly, if the observation point P is not far off-axis behind the aperture we may write

$$V_A = \frac{\exp(ik\int ndr)}{r} \tag{7.87}$$

and

$$\frac{\partial V_A}{\partial t} = -iknV_A. \tag{7.88}$$

Combining these results we obtain for I_A, the surface integral over the aperture,

$$I_A = -2ik\int_A U_A \frac{\exp(ik\int ndr)}{r} dA. \tag{7.89}$$

Let us consider next the surface integral over the small sphere surrounding P. In the limit of small radius ($\varepsilon \ll 1/k$), the only important contributor comes from the first term in (7.85),

$$\frac{\partial V_C}{\partial t} = -\frac{\exp(ik\int ndr)}{r^2}. \tag{7.90}$$

As the radius of the sphere proceeds to zero, we may take U_P and $\partial V_C/\partial t$ outside the integral sign in (7.85) and write $4\pi r^2$ for the integral over the surface area.

Thus we obtain in the limit

$$I_C = -4\pi U_P. \tag{7.91}$$

Finally, we must consider the integral over the spherical element B. We want to argue that this integral will vanish as R goes to infinity. As R becomes very large we may write in (7.85)

$$\left(\frac{\partial}{\partial t}\right)_{r=R} = -\left(\frac{\partial}{\partial r}\right)_{r=R}. \tag{7.92}$$

Furthermore, since the surface area becomes infinite as R goes to infinity, it is useful to write $ds = r^2 d\Omega$ where $d\Omega$ is an element of solid angle. Thus the integral over the spherical surface element B becomes

$$I_B = \int_B d\Omega r \left[-ikn(r)U_B + \frac{\partial U_B}{\partial r} \right] \exp(ik \int n dr). \tag{7.93}$$

This integral will vanish if

$$\lim_{r \to \infty} r \left[\frac{\partial U_B}{\partial r} - ikn(r)U_B \right] = 0. \tag{7.94}$$

This condition might be termed a modified radiation condition. It reduces to the Sommerfeld radiation condition [Ref. 7.9, p. 25] when $n(r)$ is constant. It is not difficult to see how this condition can be satisfied. For example, as r recedes to infinity we may expect U_B to appear more and more like a disturbance originating in a point source. We will show below that this leads to an expression of the form

$$U_B \simeq A \exp(i\phi) \frac{\exp(ik \int n dr)}{r}, \tag{7.95}$$

where A is the source strength and ϕ takes account of the phase difference introduced by measuring r from the point P rather than from the aperture. When this expression is substituted into (7.94) we see that the modified radiation condition is satisfied.

If we assume that the modified radiation condition can be satisfied, we obtain by combining (7.85, 89, 91, 94) the expression

$$U_P = -\frac{ik}{2\pi} \int_A U_A \frac{\exp(ik \int n dr)}{r} dA. \tag{7.96}$$

Comparing (7.58) and (7.96) we see that Huygens' principle still holds in a *medium with small variations* in the refractive index, provided that the corresponding phase variations between the aperture and the observation point are included in the superposition. This result is sometimes referred to as the pasted phase approximation. Similar results were obtained by *Kravtsov* and *Feizulin* [7.10] and also by *Lutomirski* and *Yura* [7.11].

It is convenient to expand the exponential term in (7.96) in the same way as in the preceding subsection. Proceeding along the same lines we obtain the result

$$U_P = -\frac{ik}{2\pi}\frac{\exp(ikR_0)}{R_0}$$

$$\cdot \int_A U_A \exp\left[ik\left(\frac{\xi^2}{2R_0} - \frac{x\xi}{R_0} + \frac{\eta^2}{2R_0} - \frac{y\eta}{R_0} + \int_A^P \delta n dl\right)\right] dA, \tag{7.97}$$

where δn is the difference between the local refractive index and its quiescent value, and the integral inside the exponential is to be taken along a rectilinear path between the contributing element of area in the aperture and the observation point P.

If both the refractive index disturbance and the source distribution are one dimensional, then we may integrate (7.97) over the other variables as we did earlier. We obtain after some manipulation

$$U(x) = \frac{(1-i)}{2}\sqrt{\frac{k}{\pi R_0}}\exp(ikR_0)$$

$$\cdot \int d\xi U(\xi)\exp\left[ik\left(-\frac{x\xi}{R_0} + \frac{\xi^2}{2R_0} + \int_\xi^x \delta n dl\right)\right], \tag{7.98}$$

where now the line integral of δn is confined to the $\xi - z - x$ plane illustrated in Fig. 7.1.

7.3 Fluid Mechanics

The primary objective of the fluid mechanical analysis is to calculate those fluid properties which may contribute to changes in the refractive index and therefore the propagation character of the medium. In general we may expect the refractive index to depend upon variations in temperature, pressure, density, and possibly such variables as the electric field. For air, over a wide variation in the values of pressure, temperature, and electric field strength, we find that the refractive index depends only on the density through the Gladstone-Dale relation

$$(n-1) = k\varrho, \tag{7.99}$$

where k is the Gladstone-Dale constant. For air, it has a numerical value such that $(n-1) \simeq 3 \times 10^{-4}$ under standard conditions. Because of (7.99), the following discussion will be concerned primarily with the density variations in an ideal gas which are produced when the gas is heated by a transiting laser beam.

7.3.1 Partial Differential Equation for the Density Variations in an Ideal Gas

Stationary Medium

We present here a heuristic derivation of the partial differential equation which describes the density variation in an ideal gas. The complete derivation is somewhat lengthy and would take us farther afield than is necessary at this point. A full discussion must include the fluid mechanical concepts of mass, energy, and momentum conservation. These concepts are treated by *Landau* and *Lifshitz* [7.12] and were applied to the problem of thermal blooming by *Wallace* and *Camac* [7.13]. In addition, *Hayes* [7.14] presents a fairly detailed discussion of a number of assumptions connected with the transient early time regime.

The heuristic derivation presented here complements the treatments described above. We use it because it is shorter and also because it permits more direct insight into the basic mechanisms than is possible with the more rigorous treatment. We begin by recognizing that we shall be concerned with very small disturbances and small velocities in the medium. Let us write the density in terms of the condensation $s(r, t)$ as follows:

$$\varrho = \varrho_0[1 + s(r, t)], \tag{7.100}$$

where ϱ_0 is the undisturbed density, which is assumed constant. We know that in the absence of heating by a laser beam the density variations must satisfy an acoustic wave equation such that

$$\left(\nabla^2 - \frac{1}{c_s^2} \frac{\partial^2}{\partial t^2}\right) s(r, t) = 0. \tag{7.101}$$

Now we can take the $\partial/\partial t$ of this equation and obtain

$$\frac{\partial}{\partial t}\left(\nabla^2 - \frac{1}{c_s^2} \frac{\partial^2}{\partial t^2}\right) s(r, t) = 0. \tag{7.102}$$

It turns out that the left side of (7.102) is in the correct form for the problem at hand. It remains only for us to develop the proper form for the inhomogeneous source term representing the driving forces. To develop this term, we consider the case of slow heating at constant pressure and examine the resulting density changes. By slow we mean at a rate such that acoustic waves introduced by the $(1/c_s^2)\partial^2/\partial t^2$ term in (7.102) can be neglected. After the development is completed we shall retain the $(1/c_s^2)\partial^2/\partial t^2$ term in the resulting equation. It turns out that the final result will correctly describe density variations in both the slow and fast heating regimes.

In carrying out our derivation it is convenient to use the heating rate per unit volume $h_v(r, t)$ given by

$$h_v(r, t) = \partial q_v/\partial t,$$ (7.103)

where $q_v(r, t)$ is the heat energy deposited per unit volume and is given by (7.10). By the same procedure used to obtain (7.12) we find

$$\frac{\partial}{\partial t} s(r, t) = -\frac{(\gamma - 1)}{\gamma p_0} h_v(r, t).$$ (7.104)

If we compare (7.102) and (7.104), assume that $s(r, t)$ and h_v are differentiable, and neglect the acoustic term in (7.102), we arrive at the simplest formulation of the complete problem

$$\frac{\partial}{\partial t} \left(\nabla^2 - \frac{1}{c_s^2} \frac{\partial^2}{\partial t^2} \right) s(r, t) = -\frac{(\gamma - 1)}{\gamma p_0} \nabla^2 h_v(r, t).$$ (7.105)

Equation (7.105) is in fact the correct formulation of the problem at hand.

Ordinarily the heating rate per unit volume is given in terms of the absorption of energy from the laser beam

$$h_v(r, t) = \alpha I(r, t),$$ (7.106)

where α is the power absorption per unit length and I is the beam intensity $(\mathrm{W\,m^{-2}})$. There are special cases however in which the absorbed energy is temporarily stored in excited states before being transformed to kinetic energy of the molecules. In this case (7.106) must be modified. This case, which may be important for short pulses, will be discussed later.

Uniformly Flowing Medium

We consider now the modifications to (7.105) which are required for the case of a laser beam propagating through a medium moving with a constant velocity. It is clear that some modification is required because (7.9) contains the assumption that the density changes at each point are the exclusive result of the total thermal energy deposited at that point and are not influenced by the influx of gas heated at other locations. Of course a time variation in the heating at any one point is permitted in (7.105).

Let us consider the specific case of a laser beam traveling in the z direction through a uniformly flowing gas having constant velocity v_0 in the x direction. Equation (7.105) is still valid in a coordinate system fixed in the uniformly moving gas. We introduce a new cartesian coordinate system u, v, w in which the laser beam is stationary through the transformations

$$u = x + v_0 t, \quad v = y, \quad w = z.$$ (7.107)

Let us construct a new function $S(u, v, w, t)$ through the relation

$$S(u, v, w, t) \equiv s(u - v_0 t, v, w, t). \tag{7.108}$$

Now it is easily shown that

$$\frac{\partial}{\partial t} [s(x, y, z, t)]_{x, y, z} = \left(\frac{\partial S}{\partial t}\right)_{u, v, w} + v_0 \left(\frac{\partial S}{\partial u}\right)_{v, w, t}. \tag{7.109}$$

Thus to write (7.105) in the reference frame in which the laser beam is stationary, we need only replace $s(x, y, z, t)$ by $S(u, v, w, t)$ and $\partial/\partial t$ by the expression

$$\frac{\partial}{\partial t} \rightarrow \frac{\partial}{\partial t} + v_0 \frac{\partial}{\partial u}. \tag{7.110}$$

If we now relabel the new variables S, u, v, w using the old variable names, s, x, y, z, we can write for the condensation in a moving medium

$$\left(\frac{\partial}{\partial t} + v_0 \frac{\partial}{\partial x}\right)\left[\nabla^2 - \frac{1}{c_s^2}\left(\frac{\partial}{\partial t} + v_0 \frac{\partial}{\partial x}\right)^2\right] s(x, y, z, t)$$

$$= -\frac{(\gamma - 1)}{\gamma p_0} \nabla^2 h_v(x, y, z, t), \tag{7.111}$$

where now x, y, and z are fixed with respect to the laser beam. As v_0 goes to zero this equation reduces to (7.105) as we expect.

7.3.2 Methods of Solution for the Density

In this subsection we summarize briefly our assumptions and methods of approach to the solution of (7.105) and (7.111). The basic assumption is that the spatial dependence of the heating profile does not vary with time as the heating proceeds. Of course we may expect such changes to develop since the refractive index varies throughout the medium. However a self-consistent treatment of the complete problem appears intractable at the present time except by computer methods. Although our basic assumption may appear unrealistic, it does permit insight into many of the phenomena of thermal blooming. It also permits analytical calculations whose predictions are very close to the computer results in a number of cases.

There are three principal methods employed in the solution of the fluid mechanical problems considered here. The first is a Taylor expansion in time with the coordinate variables regarded as parameters in the expansion. The second method is the use of an integral transform technique appropriate to the geometry of the problem. For one-dimensional problems in cartesian coordinates the Fourier transform is used, while for problems with cylindrical symmetry the Hankel transform is most appropriate. The properties of these integral transforms which are of importance to our discussion are summarized in the Appendix. The third method is the use of a Green's function technique based on a solution for a line heating source.

The Taylor expansion method is most suited to steady heating profiles turned on at, say, $t=0$. When it is valid, it is the easiest method to apply. It has the major disadvantage that a Taylor expansion around $t=0$ may not exist unless the heating profile and its derivatives at all points are continuous up to and including second (and sometimes higher) order. The Taylor expansion method also has the disadvantage that it is difficult to apply throughout the full time development of the density changes.

The integral transform methods, on the other hand, are much more tolerant of singularities in the heating profile and will yield correct results for all heating profiles of practical interest. The disadvantage with transform methods is that certain of the definite integrals cannot be evaluated in terms of simple functions for all values of the parameters, even for relatively simple heating profiles.

The Green's function technique complements the integral transform method but may encounter similar difficulties with the evaluation of definite integrals.

Taylor Expansion for Early Time Behavior

We consider the case of an initially quiescent medium in which $s=0$ everywhere for $t<0$ and the laser beam $I(r)$ is turned on at $t=0$. We shall examine the early time behavior by constructing a Taylor expansion valid for $t>0$. Our expression will be valid for arbitrary flow velocity. We also assume that all energy absorbed is deposited immediately as kinetic energy so that (7.106) holds.

We shall use the differential equations (7.105) and (7.111) to construct the time derivatives required for the Taylor expansion and assume that the heating profile is sufficiently well behaved so that $V^2 I$ is bounded everywhere. We wish to show that the t^3 term is the lowest order nonvanishing term in the Taylor expansion. If any of the lower order derivatives were non zero at $t=0+$, then the third-order time derivative appearing in either (7.105) or (7.111) would behave like a delta function and the differential equations would not be satisfied. To show this result analytically let us consider the integration of (7.105) with respect to time over a small interval from $t=0-$ to $t=\tau>0$. We obtain

$$\left(\frac{\partial^2 s}{\partial t^2}\right)_{t=\tau} - c_s^2 V^2(s)_{t=\tau} = \text{terms of order } \tau \tag{7.112}$$

which can be regarded as vanishing in the limit $\tau \to 0$. By repeating the integration of (7.112) across the small interval bracketing $t=0$ we easily establish that

$$s|_{t=0+} = \frac{\partial s}{\partial t}\Big|_{t=0+} = \frac{\partial^2 s}{\partial t^2}\Big|_{t=0+} = 0 \tag{7.113}$$

everywhere, with the possible exception of those points where $\nabla^2 I$ is singular.

The procedure described above yields the same results on (7.111) so that the initial conditions given by (7.113) hold also for heating in a flowing medium.

Using (7.113) we find from either (7.105) or (7.111)

$$\frac{\partial^3 s}{\partial t^3}\Big|_{t=0+} = \frac{\gamma-1}{\gamma p_0} c_s^2 \alpha \nabla^2 I(r). \tag{7.114}$$

The early time behavior given by the lowest order term in the Taylor expansion is then simply

$$s(r,t) = \frac{t^3}{3!} \frac{\gamma-1}{\gamma p_0} c_s^2 \alpha \nabla^2 I. \tag{7.115}$$

The next term in the Taylor expansion is easily computed. In the flowing medium we shall find a fourth-order term. Let us consider a stationary medium for which the next term is of fifth order. If $I(r)$ is sufficiently well behaved we may differentiate (7.105) once again with respect to time. Evaluating the result at $t=0+$ we find

$$\left(\nabla^2 \frac{\partial^2 s}{\partial t^2}\right)_{t=0+} - \frac{1}{c_s^2}\left(\frac{\partial^4 s}{\partial t^4}\right)_{t=0+} = 0. \tag{7.116}$$

Because of (7.113), $(\partial^4 s/\partial t^4)_{t=0+}$ vanishes. If we differentiate (7.105) once again with respect to time and use (7.114), we find

$$\left(\frac{\partial^5 s}{\partial t^5}\right)_{t=0+} = \frac{\gamma-1}{\gamma p_0} c_s^4 \alpha \nabla^4 I. \tag{7.117}$$

This process may be repeated indefinitely to obtain higher order terms. Thus, for well-behaved $I(r)$, the Taylor series for the condensation can be written for the non flowing medium as

$$s(r,t) = \frac{\gamma-1}{\gamma p_0} \alpha \sum_{n=1}^{\infty} \frac{t^{2n+1}}{(2n+1)!} c_s^{2n} \nabla^{2n} I(r). \tag{7.118}$$

It is instructive to compare the relative magnitudes of the first two terms in (7.118). If we denote these terms by S_5 and S_3 we obtain

$$\frac{S_5}{S_3} = \frac{t^2}{20} c_s^2 \frac{\nabla^4 I}{\nabla^2 I}. \tag{7.119}$$

We want to make an order-of-magnitude estimate of this ratio. If the intensity distribution is that corresponding to a reasonably well-collimated beam having a peak intensity I_0 and a characteristic transverse dimension a, then we can expect that

$$\nabla^{2n} I(r) \simeq \frac{I_0}{a^{2n}} \tag{7.120}$$

so that

$$\frac{S_5}{S_3} \simeq \frac{c_s^2 t^2}{a^2} = \frac{t^2}{t_0^2}, \tag{7.121}$$

where t_0 is the acoustic transit time of the beam. Thus for $t \ll t_0$, the higher order terms in (7.118) can be neglected and (7.115) should give a good approximation for $s(r, t)$.

Late Time Behavior

As we shall show presently, late time refers to times which are greater than several acoustic transit times for steady heating. In Sections 7.1.1 and 7.3.1 we used classical thermodynamic considerations in the late time regime to develop the basic differential equation for density disturbances. Thus for no-flow steady heating at late times we find

$$s(r, t) = -\frac{\gamma - 1}{\gamma p_0} \alpha I(r) t, \tag{7.122}$$

which shows a steadily increasing negative value of the condensation. The expression remains valid so long as s remains small compared to unity or until some other phenomenon such as convective cooling sets in.

Of course we expect that (7.118) would go over into (7.122) at late times so long as the Taylor expansion remained valid.

Uniform Transverse Flow at Late Times

In the presence of flow we must consider (7.111). Physically, we expect to find the gas heated as it passes through our laser beam. To examine the density changes after initial transients have died out we set $\partial/\partial t = 0$. We also assume that our beam is propagating in the z direction with variations in z which are slow enough so that they can be neglected. Making use of (7.106), (7.111) can be written

$$v_0 \frac{\partial}{\partial x} \left[(1 - M^2) \frac{\partial^2}{\partial x^2} + \frac{\partial^2}{\partial y^2} \right] s(x, y) = -\frac{\gamma - 1}{\gamma p_0} \alpha \nabla^2 I(x, y), \tag{7.123}$$

where M is the free stream Mach number given by

$$M = v_0/c_s. \tag{7.124}$$

The character of the solutions to (7.123) are radically different for $M > 1$ or $M < 1$. We shall consider only the case of low velocity flow so that we can set $M = 0$ in (7.123). It should be noted that this does not mean that we have returned to the case of no flow. Equation (7.123) still contains the flow velocity, and setting $M^2 = 0$ amounts only to dropping a second-order correction. Under these conditions (7.123) can be written

$$v_0 \frac{\partial}{\partial x} \nabla^2 s(x, y) = -\frac{\gamma - 1}{\gamma p_0} \alpha \nabla^2 I(x, y). \tag{7.125}$$

For a localized beam profile and an initial requirement that $s = 0$ before the flow enters the beam, we can drop the ∇^2 operators on both sides of (7.125). We obtain the condensation by direct integration

$$s(x, y) = -\frac{(\gamma - 1)}{\gamma p_0 v_0} \alpha \int_{-\infty}^{x} I(x', y) dx'. \tag{7.126}$$

Physically this result is identical to (7.122) if in that relation we replace t by dt, set $dt = dx/v_0$, and carry out an integration on x to sum the heating effects on a parcel of air as it transits the beam profile.

7.3.3 Formal Solution for a One-Dimensional Heating Profile

In this subsection, we obtain the formal solution for two one-dimensional problems by Fourier transform techniques. By formal solution we mean that the solution is reduced to the evaluation of a definite integral. In Section 7.3.5 the integrals will be evaluated and specific results presented for several cases of interest. The two heating profiles which we consider are a steady heating profile turned on at $t = 0$ and a heating profile with an arbitrary multiplicative time dependence also turned on at $t = 0$.

Steady Heating Turned on at $t = 0$

Let us consider the case of a slab beam propagating in the z direction but with no variation in either the y or z direction. We shall assume instantaneous heating and no flow so that (7.105) reduces to

$$\left(\frac{\partial^2}{\partial x^2} - \frac{1}{c_s^2} \frac{\partial^2}{\partial t^2} \right) \psi(x, t) = -\frac{(\gamma - 1)}{\gamma p_0} \alpha \frac{\partial^2}{\partial x^2} I(x, t), \tag{7.127}$$

where we have set

$$\frac{\partial s(x, t)}{\partial t} = \psi(x, t).$$
(7.128)

We wish to consider the solution of these equations in the transition region between the short time and long time regimes. It will be useful to begin with the case for which $I(x, t)$ has the form

$$I(x, t) = K(x)u(t),$$
(7.129)

where $K(x)$ describes the spatial dependence and $u(t)$ is the unit step function defined by the relation

$$u(t) = \begin{cases} 0 & t < 0 \\ 1 & t > 0. \end{cases}$$
(7.130)

We begin our solution of (7.127) by noting that any arbitrary function of the variable $\zeta = x - c_s t$ will satisfy the homogeneous part of the equation. This is also true for any function of the variable $\eta = x + c_s t$. In addition we note that the entire equation has the particular solution

$$\psi(x) = -\frac{(\gamma - 1)}{\gamma p_0} \alpha K(x), \quad t > 0.$$
(7.131)

Thus a complete solution for (7.127) can be written

$$\psi(x, t) = f(x - c_s t) + g(x + c_s t) - \frac{(\gamma - 1)}{\gamma p_0} \alpha K(x), \quad t > 0.$$
(7.132)

It remains only to determine the form of the functions f and g. This can be done most easily from the initial conditions which are obtained from (7.109), i.e.,

$$\psi(x, t)\bigg|_{t = 0+} = \frac{\partial}{\partial t} \psi(x, t)\bigg|_{t = 0+} = 0.$$
(7.133)

From (7.132) we obtain

$$f(x) + g(x) - \frac{(\gamma - 1)}{\gamma p_0} \alpha K(x) = 0$$
(7.134)

and

$$-f'(x) + g'(x) = 0,$$
(7.135)

both of these relations holding for all values of x. From (7.135) we find the simplest solution is $f(x)=g(x)$ which when substituted into (7.134) gives the relation

$$f(x)=g(x)=\frac{(\gamma-1)}{2\gamma p_0}\alpha K(x).\tag{7.136}$$

Thus the complete solution which satisfies the initial condition can be written

$$\psi(x,t)=\frac{(\gamma-1)}{2\gamma p_0}\alpha[K(x-c_s t)+K(x+c_s t)-2K(x)].\tag{7.137}$$

Finally the condensation is obtained by direct integration

$$s(x,t)=\frac{\gamma-1}{2\gamma p_0}\alpha\left[-2K(x)t+\int_0^t K(x-c_s t')dt'\right.$$
$$\left.+\int_0^t K(x+c_s t')dt'\right].\tag{7.138}$$

The form (7.137) is easy to interpret. The three terms on the right side consist of a pair of outgoing pulses superposed on the late time heating profile which was discussed earlier. The pulses have the same shape as the heating profile and travel outward on the positive and negative x axes at the speed of sound. After several acoustic transit times in the heated region they are well out of the way and their contribution to ψ fades away. Their contribution to further changes in the condensation also dies away but there is a small residual which is of the order of the change produced during an acoustic transit time.

Heating with a Multiplicative Time Dependence

Let us turn now to the case where the intensity has a time dependence of the form

$$I(x,t)=K(x)T(t),\tag{7.139}$$

so that (7.127) takes the explicit form

$$\left(\frac{\partial^2}{\partial x^2}-\frac{1}{c_s^2}\frac{\partial^2}{\partial t^2}\right)\psi(x,t)=-\frac{\gamma-1}{\gamma p_0}\alpha T(t)\frac{d^2}{dx^2}K(x).\tag{7.140}$$

If we take the Fourier transform (cf. Appendix) on the x coordinate of the entire (7.140), carrying the time along as a parameter, we obtain

$$(-i\lambda)^2\bar{\psi}(\lambda,t)-\frac{1}{c_s^2}\frac{d^2\bar{\psi}}{dt^2}=-\frac{(\gamma-1)}{\gamma p_0}\alpha(i\lambda)^2\bar{K}(\lambda)T(t),\tag{7.141}$$

which we treat as an ordinary differential equation with λ as a parameter and t as the independent variable. We shall obtain the solution of this differential equation for $\bar{\psi}(\lambda, t)$ and then apply the inversion theorem to obtain $\psi(x, t)$.

If we take as our initial conditions

$$\psi(x, t)\Big|_{t=0+} = \frac{\partial \psi}{\partial t}\Big|_{t=0+} = 0, \tag{7.142}$$

the same relations hold for the Fourier transforms of these quantities. Thus we require

$$[\bar{\psi}(\lambda, t)]_{t=0+} = \frac{d\bar{\psi}}{dt}\Big|_{t=0+} = 0. \tag{7.143}$$

The solution of (7.141) subject to these conditions is easily found by the method of variation of parameters [7.15]. The result is

$$\bar{\psi}(\lambda, t) = \frac{\gamma - 1}{\gamma p_0} \alpha c_s \lambda \bar{K}(\lambda) \left[\cos(c_s \lambda t) \int_0^t T(t') \sin(c_s \lambda t') dt' \right. $$
$$\left. - \sin(c_s \lambda t) \int_0^t T(t') \cos(c_s \lambda t') dt' \right]. \tag{7.144}$$

It will be helpful to simplify this expression by rewriting it in the form

$$\bar{\psi}(\lambda, t) = \frac{-(\gamma - 1)}{\gamma p_0} \alpha \bar{K}(\lambda) [M_1(\lambda, t) + M_2(\lambda, t)], \tag{7.145}$$

where M_1 and M_2 represent the product of $c_s \lambda$ with the individual terms appearing inside the brackets of (7.144). Thus

$$M_1 = -c_s \lambda \cos(c_s \lambda t) \int_0^t T(t') \sin(c_s \lambda t') dt'. \tag{7.146}$$

Integrating by parts we easily obtain

$$M_1 = -\cos(c_s \lambda t) \left\{ [T(0) - T(t) \cos(c_s \lambda t)] + \int_0^t \cos(c_s \lambda t') \frac{dT}{dt'} dt' \right\}, \tag{7.147}$$

where $T(0)$ is the value of the time-dependent term at $t = 0+$. If we integrate M_2 by parts and combine the results we find

$$M_1 + M_2 = -T(0) \cos(c_s \lambda t) + T(t)$$
$$- \int_0^t \cos[c_s \lambda(t - t')] \frac{dT}{dt'} dt'. \tag{7.148}$$

We have only to combine (7.145) and (7.148) and apply the inversion theorem to obtain $\psi(x, t)$. We illustrate the procedure by calculating one of the terms. For example,

$$L = \frac{1}{\sqrt{2\pi}} \int_{-\infty}^{\infty} d\lambda \, e^{-i\lambda x} \bar{K}(\lambda) \int_{0}^{t} \cos[c_s \lambda(t - t')] \frac{dT}{dt'} dt'. \tag{7.149}$$

If we interchange the order of integration and express the cosine by the relation

$$\cos z = (e^{iz} + e^{-iz})/2 \tag{7.150}$$

we obtain

$$L = \frac{1}{2} \int_{0}^{t} dt' \frac{dT}{dt'} \frac{1}{\sqrt{2\pi}} \int_{-\infty}^{\infty} d\lambda \bar{K}(\lambda)$$

$$\cdot (\exp\{-i\lambda[x - c_s(t - t')]\} + \exp\{-i\lambda[x + c_s(t - t')]\}), \tag{7.151}$$

which by the inversion theorem can be written

$$L = \frac{1}{2} \int_{0}^{t} dt' \frac{dT}{dt'} \{K[x - c_s(t - t')] + K[x + c_s(t - t')]\}. \tag{7.152}$$

Applying similar procedures to the other terms we easily obtain the result

$$\psi(x, t) = \frac{-(\gamma - 1)}{\gamma p_0} \alpha \left(T(t) K(x) - \frac{T(0)}{2} [K(x - c_s t) + K(x + c_s t)] \right.$$

$$\left. - \frac{1}{2} \int_{0}^{t} dt' \frac{dT}{dt'} \{K[x - c_s(t - t')] + K[x + c_s(t - t')]\} \right). \tag{7.153}$$

The three terms within the outermost parentheses in (7.153) have a direct physical interpretation. The first term corresponds to the results of slow heating with a variable time dependence. The second term is an acoustic pulse radiated by the initial turn-on transient. For steady heating the first two terms give a result identical to that obtained earlier in (7.137). The last term shows that dT/dt is a continuous source of outwardly traveling heating pulses. In fact the second term containing $T(0)$ is a special case of this more general term.

It is also possible to handle heating functions of the form $I(x + vt)$ by exactly the same kind of analysis we have just completed. In this way we can construct the solution for the full time development in a steady wind for the heating profile turned on at $t = 0$ with the solution carried out in a reference frame moving with the gas. It is a simple matter to transform the result to the reference frame in which the laser is stationary.

7.3.4 Formal Solutions for Heating in a Cylindrical Geometry

We consider now the case of heating in the absence of flow in a cylindrical geometry with coordinates r, θ, z where the beam propagates in the z direction with a radial profile independent of θ. Slow variations of the beam profile in the z direction are acceptable as long as they are small compared to the radial variations. If there are such variations, the resulting z dependence can be handled parametrically. We first apply the Hankel transform techniques to the solution of this problem. By formal solution we mean that we shall reduce the problem to the evaluation of a definite integral. Because of the character of wave solutions in cylindrical coordinates we shall not be able to produce simple results such as (7.137) for plane waves. Also we shall consider only the case of a uniform beam turned on at $t=0$.

Hankel Transform Solution for the Uniform Beam Turned on at $t=0$

We begin by writing (7.105) in cylindrical coordinates. We obtain

$$\left(\frac{\partial^2}{\partial r^2} + \frac{1}{r}\frac{\partial}{\partial r} - \frac{1}{c_s^2}\frac{\partial^2}{\partial t^2}\right)\psi(r,t) = -\frac{\gamma-1}{\gamma P_0}\alpha V^2 I(r), \tag{7.154}$$

where $\partial s/\partial t = \psi$. If we take the Hankel transform (cf. Appendix) of the entire equation we obtain

$$\left(\frac{d^2}{dt^2} + \lambda^2 c_s^2\right)\bar{\psi}(\lambda,t) = -\frac{\gamma-1}{\gamma P_0}\alpha c_s^2 \lambda^2 \bar{I}(\lambda), \tag{7.155}$$

where we have assumed that the operations of differentiation with respect to time and taking the Hankel transform on the radial coordinate can be interchanged. Equation (7.155) is now an ordinary differential equation whose general solution is

$$\bar{\psi}(\lambda,t) = A\cos(\lambda c_s t) + B\sin(\lambda c_s t) - \frac{\gamma-1}{\gamma P_0}\alpha \bar{I}(\lambda), \tag{7.156}$$

where A and B are constants to be determined from the initial conditions. If we have an initially quiescent medium with a steady heating pulse turned on at $t=0$ then, as before, both ψ and $\partial\psi/\partial t$ and their transforms will vanish at $t=0+$. Applying these considerations to (7.156) we obtain the relation

$$\bar{\psi}(\lambda,t) = \frac{\gamma-1}{\gamma P_0}\alpha \bar{I}(\lambda)[\cos(c_s \lambda t) - 1]. \tag{7.157}$$

Applying the inversion theorem (cf. Appendix) to this expression we obtain

$$\psi(r,t) = -\frac{\gamma-1}{\gamma P_0}\alpha\left[I(r) + \int_0^\infty \lambda\bar{I}(\lambda)\cos(c_s \lambda t)J_0(\lambda r)d\lambda\right], \tag{7.158}$$

which can be integrated with respect to time to obtain the condensation

$$s(r, t) = -\frac{\gamma - 1}{\gamma p_0} \alpha \left[I(r)t - \frac{1}{c_s} \int_0^\infty \bar{I}(\lambda) \sin(c_s \lambda t) J_0(\lambda r) d\lambda \right]. \tag{7.159}$$

This expression reduces the solution of the partial differential equation (7.154) to the evaluation of two definite integrals: the first for $\bar{I}(\lambda)$ from the initial heating profile and the second for the integral appearing in (7.159). The first integral can be evaluated in a number of cases of interest. The second integral is generally intractable except for selected values of r and t.

Equation (7.159) is easily seen to have the features we require in $s(r, t)$. For example, $s(r, t)$ and its first and second partial derivatives with time vanish at $t = 0+$. Furthermore, the integral in (7.159) can be seen to vary inversely with t for large t because of the increasingly rapid oscillation of the quantity $\sin(c_s \lambda t)$ in the integrand. This can be explicitly demonstrated by repeated partial integrations so as to develop an expansion in powers of $1/t$ valid for late times.

Green's Function for the Uniform Beam Turned on at $t = 0$

Making use of (7.159) we now proceed to construct the solution for the condensation produced at a distance r from an infinite line source turned on at $t = 0$ in an initially quiescent medium. Let us begin with the normalized intensity distribution

$$I(r) = \frac{\exp(-r^2/a^2)}{\pi a^2}. \tag{7.160}$$

If we take the Hankel transform of (7.160) we obtain [7.16]

$$\bar{I}(\lambda) = \frac{\exp(-\lambda^2 a^2/4)}{2\pi}. \tag{7.161}$$

At this point we evaluate (7.161) in the limit $a = 0$ so that (7.160) will represent a normalized line source. Inserting the result into (7.159) we find that we require the following definite integral which has been evaluated by *Watson* [Ref. 7.16, p. 405]:

$$\int_0^\infty J_0(\lambda r) \sin(\lambda c_s t) d\lambda = \begin{cases} 0 & c_s t < r \\ \infty & c_s t = r \\ \dfrac{1}{\sqrt{c_s^2 t^2 - r^2}} c_s t > r. \end{cases} \tag{7.162}$$

Combining (7.159–162) we obtain the condensation for a normalized line source distribution. This solution is our required Green's function, $G(r, t)$. We find

$$G(r,t) = -\frac{\gamma-1}{\gamma p_0}\alpha t\delta(r) \quad \text{for} \quad c_s t < r \quad \text{and} \tag{7.163}$$

$$G(r,t) = -\frac{\gamma-1}{\gamma p_0}\alpha\left[t\delta(r) - \frac{1}{2\pi c_s\sqrt{c_s^2 t^2 - r^2}}\right] \quad \text{for} \quad c_s t \geq r, \tag{7.164}$$

where we have introduced the Dirac delta function in cylindrical coordinates, $\delta(r)$, to permit recovery of the $I(r)$ term inside the brackets of (7.159). We see that the Green's function consists of a delta function representing late time heating at the source together with an outgoing cylindrical wave which is singular at $r = c_s t$.

It should be noted in (7.163) and (7.164) that since the Green's function operates on the intensity and not on its Laplacian, it is strictly speaking not a Green's function for the operator appearing on the left side of (7.105) but rather for the operator consisting of the inverse Laplacian applied to the operator on the left side of (7.105).

For an extended source distribution we may obtain the condensation by superposition. If ϱ and θ are cylindrical polar coordinates with their origin at the observation point r_0 we may write

$$s(r_0,t) = \int\limits_0^{c_s t} d\varrho\varrho \int\limits_0^{2\pi} d\theta G(\varrho,t)I(\varrho,\theta) \tag{7.165}$$

where the upper limit on the ϱ integration need not extend beyond $c_s t$ because there is no contribution from the Green's function beyond that distance.

We may give a physical interpretation to (7.165) by saying that it represents an information-gathering cylinder moving out from each observation point at the speed of sound such that each point within the cylinder makes a contribution to the condensation at the observation point determined by the Green's function and that there is no contribution from the intensity at points outside the information-gathering cylinder.

7.3.5 Representative Solutions in a Slap Geometry

Based on the general solutions presented in Section 7.3.3 we now develop explicit analytical expressions for several problems of interest.

Parabolic Slab in a Transverse Flow

Let us imagine an intensity profile of the form

$$I(x) = I_0(1 - x^2/a^2) \quad |x| < a. \tag{7.166}$$

The intensity $I(x)$ vanishes for $|x| > a$. If we put (7.166) into (7.126) and carry out the integration we obtain

$$s(x) = -\frac{(\gamma - 1)\alpha}{\gamma p_0 v_0} I_0 \left(x - \frac{x^3}{3a^2} + \frac{2a}{3}\right). \tag{7.167}$$

We shall make use of this result later in estimating blooming effects with this heating profile.

Gaussian Beam Turned on at $t = 0$

If a gaussian heating profile

$$I(x) = I_0 \exp(-x^2/a^2) \tag{7.168}$$

is applied at $t = 0$ to an initially undisturbed medium with $s = 0$, we may use (7.138) directly. The integrals occurring therein can be evaluated in terms of error functions. We obtain

$$s(x, t) = \frac{(\gamma - 1)\alpha I_0}{2\gamma p_0} \left\{ -2t \exp(-x^2/a^2) \right.$$

$$\left. + \frac{a}{c_s} \frac{\sqrt{\pi}}{2} \left[\mathrm{erf}\left(\frac{x + c_s t}{a}\right) - \mathrm{erf}\left(\frac{x - c_s t}{a}\right) \right] \right\}. \tag{7.169}$$

For times much greater than an acoustic transit time a/c_s, the sum of the error functions becomes 2 and the condensation can be written

$$s(x, t) = \frac{(\gamma - 1)\alpha I_0}{\gamma p_0} \left[-t \exp(-x^2/a^2) + \frac{a}{c_s} \frac{\sqrt{\pi}}{2} \right], \tag{7.170}$$

so that the residual disturbance in this case is independent of x, of opposite sign to the steady state growth, and of order of the density change accumulated in one acoustic transit time.

Uniform Slab Turned on at $t = 0$

Consider an initially undisturbed medium with $s = 0$ subject to the heating profile

$$I(x, t) = \begin{cases} I_0 u(t) & |x| < a \\ 0 & |x| > a. \end{cases} \tag{7.171}$$

As a preliminary observation we note that for $t > 0$, $\nabla^2 I$ vanishes everywhere except at $x = \pm a$ where it is singular. Thus (7.115) will not yield a satisfactory solution for early times because the correct solution does not have a Taylor expansion. We can obtain the correct solution, however, from (7.138), a form which is essentially equivalent to a Fourier transform solution. When (7.171) is put into (7.138) the resulting integrals are readily evaluated. As an example, consider the integral representing the right-traveling wave in the region $x > 0$. We obtain

$$R(x,t) = \int_0^t I(x - c_s t') dt', \tag{7.172}$$

$$R(x,t) = \begin{cases} I_0 t & 0 \leq t \leq (x+a)/c_s \\ I_0(x+a)/c_s & t \geq (x+a)/c_s \end{cases} \tag{7.173}$$

when $x < a$. For $x > a$ we obtain

$$R(x,t) = \begin{cases} 0 & t \leq (x-a)/c_s \\ I_0[t - (x-a)/c_s] & (x-a)/c_s \leq t \leq (x+a)/c_s \\ 2I_0 a/c_s & t \geq (x+a)/c_s. \end{cases} \tag{7.174}$$

The other terms in (7.138) are readily evaluated to obtain the final results. They are most readily stated for $0 < x < a$, although exactly symmetrical results obtain for $-a < x < 0$. For $0 < x < a$ we find three regions and three different expressions for the solution $s(x, t)$ which we designate s_1, s_2, and s_3. The regions of validity of the solutions are as follows:

$$s_1(x,t): \quad 0 \leq t \leq (-x+a)/c_s$$
$$s_2(x,t): \quad (-x+a)/c_s \leq t \leq (x+a)/c_s \tag{7.175}$$
$$s_3(x,t): \quad (x+a)/c_s \leq t < \infty.$$

The boundaries $t = (-x+a)/c_s$ and $t = (x+a)/c_s$ correspond physically to the end points of the left-going and right-going heating pulses, respectively. The solutions for the three regions are given by the relations

$$s_1(x,t) = 0$$

$$s_2(x,t) = \frac{\gamma-1}{2\gamma p_0} \alpha I_0 \left(-t + \frac{a-x}{c_s}\right) \tag{7.176}$$

$$s_3(x,t) = \frac{\gamma-1}{\gamma p_0} \alpha I_0 \left(-t + \frac{a}{c_s}\right).$$

The location of these regions in $x - t$ space is shown in Fig. 7.3. As we might expect on physical grounds, these solutions are continuous at their common boundaries.

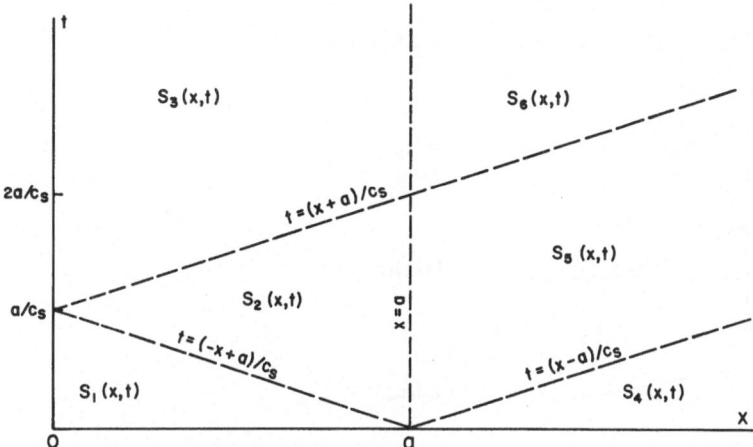

Fig. 7.3. Condensation as a function of x and t in various parts of $x-t$ space for one-dimensional heating

In the region $x > a$, only the right-traveling pulse contributes. We again find three regions and denote the solutions $s_4(x,t)$, $s_5(x,t)$, and $s_6(x,t)$ with the boundaries

$$s_4(x,t): \quad 0 \leq t \leq (x-a)/c_s$$

$$s_5(x,t): \quad (x-a)/c_s \leq t \leq (x+a)/c_s \tag{7.177}$$

$$s_6(x,t): \quad (x+a)/c_s \leq t < \infty.$$

These regions are also illustrated in Fig. 7.3. The condensation functions are readily found and given by the relations

$$s_4(x,t) = 0$$

$$s_5(x,t) = \frac{\gamma-1}{2\gamma p_0} \alpha I_0 \left(t - \frac{x-a}{c_s} \right) \tag{7.178}$$

$$s_6(x,t) = \frac{\gamma-1}{\gamma p_0} \alpha I_0 \frac{a}{c_s}.$$

Once again the functions s_4, s_5, and s_6 are continuous at their common boundaries.

For the case of the heating profile given by (7.171) we see that at $t=0$, all partial derivatives $(\partial^n s/\partial t^n)_{t=0}$ vanish everywhere except at $x=a$, which precludes the possibility of constructing a Taylor expansion for early times. We notice also that at the edge of the beam $x=a$ there is a discontinuity in $s(x,t)$

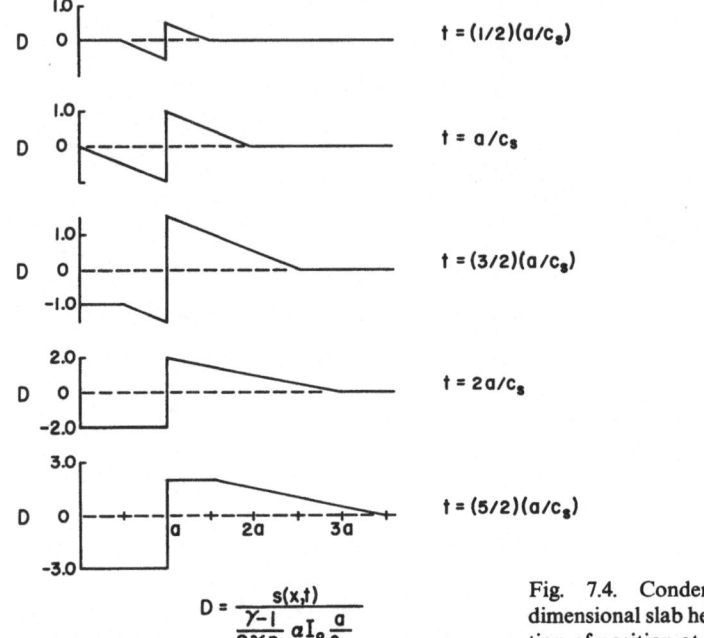

$$D = \dfrac{s(x,t)}{\dfrac{\gamma - 1}{2\gamma P_0}\, \alpha I_o\, \dfrac{a}{c_s}}$$

Fig. 7.4. Condensation for a one-dimensional slab heating profile as a function of position at various times

whose form is the same for all $t > 0$, namely

$$\Delta s(a, t) = s_2(a, t) - s_5(a, t) = s_3(a, t) - s_6(a, t) = -\frac{\gamma - 1}{\gamma P_0} \alpha I_0 t. \tag{7.179}$$

Figure 7.4 shows $s(x, t)$ as a function of x for various times.

Parabolic Slab Turned on at $t = 0$

The parabolic heating profile even in one dimension is a case of important practical interest. Consider the heating profile given by the relation

$$I(x, t) = \begin{cases} I_0(1 - x^2/a^2)u(t) & |x| < a \\ 0 & |x| > a. \end{cases} \tag{7.180}$$

The calculations are more tedious but they can be carried out along exactly the same lines as described above. We again find six regions in the $x - t$ plane, three within the heated region $|x| < a$ with the same boundaries as shown in Fig. 7.3. Let us introduce the symbol R defined by the expression

$$R = \frac{\gamma - 1}{\gamma P_0} \alpha I_0. \tag{7.181}$$

The expressions for the condensation in the various regions are given below. In the interval $0 \leq x \leq a$ we find

$$s_1(x, t) = - R \frac{c_s^2}{a^2} \frac{t^3}{3} \tag{7.182}$$

when $0 \leq t \leq (a - x)/c_s$;

$$s_2(x, t) = \frac{R}{2} \left[\frac{(a - x)^2 (2a + x)}{3a^2 c_s} - \left(1 - \frac{x^2}{a^2}\right) t \right.$$
$$\left. + \frac{x c_s}{a^2} t^2 - \frac{c_s^2}{a^2} \frac{t^3}{3} \right] \tag{7.183}$$

when $(a - x)/c_s \leq t \leq (a + x)/c_s$; and

$$s_3(x, t) = R \left[-\left(1 - \frac{x^2}{a^2}\right) t + \frac{2a}{3 c_s} \right] \tag{7.184}$$

when $t \geq (a + x)/c_s$. In the region $x \geq a$, we find

$$s_4(x, t) = 0 \tag{7.185}$$

for $0 \leq t \leq (x - a)/c_s$;

$$s_5(x, t) = - \frac{R}{6} \frac{c_s^2}{a^2} \left[\left(t - \frac{x - a}{c_s}\right)^3 - \frac{3a}{c_s} \left(t - \frac{x - a}{c_s}\right)^2 \right] \tag{7.186}$$

for $(x - a)/c_s \leq t \leq (x + a)/c_s$; and finally

$$s_6(x, t) = \frac{2}{3} R \frac{a}{c_s} \tag{7.187}$$

when $t \geq (x + a)/c_s$.

By examination of (7.182–187) we can see that $s(x, t)$ and $\partial s / \partial t$ are continuous at their common boundaries in the x, t plane. Physically the first condition arises from the fact that the heating profile (7.180) is continuous, and the second condition corresponds to the absence of impulsive forces when crossing the boundaries.

The point $x = a$ where the source terms in the partial differential equation for the condensation are singular deserves special attention. We find from either s_2 or s_5

$$s(a, t) = \frac{R c_s}{2a} \left(t^2 - \frac{c_s}{3a} t^3\right) \tag{7.188}$$

for the region $0 < t \leqq (x+a)/c_s$. At later times $t \geqq (x+a)/c_s$

$$s(a, t) = \frac{2}{3} Ra/c_s. \tag{7.189}$$

The t^2 term in (7.188) will introduce a singularity in $\partial^3 s/\partial t^3$ computed at $x = a$ for $t = 0+$, and this explains our difficulty in constructing a Taylor expansion for early times at the point $x = a$. On the other hand, we see that both $s_1(x, t)$ and $s_4(x, t)$ are consistent with the Taylor expansions generated by either (7.115) or (7.118), and the limits of validity of these expansions in time are the same as those for the validity of the expressions for $s_1(x, t)$ or $s_4(x, t)$.

We can get fairly simple physical picture of the region of validity of the t^3 term in the early time Taylor expansion for the parabolic one-dimensional heating profile. The expression for $\partial s/\partial t$, (7.138), contains three terms: a stationary image, and a left-traveling and right-traveling image of the heating profile. For times such that all three terms are different from zero, we have the early time t^3 dependence. When one of these terms is not present, a t^3 term will not describe the early time behavior. At $x = a$, for example, the left-traveling image of the heating profile departs as soon as the heating profile is turned on and thus the time interval for the validity of a t^3 dependence vanishes.

7.3.6 Representative Solutions in a Cylindrical Geometry

We consider now some representative solutions for specific cylindrical heating profiles.

Gaussian Beam-Early Time

Let us imagine a medium with $s = 0$ initially which is exposed to the gaussian heating profile

$$I = I_0 \exp(-r^2/a^2) \tag{7.190}$$

for all $t > 0$. The beam is considered to extend to $\pm \infty$ in the z direction and to have no z dependence. The Laplacian of (7.190) is given by

$$\nabla^2 I = \frac{4I_0}{a^2} \left(\frac{r^2}{a^2} - 1 \right) \exp(-r^2/a^2). \tag{7.191}$$

From (7.115) we see that the condensation at early times is simply

$$s(r, t) = \frac{2}{3} \frac{\gamma - 1}{\gamma p_0} \alpha I_0 \exp(-r^2/a^2) \frac{c_s^2 t^3}{a^2} \left(\frac{r^2}{a^2} - 1 \right). \tag{7.192}$$

From the consideration which entered the development of the differential equation for $s(r, t)$ we see that (7.192) is of the same order as the late time expression (7.122) multiplied by $(t/t_0)^2$ and may be of opposite sign. This is generally true for all well-behaved intensity profiles.

Gaussian Beam in a Transverse Flow

We consider now the case of a gaussian beam in a steady transverse flow at late times. Thus the initial transients have died out and (7.126) will describe the density variations. It is convenient to write the cylindrical gaussian beam in the form

$$I(x, y) = I_0 \exp[-(x^2 + y^2)/a^2].$$ (7.193)

Substituting this expression into (7.126) and carrying out the integration we obtain

$$s(x, y) = -\frac{\gamma - 1}{\gamma p_0 v_0} \alpha I_0 \exp(-y^2/a^2) \frac{a\sqrt{\pi}}{2} [1 + \text{erf}(x/a)],$$ (7.194)

where $\text{erf}(x/a)$ is the error function. From (7.194) we see that the y dependence of the intensity distribution goes over directly into the condensation profile. From (7.126) we see that this behavior is expected for any intensity distribution which is separable in x and y.

Gaussian Beam Turned on at $t = 0$

We turn now to a consideration of the Hankel transform solution for the gaussian heating profile given in (7.190). If we take the Hankel transform (cf. Appendix) of (7.190) we obtain [7.16]

$$\bar{I}(\lambda) = \frac{I_0 a^2}{2} \exp(-\lambda^2 a^2/4).$$ (7.195)

Inserting this expression into (7.159) we obtain

$$s(r, t) = -\frac{\gamma - 1}{\gamma p_0} \alpha \left[I(r)t - \frac{I_0 a^2}{2c_s} \right.$$
$$\left. \int_0^\infty \exp(-\lambda^2 a^2/4) \sin(c_s \lambda t) J_0(\lambda r) d\lambda \right].$$ (7.196)

This integral does not appear to be tractable in terms of simple functions for arbitrary r and t. However, on the axis $r = 0$ the integral can be expressed in terms of Dawson's integral $F(x)$ which is a tabulated function [7.17] defined by the expression

$$F(x) = \exp(-x^2) \int_0^x \exp(t^2) dt.$$ (7.197)

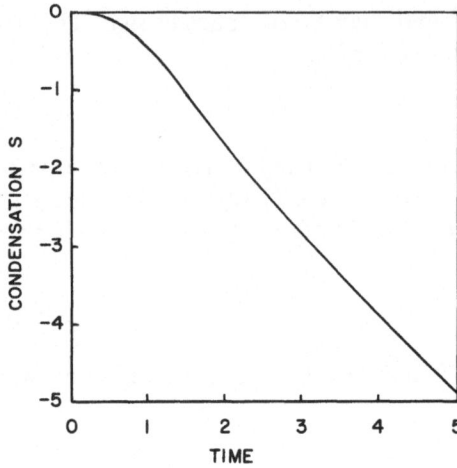

Fig. 7.5. Density variation on axis for a gaussian beam turned on at $t=0$ [from (7.203)]: abscissa is in units of the acoustic transit time; ordinate is in units of the late time condensation change per unit acoustic transit time

Making use of the relation [Ref. 7.17, p. 302]

$$\int_0^\infty \exp(-a^2\lambda^2/4)\sin(c_s t\lambda)d\lambda = \frac{2}{a}F(c_s t/a) \tag{7.198}$$

we find that on the axis

$$s(0,t) = -\frac{(\gamma-1)}{\gamma p_0}\alpha I_0 \frac{a}{c_s}\left[\frac{c_s t}{a} - F(c_s t/a)\right]. \tag{7.199}$$

This expression describes the circularly symmetric gaussian beam on-axis for all time regimes. It is of interest to consider the late time regime behavior of $F(c_s t/a)$. For large values of x

$$F(x) \to 1/(2x), \tag{7.200}$$

so that for late times

$$s(0,t) = -\frac{\gamma-1}{\gamma p_0}\alpha I_0\left[t - \frac{(a/c_s)^2}{2t}\right]. \tag{7.201}$$

We see that after only a few acoustic transit times the condensation corresponding to slow heating is achieved. By repeated partial integration of (7.196), it can be shown that the $1/t$ term in (7.201) is valid for all values of r. It is in fact the first term in an expansion of the integral in (7.196) in a series of inverse powers of t.

If we define

$$t_0 = a/c_s \tag{7.202}$$

as the acoustic transit time for the beam, we can write (7.199) in the form

$$s(0,t) = -s_0[(t/t_0) - F(t/t_0)],\tag{7.203}$$

where s_0 is the late time change in the condensation in one acoustic transit time. This variation is shown in Fig. 7.5.

Uniform Beam of Radius a Turned on at $t=0$

We examine next the intensity distribution given by

$$I(r,t) = \begin{cases} I_0 u(t) & 0 < r < a \\ 0 & r > a. \end{cases}\tag{7.204}$$

We note that $\nabla^2 I$ vanishes everywhere except at $r=a$ where it has a singular behavior. If we take the Hankel transform of this profile we obtain

$$\bar{I}(\lambda) = I_0 \int_0^a r J_0(\lambda r) dr = \frac{I_0 a}{\lambda} J_1(a\lambda).\tag{7.205}$$

To obtain the condensation we insert this result into (7.159). Once again the problem has been reduced to the evaluation of a single definite integral. The integrand in this case involves the product of two Bessel functions and a trigonometric function. Integrals of this type have been discussed by *Watson* [Ref. 7.16, Chap. XIII]. Their evaluation in the general case is quite difficult. We shall consider only the on-axis case $r=0$, for which we can obtain interesting results rather easily. At $r=0$, $J_0(0)=1$ and (7.159) involves the integral

$$L(t,a) = \int_0^\infty J_1(a\lambda) \sin(c_s t \lambda) \frac{d\lambda}{\lambda}.\tag{7.206}$$

From *Watson* [Ref. 7.16, p. 405] we find

$$L(t,a) = \begin{cases} c_s t/a & a > c_s t > 0 \\ 1 & a = c_s t \neq 0 \\ \dfrac{a}{c_s t + \sqrt{c_s^2 t^2 - a^2}} & 0 < a < c_s t. \end{cases}\tag{7.207}$$

Combining (7.159) and (7.207) we find that the condensation on-axis can be written

$$s(0,t) = 0 \quad c_s t < a\tag{7.208}$$

and

$$s(0,t) = -\frac{\gamma-1}{\gamma p_0} \alpha I_0 \left(t - \frac{a^2/c_s}{c_s t + \sqrt{c_s^2 t^2 - a^2}} \right) \quad c_s t \geq a.\tag{7.209}$$

If we examine $\partial s/\partial t$ we find that it has a negative infinity at $t=a/c_s$ and increases monotonically to the negative value corresponding to steady heating at late times. On physical grounds we expect that this behavior is due to the constructive interference at $r=0$ and $t=a/c_s$ of all the impulse disturbances launched from $r=a$ at $t=0$ by the singularities in the function $\nabla^2 I$. We should not expect to find this behavior at other points $r \neq 0$.

The Green's function method is an alternative approach to this problem. Let us examine the condensation at an arbitrary radius r_0 for times

$$t \leq t_0 = (a - r_0)/c_s, \tag{7.210}$$

where t_0 is the time at which the information-gathering cylinder centered on r_0 breaks through the boundary of the cylinder $r=a$. If we combine (7.204) with (7.163–165) we obtain

$$s(r_0, t) = -\frac{\gamma-1}{\gamma p_0} \alpha \left\{ I_0 t - \frac{I_0}{c_s} \int_0^{c_s t} \frac{\varrho\, d\varrho}{\sqrt{c_s^2 t^2 - \varrho^2}} \right\}. \tag{7.211}$$

The integral is readily evaluated with the result

$$s(r_0, t) = 0 \quad t < t_0, \tag{7.212}$$

so that the condensation does not change for the heating profile given by (7.204) until the information-gathering cylinder breaks through the boundary of the heating profile. After breakthrough, the double integration in (7.165) becomes much more complex. The final result appears in the form of elliptic integrals, but will not be presented here.

Parabolic Heating Profile Turned on at $t=0$

Finally, we consider an initially quiescent medium to which we apply a parabolic intensity given by

$$I(r, t) = \begin{cases} I_0(1 - r^2/a^2)u(t) & r < a \\ 0 & r > a. \end{cases} \tag{7.213}$$

For this function we find

$$\nabla^2 I = -4 I_0/a^2 \tag{7.214}$$

for $r < a$ and $t > 0$. If we take the Hankel transform of this heating profile we find from *Sneddon* [7.18]

$$\bar{I}(\lambda) = \frac{2 I_0}{\lambda^2} \left[\frac{2a}{\lambda} J_1(\lambda a) - J_0(\lambda a) \right]. \tag{7.215}$$

The individual terms in (7.215) have opposite singularities at $\lambda = 0$. To avoid this difficulty we combine them using the recursion relations for the Bessel functions. We obtain finally

$$\bar{I}(\lambda) = \frac{2I_0}{\lambda^2} J_2(a\lambda), \tag{7.216}$$

which is properly behaved in the limit $\lambda \to 0$. The evaluation of $s(r,t)$ is now reduced to the evaluation of a single definite integral of the same type discussed above. The required integrations can be carried out if we limit our considerations to $s(0,t)$. We find two different forms depending on whether $c_s t$ is less than or greater than a. In conformity with the regions shown in Fig. 7.3, we designate these solutions as s_1 and s_3, respectively. Thus we find for s_1 and s_3

$$s_1(0,t) = -\frac{2}{3}\frac{\gamma-1}{\gamma p_0}\alpha I_0 (c_s/a)^2 t^3 \tag{7.217}$$

for $c_s t < a$, and

$$s_3(0,T) = -\frac{\gamma-1}{\gamma p_0}\alpha I_0 \frac{a}{c_s}\left[T - \frac{T^2 + T\sqrt{T^2-1} - 2/3}{(T+\sqrt{T^2-1})^3}\right] \tag{7.218}$$

for $c_s t > a$, where we have written $T = c_s t/a$. As we expect, the expression for s_1 is identical with the early time Taylor expansion. At late times the function $s_3(0,t)$ goes over into the form

$$s_3(0,T) = -\frac{\gamma-1}{\gamma p_0}\alpha I_0 \frac{a}{c_s}\left(T - \frac{1}{4T}\right), \tag{7.219}$$

the first term of which is identical to that expected for slow steady heating.

Let us consider now the Green's function method applied to this problem. First we express (7.213) in terms of the polar coordinates ϱ and θ whose origin is at the observation point r_0. Thus we find

$$I(\varrho,\theta) = I_0\left(1 - \frac{r_0^2 + \varrho^2 + 2r_0\varrho\cos\theta}{a^2}\right). \tag{7.220}$$

Combining (7.220) and (7.163)–(7.165) we obtain

$$s(r_0,t) = -\frac{\gamma-1}{\gamma p_0}\alpha\left[I_0 t\left(1 - \frac{r_0^2}{a^2}\right)\right.$$
$$\left. -\frac{I_0}{2\pi c_s}\int_0^{2\pi} d\theta \int_0^{c_s t} \frac{\varrho d\varrho}{\sqrt{c_s^2 t^2 - \varrho^2}}\left(1 - \frac{r_0^2 + \varrho^2 + 2r_0\varrho\cos\theta}{a^2}\right)\right]. \tag{7.221}$$

After evaluating the integrals we obtain

$$s(r_0, t) = -\frac{2}{3}\frac{\gamma - 1}{\gamma p_0}\alpha I_0 t(c_s t/a)^2 \tag{7.222}$$

for $t \leq t_0$, with t_0 given by (7.210). Thus we have obtained an expression for the condensation off the axis and established a limit for the early time behavior. Just as in the one-dimensional case, we see that the early time t^3 dependence holds exactly until the information-gathering cylinder passes through the boundary of the heating profile. Of course this does not hold true for an arbitrary heating profile.

7.3.7 Repetitive Pulses

Thus far in our discussion we have considered primarily the case of a heating function which is turned on at $t = 0$ and left on for all time. In this section we examine briefly the effects produced by a repetitively pulsed beam.

The first pulse in a multiple pulse train will see the same initial conditions that we have used for most of our solutions thus far. Succeeding pulses, however, will encounter different initial conditions for $s(r, t)$ and its time derivatives than the zero values which we have assumed up to now. Let us consider a single pulse of duration t_1 and examine the condensation after it has been turned off. Let $s_0(r, t)$ be the condensation produced by a beam $I(r)$ turned on at $t = 0$ and left on indefinitely. If the beam is actually turned off at $t = t_1$, then the condensation $s(r, t)$ can be obtained by the principle of superposition,

$$s(r, t) = s_0(r, t) - s_0(r, t - t_1) \tag{7.223}$$

for $t > t_1$. If the pulses are spaced at an interval t_0 and of duration t_1, then the condensation during the nth pulse is simply

$$s(r, t, n) = \sum_{j=1}^{n} s_0[r, t - (j-1)t_0] - \sum_{j=2}^{n} s_0[r, t - (j-2)t_0 - t_1]. \tag{7.224}$$

There are some special cases for which the results can be given very simply. In the presence of flow, for example, one may find a pulse spacing t_0 which exceeds the flow time across the beam. In this case each pulse starts with an undisturbed medium and the appropriate expressions developed previously can be used.

In the absence of flow, we may expect the late time contributions from each heating pulse to be cumulative. In addition, there will be transients associated with heating pulses which have occurred within the last several acoustic transit times.

Finally, if the pulse duration t_1 is short compared to an acoustic transit time and the functions are sufficiently well behaved, we may expand the second term in (7.223) in a Taylor series about the time t

$$s_0(r, t - t_1) = s_0(r, t) + \left(\frac{\partial s_0}{\partial t}\right)_t (-t_1).$$
(7.225)

If we combine (7.223) and (7.225) we obtain for the condensation after a single short pulse started at $t = 0$

$$s(r, t) = \left(\frac{\partial s_0}{\partial t}\right) t_1 = \psi_0(r, t) t_1.$$
(7.226)

For repeated pulses these changes may be superposed in a manner similar to (7.224).

7.3.8 Absorption with a Time Delay in Transfer to Kinetic Energy

In our treatment thus far we have assumed that when energy is absorbed from the laser beam it is transferred instantaneously to the kinetic energy reservoir and that, in turn, it produces an instantaneous change in the kinetic temperature. It may happen, however, that the absorption produces electronic or vibrational excitation which decays to the kinetic temperature reservoir in a nonnegligible time t_1. This phenomenon is an ingredient in the process of kinetic cooling [7.19] which occurs for example in the absorption of 10.6 µm laser radiation in the atmosphere. We shall not treat the complete process of kinetic cooling here, but shall consider only the pure case of intermediate energy storage with a decay time.

Let $E(r, t)$ represent the energy stored per unit volume outside the kinetic energy reservoir. This energy is supplied by the laser beam at a rate $\alpha I(r, t)$ and it decays to the kinetic temperature reservoir at a rate E/t_1. Thus we may write

$$\frac{d}{dt} E(r, t) = \alpha I(r, t) - \frac{E(r, t)}{t_1}$$
(7.227)

and

$$h_v(r, t) = \frac{E(r, t)}{t_1}.$$
(7.228)

If this reservoir is empty at $t = 0$ and the laser beam $I(r, t)$ is turned on at $t = 0$, (7.227) and (7.228) yield the result

$$h_v(r, t) = \frac{\alpha}{t_1} \exp(-t/t_1) \int_0^t I(r, t') \exp(t'/t_1) dt'.$$
(7.229)

For the special case of a beam $I_0(r)$ which is independent of time after it is turned on at $t=0$, (7.229) becomes

$$h_v(r,t) = \alpha I_0(r)[1 - \exp(-t/t_1)]. \tag{7.230}$$

This phenomenon has its most significant impact on the early time solution for the condensation. From (7.230) we note that $h_v(r,0)=0$ and therefore the right side of (7.114) vanishes at $t=0+$. For the case of $s(r,0-)=0$ everywhere and sufficiently well-behaved $I(r)$, we can show from (7.105) and (7.230) using the same procedure as is described in Section 7.3.2 that

$$s(r,0)\Big|_{t=0+} = \frac{\partial s}{\partial t}\Big|_{t=0+} = \frac{\partial^2 s}{\partial t^2}\Big|_{t=0+} = \frac{\partial^3 s}{\partial t^3}\Big|_{t=0+} = 0. \tag{7.231}$$

In order to construct the solution we take the partial derivatives with respect to time of (7.105) and (7.230). We obtain

$$\frac{\partial^4 s}{\partial t^4}\Big|_{t=0+} = \frac{\gamma-1}{\gamma p_0} \frac{c_s^2 \alpha}{t_1} \nabla^2 I_0(r). \tag{7.232}$$

Thus the lowest order term in the Taylor expansion for $s(r,t)$ can be written

$$s(r,t) = \frac{\gamma-1}{\gamma p_0} \frac{c_s^2 \alpha}{t_1} \frac{t^4}{4!} \nabla^2 I_0(r). \tag{7.233}$$

Thus for one intermediate reservoir, the lowest order nonvanishing term is one order higher than for the case of instantaneous deposition in the kinetic temperature reservoir. The late time dependence is very similar to the results which have been developed previously except that an amount of energy $\alpha I_0 t_1$ is permanently stored in the intermediate reservoir.

The generalization to the case of a sequential flow of energy through n intermediate storage locations is easily carried out. For n reservoirs we may write

$$\frac{dE_1}{dt} = I_0 - \frac{E_1}{t_1}, \quad \frac{dE_2}{dt} = \frac{E_1}{t_1} - \frac{E_2}{t_2},$$

$$\frac{dE_n}{dt} = \frac{E_{n-1}}{t_{n-1}} - \frac{E_n}{t_n}, \quad h_v = \frac{E_n}{t_n}. \tag{7.234}$$

We wish to calculate the lowest order time-dependent term in the Taylor expansion of h_v. For the case of the steady beam $I_0(r)$ turned on at $t=0$ with all of the reservoirs initially empty, we can write

$$\frac{dE_1}{dt}\Big|_{t=0} = \alpha I_0. \tag{7.235}$$

Differentiating the second of (7.234) with respect to time and noting that dE_2/dt vanishes at $t=0$ we obtain

$$\frac{d^2E_2}{dt^2}\bigg|_{t=0} = \frac{1}{t_1}\frac{dE_1}{dt}\bigg|_{t=0} = \frac{\alpha I_0}{t_1}. \tag{7.236}$$

Repeating the process n times we obtain finally

$$\frac{d^n h_v(r)}{dt^n}\bigg|_{t=0} = \frac{\alpha I_0(r)}{t_1\cdots t_n}. \tag{7.237}$$

To find the early time dependence of $s(r,t)$ we proceed in the same manner as before. The lowest order nonvanishing time derivative of $s(r,t)$ is given by the relation

$$\frac{\partial^{n+3}}{\partial t^{n+3}}s(r,t) = \frac{\gamma-1}{\gamma p_0}c_s^2 \nabla^2 \frac{\partial^n}{\partial t^n}h_v, \tag{7.238}$$

so that for properly behaved $I(r)$ at early times

$$s(r,t) = \frac{\gamma-1}{\gamma p_0}\frac{\alpha c_s^2}{t_1\cdots t_n}\frac{t^{n+3}}{(n+3)!}\nabla^2 I(r). \tag{7.239}$$

Thus we see that even for a few intermediate reservoirs the early time development can be slowed considerably. It must be remembered that this relation holds only for times short compared to each of the storage times t_1 through t_n as well as the acoustic transit time. It may happen that one or several of the decay times are very much shorter than the others, thus severely restricting the range of applicability of (7.239). Let us imagine that t_k is very much shorter than the other storage times. As we might expect on physical grounds it is easy to show that for $t \gg t_k$ (7.239) is still valid if n is replaced by $n-1$ and t_k is dropped from the denominator, provided of course that t remains small compared to the remaining t_n and the acoustic transit time. The same conclusion applies when there are several very short relaxation times, provided that t is much greater than the largest of these short relaxation times.

7.3.9 Electrostrictive Effects

In our discussion so far we have been concerned exclusively with density changes associated with the heating of the medium by laser beam absorption. For very short pulsed Q-switched lasers, however, very strong electric fields may be produced but the time duration of the pulse is so short as to preclude the possibility of substantial density changes due to heating. The electric fields do produce forces directly on the gas molecules and it is these forces we wish to consider now.

When an electric dipole is placed in an electric field whose strength varies with position there is a net force on the dipole given by Ref. 7.4, p. 149

$$F = (p \cdot V)E. \tag{7.240}$$

If we have a gas like air in which p is also proportional to E then we can write

$$F = \mu(E \cdot V)E. \tag{7.241}$$

This force will always be in the direction of the local electric field. In the electrostatic case this term can be written in terms of $V(E^2)$, but in the electrodynamic case this is only true if our light beam is circularly polarized. For a collimated paraxial beam of circularly polarized light it is fairly easy to show that the force per unit volume produced by a transverse gradient in the intensity distribution is given by the relation

$$f_v = \frac{n-1}{2c} VI, \tag{7.242}$$

where n is the refractive index of the medium, c is the speed of light, and f_v is the result of averaging over the period and polarization direction of the circularly polarized light beam.

We wish now to develop a heuristic differential equation for the condensation in the same spirit as the treatment of Section 7.3.1. We begin by recognizing that if we make slow variations in the intensity distribution, the inertial terms can be neglected and the gas will develop a pressure field which will maintain equilibrium with the volume forces. The condition for equilibrium is simply

$$f_v - Vp = 0. \tag{7.243}$$

We must now express the pressure field in terms of the condensation. For adiabatic variations in an ideal gas we have the relation

$$p = \text{const} \times \varrho^\gamma, \tag{7.244}$$

which for small variations in p and ϱ leads to the expression

$$\delta p = \gamma p_0 s. \tag{7.245}$$

From (7.243) we can write

$$f_v - \gamma p_0 Vs = 0. \tag{7.246}$$

Comparing this relation with (7.242) we obtain finally

$$s(r) = \frac{n-1}{2\gamma p_0 c} I(r) \tag{7.247}$$

for the equilibrium condensation associated with the intensity distribution $I(r)$.

Now $s(r, t)$ must also satisfy a differential equation which accomodates sound waves. Proceeding in the same manner as in Section 7.3.1 we can write the heuristic differential equation for the condensation when driven by electrostrictive effects

$$\left(\nabla^2 - \frac{1}{c_s^2} \frac{\partial^2}{\partial t^2} \right) s(r, t) = \frac{n-1}{2\gamma p_0 c} \nabla^2 I(r, t). \tag{7.248}$$

Comparing (7.105) and (7.248) we see that electrostrictive effects could be introduced simply by taking $\partial/\partial t$ of the right side of (7.248) and adding the result to the right side of (7.105). It is not necessary to do this, however. The solutions obtained individually in (7.105) and (7.248) can simply be added, provided that we take proper account of the initial conditions.

Let us consider the early time dependence of $s(r, t)$ from (7.248). Let us assume that the laser beam is turned on at $t = 0$ and that both $s(r, t)$ and $\partial s/\partial t$ vanish at $t = 0+$. The early time Taylor expansion around points where $\nabla^2 I$ is properly behaved is given by

$$s(r, t) = -\frac{n-1}{2c} \frac{c_s^2}{\gamma p_0} \nabla^2 I(r) \frac{t^2}{2}. \tag{7.249}$$

We notice that the condensation has a t^2 early time dependence instead of the t^3 dependence characteristic of thermal blooming. This is a natural consequence of Newton's second law in which displacements appear in a Taylor expansion with a time dependence two powers of t higher than the forces which drive them. Since electrostrictive forces appear full strength at the instant the laser beam is turned on, the displacements follow with a t^2 time dependence. In the case of heating, the driving force depends on the energy deposited per unit volume which grows linearly in time, and the displacements follow a t^3 time dependence.

Comparing (7.115) and (7.249) we see that they have the same spatial dependence and are of opposite sign. Thus we can find the time at which the two effects exactly cancel. Setting these two expressions equal to one another we obtain

$$\alpha t = \frac{3}{2} \frac{n-1}{c(\gamma - 1)}, \tag{7.250}$$

a result first suggested by Townes. For air under standard conditions where $n - 1 \simeq 3 \times 10^{-4}$ and $\gamma = 1.4$ we obtain

$$\alpha t = 3.8 \times 10^{-14} \, \text{s cm}^{-1} \tag{7.251}$$

for exact cancellation. For later times, thermal blooming takes over and dominates the problem.

We can make the same comparison for the case of heating with a time delay. Let us consider the case of a single reservoir with time delay t_1. If we set (7.233) and (7.249) equal to one another we find

$$\frac{\alpha t^2}{t_1} = \frac{6(n-1)}{c(\gamma-1)}. \tag{7.252}$$

For air under standard conditions we obtain

$$\frac{\alpha t^2}{t_1} \simeq 1.5 \times 10^{-13}\,\mathrm{s\,cm^{-1}}. \tag{7.253}$$

This expression will yield a crossover point at a later time than (7.251) because thermal blooming will take longer to catch up with the electrostrictive effects.

7.4 Approximate Analytical Solutions

Because of its complexity, the problem of the passage of a high-energy laser beam through an absorbing gas has not proven amenable to exact analytic solution. A standard technique which has been successfully applied to many thermal blooming problems is a form of perturbation theory. In this approach, the laser beam is initially assumed to propagate in the quiescent medium without disturbing it. The resulting intensity distribution throughout the medium multiplied by the absorption coefficient is used as a local heat source. This heating gives rise in turn to density and refractive index changes. The propagation of the laser beam in the modified refractive index profile is then calculated to first order. This model can be viewed as the first step of an iterative process which, we assume, if continued ad infinitum would converge to the correct, self-consistent description of the laser beam everywhere for all time. Higher order corrections are extremely difficult to compute in closed form. Indeed, one is often frustrated in an attempt to produce closed-form solutions for first-order approximations even where the laser beam is assumed to have simple space-time characteristics. We present in this section some of the cases for which analytic solutions have been successfully derived.

In our discussion, we consider three complementary approaches to the propagation problem. The first of these is through geometric optics for which the calculations are generally simpler and a greater variety of blooming problems are tractable. The second approach is that of perturbation theory applied directly to the wave equation for paraxial beams. This method, which is simple in concept, leads to analytic expressions which are tractable only in a limited number of special cases. On the other hand, it is expected to give useful results in regions near a focus where geometric optics solutions exhibit singular behavior.

The third approach employs an extended Huygens-Fresnel method called the pasted phase approximation to calculate beam disturbance. It is a wave optics approach formulated by means of a diffraction integral rather than through the wave equation. This method appears analytically tractable only in a limited number of cases. On the other hand, the diffraction integrals can be evaluated fairly simply by computer calculation.

7.4.1 General Solutions in the Geometric Optics Approximation

In this subsection, we take the results of Section 7.2.4 and apply them to the calculation of ray trajectories and intensity variation for paraxial beams. Our treatment parallels that given by *Keller* [7.20] and later by *Avizonis* et al. [7.21] for the ray trajectories and by *Bergmann* [7.22] for the intensity variation along the rays. We modify Bergmann's treatment to include extinction along the ray.

Ray Trajectories for Paraxial Beams

Let us write the refractive index in the form

$$n(r, t) = n_0 + \lambda n_1(r, t), \tag{7.254}$$

where n_0 is the quiescent value, $n_1(r, t)$ is the disturbance produced by thermal blooming, and λ is a parameter in the expansion. We wish to determine the ray trajectory $r(s, t, R_0, \lambda)$ which is a function of the initial position R_0, the arclength along the ray s, the time t, and λ, the parameter of the expansion. We assume the ray trajectory can be written as a Taylor series in λ which to first order is given by

$$r(s, t, R_0, \lambda) = r_0(s, t, R_0) + \lambda r_1(s, t, R_0). \tag{7.255}$$

If we substitute (7.254) and (7.255) into (7.51) we obtain

$$\frac{d}{ds}\left[(n_0 + \lambda n_1)\frac{d}{ds}(r_0 + \lambda r_1) \right] = V(n_0 + \lambda n_1). \tag{7.256}$$

Setting $\lambda = 0$ in (7.256) and recognizing that n_0 is a constant we obtain

$$(d^2/ds^2)(r_0) = 0, \tag{7.257}$$

which has the solution

$$r_0(s, t, R_0) = R_0 + su(R_0), \tag{7.258}$$

where $u(R_0)$ is a unit vector tangent to the ray at position R_0. Thus r_0 is a straight-line trajectory from the starting point. We next take the derivative of (7.256) with respect to λ and again set λ equal to zero to obtain

$$\frac{d}{ds}\left(n_0\frac{dr_1}{ds} + n_1\frac{dr_0}{ds}\right) = \nabla n_1. \tag{7.259}$$

Making use of (7.258), this equation can be written after some manipulation

$$n_0\frac{d^2r_1}{ds^2} = \nabla_T n_1(R_0 + us), \tag{7.260}$$

where $n_1(R_0 + us)$ is the disturbance in the refractive index calculated along the unperturbed trajectory and ∇_T is the transverse gradient operator. For paraxial beams the transverse coordinates will be taken as normal to the beam axis. Equation (7.260) can be solved by integrating twice with respect to s. Thus we obtain

$$r_1 = \frac{1}{n_0}\int_0^s d\varrho \int_0^\varrho d\sigma \nabla_T n_1(R_0 + u\sigma) \tag{7.261}$$

which can be integrated by parts to obtain finally

$$r_1(s,t,R_0) = \frac{1}{n_0}\int_0^s (s-\sigma)\nabla_T n_1(R_0 + u\sigma)d\sigma. \tag{7.262}$$

This is the basic expression for the calculation of ray trajectories.

Intensity Variation Along the Rays

There are several approaches by which we may examine the intensity variation along a ray trajectory. *Born* and *Wolf* [Ref. 7.6, p. 117] express the ratio of the intensity at two points along the ray in terms of a path integral of the Laplacian of the optical path function S. *Kline* [7.23] gives a similar expression involving the principal radii of curvature at each point along the path. Both of these approaches give exact results within the approximations of geometric optics.

In the thermal blooming problem, we generally do not have sufficient information to carry out the integrations described above. Since the refractive index disturbances produced by thermal blooming are small, we shall proceed along the lines of perturbation theory. Our approach parallels that given by *Bergmann* [7.22].

The basic equations are (7.48), which we rewrite here,

$$(\nabla S)^2 = n^2, \tag{7.263}$$

and (7.56) which can be rewritten using (7.50) as

$$\boldsymbol{V} \cdot [(I/n)\boldsymbol{V}S] + \alpha I = 0. \tag{7.264}$$

If we make the substitution

$$L = \ln(I/n), \tag{7.265}$$

(7.264) becomes

$$\boldsymbol{V}L \cdot \boldsymbol{V}S + \boldsymbol{V}^2 S + n\alpha = 0. \tag{7.266}$$

Equations (7.263) and (7.266) are the basic equations of the problem. We proceed to develop the perturbation theory in the same way as above, i.e., by setting

$$n = n_0 + \lambda n_1. \tag{7.267}$$

We next assume that the functions S and L can be expanded in a Taylor series in λ. Writing only terms to first order we have

$$S = S_0 + \lambda S_1,$$
$$L = L_0 + \lambda L_1. \tag{7.268}$$

Substituting (7.267) and (7.268) into (7.263) and (7.266), we may set $\lambda = 0$ to obtain the lowest order terms. Next we may take the derivative with respect to λ and then set $\lambda = 0$ to obtain the first-order equations. The lowest order equations are

$$(\boldsymbol{V}S_0)^2 = n_0^2 \tag{7.269}$$

and

$$\boldsymbol{V}L_0 \cdot \boldsymbol{V}S_0 + \boldsymbol{V}^2 S_0 + n_0\alpha = 0. \tag{7.270}$$

Since n_0 is a constant, we see by comparing these expressions with (7.263) and (7.266) that the zeroth-order problem corresponds to propagation in a uniform medium. From the discussion given earlier, the rays are simply a series of straight lines. Thus the solution of (7.269) can be written

$$\boldsymbol{V}S_0 = n_0 \boldsymbol{u}, \tag{7.271}$$

where $\boldsymbol{u}$ is a unit vector whose orientation may vary between rays but remains fixed along a given ray. If (7.271) is put into (7.270), we can write

$$\frac{dL_0}{ds} + \frac{\boldsymbol{V}^2 S_0}{n_0} + \alpha = 0, \tag{7.272}$$

where now s is a scalar which measures distance along a ray.

The first-order equations are easily obtained. We find

$$\boldsymbol{u} \cdot \boldsymbol{V} S_1 = n_1 \tag{7.273}$$

and

$$\boldsymbol{u} \cdot \boldsymbol{V} L_1 = -\frac{n_1}{n_0} \alpha - \boldsymbol{V} L_0 \cdot \frac{\boldsymbol{V} S_1}{n_0} - \frac{\boldsymbol{V}^2 S_1}{n_0}, \tag{7.274}$$

where $\boldsymbol{u}$ is the unit vector along the unperturbed trajectory. The three terms on the right side of (7.274) have a direct physical interpretation. The first term is simply extinction along the ray. It can generally be neglected in this order. The second term describes intensity variation along the unperturbed trajectory due to lateral displacement of adjacent rays. The third term describes the intensity variation due to lensing effects produced by the perturbation n_1. The terms in the Laplacian of S appearing in (7.270) or (7.274) are the same as those appearing in the treatment of *Born* and *Wolf* [Ref. 7.6, p. 117]. The advantages of the present formulation are that the perturbation theory is developed systematically and that the displacements of nearby rays are automatically taken into account.

Fortunately, (7.273) and (7.274) can be solved separately. The solution to (7.273) can be written

$$S_1(r) = \int_0^r n_1(s) ds, \tag{7.275}$$

where s is the path length along the unperturbed trajectory and $s=0$ at the aperture or other location where the initial conditions are specified. The position vector r indicates an arbitrary point on the unperturbed trajectory. Before proceeding further, we shall consider a simple example..

7.4.2 Example of a Gaussian Slab

Consider a collimated beam with intensity distribution

$$I(x) = I_0 \exp(-x^2/a^2) \tag{7.276}$$

which propagates in the positive z direction from an aperture at $z=0$. To simplify matters, we let the beam extended to infinity in the $\pm y$ directions and neglect the decrease in intensity in the z direction due to absorption. In addition, we introduce a transverse flow with velocity v_0 in the positive x direction. The condensation for this example is given by (7.126). Carrying out the integration we find

$$s(x) = -\frac{(\gamma - 1)}{\gamma p_0 v_0} \alpha I_0 \frac{a\sqrt{\pi}}{2} [1 + \mathrm{erf}(x/a)]. \tag{7.277}$$

Before proceeding further, we require a relation between the refractive index and the condensation. We shall assume for all our calculations that $(n-1)$ is proportional to the density of the medium. Thus we write

$$(n-1)=(n_0-1)(1+s), \tag{7.278}$$

where n_0 is the quiescent refractive index for $s=0$. Then

$$n_1=(n-n_0)=(n_0-1)s. \tag{7.279}$$

We turn first to a consideration of the ray trajectories. In this problem the ray trajectories can be described by a single scalar function of z, $u(z, x_0)$ which is the displacement of a ray beginning at $x=x_0$, $z=0$. The unperturbed trajectories u_0 of the collimated beam are given simply by

$$u_0(z, x_0)=x_0. \tag{7.280}$$

To find the perturbed trajectories, we make use of (7.262), but we require an expression for $\partial n_1/\partial x$. From (7.277) and (7.279) we find

$$\frac{1}{n_0}\frac{\partial n_1}{\partial x}=-\frac{(n_0-1)}{n_0}\frac{(\gamma-1)}{\gamma p_0 v_0}\alpha I_0\exp(-x^2/a^2). \tag{7.281}$$

Since n_1 is independent of distance along the ray trajectory in this example, the integration in (7.262) is easily carried out. We obtain finally

$$u(z, x_0)=x_0-\frac{z^2}{2}\frac{n_0-1}{n_0}\frac{(\gamma-1)}{\gamma p_0 v_0}\alpha I_0\exp(-x_0^2/a^2). \tag{7.282}$$

In this approximation, we find that all the rays curve upstream to the flow by an amount proportional to the beam intensity at x_0 and the square of the distance from the aperture. For the perturbation theory to be valid, we require that the deflection $(u-x_0)$ be small compared to a, the characteristic dimension of the unperturbed beam.

If we had included extinction in (7.276) we would find that the rays deflect upstream by an amount proportional to z^2 for small z but linear in z for z greater than several extinction lengths. The linear portion of the trajectory exhibits an angular deflection upwind given by

$$\theta=\tan^{-1}\left[\frac{n_0-1}{n_0}\frac{\gamma-1}{\gamma p_0 v_0}I_0\exp(-x_0^2/a^2)\right], \tag{7.283}$$

subject also to the requirement that the lateral displacement be small compared to a.

We turn now to the question of ray crossings which is an important limit to the range of validity of our geometric optics calculations. From (7.282), for example, we see that the rays from $x_0 = 0$ curve toward the negative x direction more rapidly than rays from any negative x_0. Thus we may anticipate ray crossings for $x_0 < 0$ in this example. In particular, consider

$$\frac{\partial u}{\partial x_0}\bigg|_z = 1 + \frac{x_0 z^2}{a^2} \frac{n_0 - 1}{n_0} \frac{\gamma - 1}{\gamma p_0 v_0} \alpha I_0 \exp(-x_0^2/a^2), \tag{7.284}$$

which represents the ratio of the change in ray deflection at fixed z to the value of a small change in x_0. If this quantity vanishes, we may conclude that adjacent rays from the vicinity of x_0 have intersected. From (7.284) we see that for any negative x_0, we can always find a value of z for which adjacent rays have intersected. The intensity distribution derived for these values of the coordinates must be considered suspect.

The calculation of the intensity distribution for this example is straightforward. Because the beam is collimated, the unit vector u in (7.271) is everywhere parallel to the z axis. The solution to (7.271) is therefore

$$S_0(x, z) = n_0 z. \tag{7.285}$$

The solution to (7.272) which satisfies the initial conditions is simply

$$L_0(x, z) = \ln(I_0/n_0) - x^2/a^2. \tag{7.286}$$

From (7.273) we obtain

$$S_1(x, z) = \int_0^z n_1(x, z') dz'. \tag{7.287}$$

From (7.277) and (7.279), this expression becomes

$$S_1(x, z) = -(n_0 - 1) \frac{\gamma - 1}{\gamma p_0 v_0} \alpha I_0 a \frac{\sqrt{\pi}}{2} [1 + \mathrm{erf}(x/a)] z. \tag{7.288}$$

Using (7.286) and (7.288), (7.274) becomes, after dropping the term in $n_1 \alpha$,

$$\frac{\partial L_1}{\partial z} = -4 \frac{(n_0 - 1)}{n_0} \frac{(\gamma - 1)}{\gamma p_0 v_0} \alpha I_0 \frac{xz}{a^2} \exp(-x^2/a^2). \tag{7.289}$$

We may now integrate this expression from $z = 0$ adding an arbitrary $f(x)$ as a constant of integration. Finally the intensity distribution is recovered from (7.265) by the relation

$$I = (n_0 + n_1) \exp(L_0) \exp(L_1). \tag{7.290}$$

Integrating (7.289) and combining the results with (7.286) and (7.290), we obtain

$$I(x, z) = I_0 \exp(-x^2/a^2)$$
$$\cdot \exp\left[-2\frac{(n_0-1)}{n_0}\frac{(\gamma-1)}{\gamma p_0 v_0} \alpha I_0 \frac{xz^2}{a^2} \exp(-x^2/a^2)\right], \tag{7.291}$$

where we have defined $f(x)$ by the expression

$$1 + n_1(x)/n_0 = \exp[-f(x)] \tag{7.292}$$

in order to satisfy (7.276) at $z=0$.

Since the success of the perturbation theory requires L_1 to be small compared to L_0, we may expand the second exponential term in (7.291) in a Taylor series to obtain finally

$$I(x, z) = I_0 \exp(-x^2/a^2)$$
$$\cdot \left[1 - 2\frac{(n_0-1)}{n_0}\frac{(\gamma-1)}{\gamma p_0 v_0} \alpha I_0 \frac{xz^2}{a^2} \exp(-x^2/a^2)\right]. \tag{7.293}$$

We note from (7.293) that the perturbed intensity in this approximation increases for negative x, although we must be alert to the possibility of ray crossing for sufficiently large z in this region. We note also that the perturbation vanishes at $x=0$. This occurs because both $\nabla^2 S_1$ and ∇L_0 in (7.274) vanish at $x=0$. The physical explanation is that there are no focusing effects because the refractive index has an inflection point at $x=0$. Furthermore, since the gradient of L_0 also vanishes at $x=0$, beam steering will have no effect on the intensity perturbation.

7.4.3 Intensity Variation for Collimated Beams

The general perturbation theory for the intensity variation along the ray trajectories can be simplified considerably for the case of collimated beams. Let us imagine a paraxial beam launched in the positive z direction from an aperture at $z=0$. The initial intensity distribution $I(x, y, z, t)$ will be specified by L_0 given by

$$L_0(x, y, 0, t) = \ln[I_0(x, y, 0, t)/n_0]. \tag{7.294}$$

Equations (7.271) and (7.272) are readily solved to give

$$S_0 = n_0 z \tag{7.295}$$

and

$$L_0(x, y, z, t) = L_0(x, y, 0, t) - \alpha z. \tag{7.296}$$

Next, we obtain $n_1(x, y, z, t)$ from (7.279) and the condensation which may be found by the methods of Section 7.3. Integrating (7.273) for a collimated beam we obtain

$$S_1(x, y, z, t) = \int_0^z n_1(x, y, z', t) dz' . \tag{7.297}$$

We now require both the gradient and the Laplacian of this expression for the calculation of L_1. If we assume that the extinction length $(1/\alpha)$ is much larger than the characteristic transverse dimension of the beam, then we may neglect the derivatives with respect to z. With this assumption we have from (7.297)

$$\boldsymbol{V}S_1 = \int_0^z (\boldsymbol{V}_T n_1) dz' \tag{7.298}$$

and

$$V^2 S_1 = \int_0^z (V_T^2 n_1) dz' . \tag{7.299}$$

Substituting these expressions into (7.274) and dropping $n_1 \alpha / n_0$, we obtain

$$\frac{dL_1}{dz} = - \frac{\boldsymbol{V} I_0(x, y, 0, t)}{n_0 I_0(x, y, 0, t)} \cdot \int_0^z \boldsymbol{V}_T n_1 dz' - \frac{1}{n_0} \int_0^z V_T^2 n_1 dz' . \tag{7.300}$$

If we integrate this expression with respect to z we obtain a pair of double integrals plus an arbitrary function $f(x)$ which is needed to satisfy the initial conditions. The double integrals are readily evaluated for collimated beams, assuming n_1 falls as $\exp(-\alpha z)$ away from the aperture. Finally we obtain from (7.294, 296, 300)

$$I(x, y, z, t) = I_A(x, y, t) \exp(-\alpha z)$$
$$\cdot \exp\{[1 - \alpha z - \exp(-\alpha z)](\boldsymbol{V} I_A \cdot \boldsymbol{V}_T n_A + V^2 n_A)/(\alpha n_0)\} , \tag{7.301}$$

where in this expression n_A and $I_A(x, y, t)$ are the refractive index and intensity profiles, respectively, at $z = 0$. This expression presents the intensity distribution in terms of the refractive index perturbation directly.

7.4.4 Geometric Optics Treatment of Focused Gaussian Beams

In the geometric optics approximation, the beam intensity goes to infinity as we approach a real focal point. Of course the methods of Section 7.4.1 may be applied upstream of the focus until the beam begins to approach the

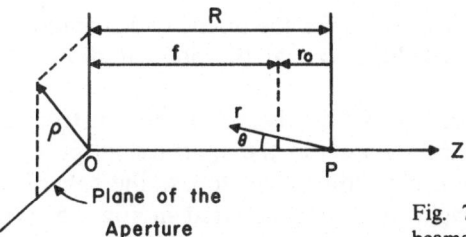

Fig. 7.6. Ad hoc treatment of focused gaussian beams: the straight-line ray trajectories converge toward P from the plane of the aperture

diffraction-limited spot size. In this subsection we develop an ad hoc method for the geometric optics treatment of thermal blooming effects at the diffraction focus of a gaussian beam. As before, we consider a beam propagating in the positive z direction from an aperture located at $z=0$. Let ϱ designate a radial cylindrical coordinate as illustrated in Fig. 7.6.

We begin by noting from (7.45) that the intensity distribution of an initially gaussian beam remains gaussian along its entire length. Of course the normals to the surfaces of constant phase given by (7.43) change direction as we move along the beam. In our ad hoc treatment, we replace the converging wave optics beam given in (7.45) by a geometric optics beam with straight-line ray trajectories converging to a point P located downstream from the diffraction focus as shown in Fig. 7.6. The geometric optics beam parameters are chosen to give the same radial intensity distribution at both the aperture and the diffraction focus as given by (7.45). In (7.45) it is convenient to write

$$a^2(z)=w^2(z)/2 \quad \text{and} \quad a_0^2=r_0^2/2, \tag{7.302}$$

where $a(z)$ is the $1/e$ radius of the gaussian intensity distribution as a function of z and a_0 is the value at the aperture. The actual profile we wish to simulate with our ad hoc distribution is given by

$$I(\varrho,0)=I_0\exp(-\varrho^2/a_0^2) \tag{7.303}$$

at the aperture and by

$$I(\varrho,f)=I_0(k^2a_0^4/f^2)\exp\left[\frac{-\varrho^2}{f^2/(k^2a_0^2)}\right] \tag{7.304}$$

at $z=f$. Let us consider now the ray trajectory $b(z)$ given by

$$b(z)=a_0(1-z/R), \tag{7.305}$$

where R is shown in Fig. 7.6 and is given by the expression

$$R=\frac{f}{1-f/(ka_0^2)}. \tag{7.306}$$

This ray trajectory passes through the 1/e points for the intensity distributions given in (7.303) and (7.304). It intersects the z axis at the point P shown in Fig. 7.6.

We now proceed to demonstrate that our ad hoc beam can be constructed using a distribution of rays from arbitrary points in the aperture, all passing through the point P. In our discussion, it is convenient to use the spherical polar coordinate system r, θ with origin at P as illustrated in Fig. 7.6. We proceed in the same manner as in Section 7.4.2. The solution of (7.269) for this case is simply

$$S_0 = n_0(R-r), \tag{7.307}$$

where we have arbitrarily chosen $S_0 = 0$ at the center of the aperture. The sign is chosen so that S_0 increases along the ray as required by (7.51). Noting that in this case $dL/ds = -dL/dr$, we easily find that the solution for (7.272) is

$$L_0(r,\theta) = \alpha r - 2\ln r + A(\theta), \tag{7.308}$$

where $A(\theta)$ is an arbitrary function. If we assume a paraxial beam, we may use the approximation

$$\varrho = R\theta \tag{7.309}$$

in (7.303) and (7.304). The function $A(\theta)$ is determined by the boundary conditions in the aperture given by (7.303). Using (7.309) and (7.265) we find

$$L_0(r,\theta) = \ln(I_0/n_0) + \alpha(r-R) + 2\ln(R/r) - R^2\theta^2/a_0^2. \tag{7.310}$$

This function also reproduces the required distribution (7.304) at the plane $z = f$.

Case of Steady Heating

To proceed further, let us consider the case of steady heating given by (7.122). Using (7.279) we can write

$$n_1(r,\theta) = -(n_0-1)\frac{\gamma-1}{\gamma p_0}\alpha I_0 t \exp[\alpha(r-R)]$$
$$\cdot (R/r)^2 \exp[-(R\theta/a_0)^2]. \tag{7.311}$$

We now obtain S_1 by carrying out the integration indicated in (7.275),

$$S_1(r,\theta) = \int_R^r n_1(r',\theta)dr'. \tag{7.312}$$

To simplify matters we now restrict our attention to the blooming effects on the axis at $z = f$ $(r = r_0)$. On the axis the transverse gradient of L_0 vanishes so that from (7.274) we need only compute the Laplacian of S_1 for $\theta = 0$. We obtain

$$(\nabla^2 S_1)_{\theta = 0} = (n_0 - 1)\frac{\gamma - 1}{\gamma p_0}\alpha I_0 t \exp\left[\alpha(r - R)\right]R(R/r - 1)4R^2/(a_0 r)^2 . \qquad (7.313)$$

This expression may be substituted into (7.274) and the latter integrated to obtain the axial distribution $L_1(r, 0)$. At $r = r_0$ we obtain finally

$$I(r_0, 0) = I_0 \exp(-\alpha z)(R/r_0)^2$$
$$\cdot \exp\left[-2\frac{(n_0 - 1)}{n_0}\frac{(\gamma - 1)}{\gamma p_0}\alpha I_0 t k^2 a_0^2\right]. \qquad (7.314)$$

To examine the limit of no blooming we may set $(n_0 - 1) = 0$. In this case (7.314) is the same as (7.304) with the term $\exp(-\alpha z)$ added. As always the validity of our approximations requires that the argument of the second exponential be small compared to unity. It is interesting to express the argument of the second exponential in terms of the optical phase shift δN of the unperturbed beam L_0 due to the perturbation $n_1(r, 0)$. Using (7.2) we find that the phase shift in radians can be written

$$\delta N = (k/n_0)\int_R^{r_0} n_1(r', 0)(-dr'), \qquad (7.315)$$

where k is the wave number in the medium. Using (7.311) we obtain for the axial phase shift

$$\delta N = -\frac{(n_0 - 1)}{n_0}\frac{(\gamma - 1)}{\gamma p_0}\alpha I_0 t (k a_0)^2 . \qquad (7.316)$$

Putting this relation into (7.314) we find that the axial intensity at the diffraction focus can be written

$$I(r_0, 0) = I_0 \exp(-\alpha z)(R/r_0)^2 \exp(2\delta N). \qquad (7.317)$$

This result is compatible with our assumptions in Section 7.1.1 that the threshold of thermal blooming could be estimated by setting $\delta N \simeq -1$ rad.

Early Time Dependence

The analysis of the early time dependence of our ad hoc gaussian beam proceeds in nearly the same manner as given above. The condensation is computed from the ad hoc intensity distribution using (7.310) and (7.115). We then calculate in turn $n_1(r, \theta)$, $S_1(r, \theta)$, $L_1(r, 0)$, and finally $I(r_0, 0)$. The calculations are fairly straightforward and will not be presented here. The final result for the axial intensity at the plane $z = f$ can be written

$$I(r_0, 0) = I_0 \exp(-\alpha z)(R/r_0)^2 \exp[(4/9)\delta N(t/t_0)^2], \tag{7.318}$$

where

$$t_0 = \frac{f}{ka_0 c} \tag{7.319}$$

is the acoustic transit time across the spot radius to the $1/e$ point measured at the diffraction focal point $r = r_0$. It should be noted that δN is always negative in (7.317) and (7.318). Results equivalent to (7.317) and (7.318) were presented by *Aitken* et al. [7.24].

7.4.5 Wave Optics Perturbation Theory

Basic Formulation

We turn now to the development of a perturbation theory applied directly to the wave equation. We begin with (7.34) for paraxial beams. As before, we assume that the perturbation can be developed in a Taylor series in a parameter λ. Keeping only first-order terms, we may write

$$n(r, t) = n_0 + \lambda n_1(r, t) \tag{7.320}$$

and

$$\psi(r, t) = \psi_0(r, t) + \lambda \psi_1(r, t). \tag{7.321}$$

Substituting these expressions into (7.35) and setting $\lambda = 0$ we obtain

$$\nabla_T^2 \psi_0 + 2ik(\partial \psi_0/\partial z) = 0. \tag{7.322}$$

We may obtain the first-order contribution by putting (7.320) and (7.321) into (7.34) and taking the derivative with respect to λ before setting $\lambda = 0$. We find

$$\nabla_T^2 \psi_1 + 2ik(\partial \psi_1/\partial z) + (2k^2/n_0)n_1(r, t)\psi_0 = 0. \tag{7.323}$$

Equations (7.322) and (7.323) constitute the basic formulation of the problem. Equation (7.322) for the propagation in a uniform medium should be solved first with proper attention to the boundary conditions. Then the condensation associated with this unperturbed intensity distribution may be determined by the methods of Section 7.3. The resulting $n_1(r, t)$ may be obtained from (7.279). Given these values for $n_1(r, t)$ and $\psi_0(r, t)$, (7.323) may be solved for $\psi_1(r, t)$

subject to the appropriate boundary conditions. Thus ψ_1 should vanish at the aperture where ψ_0 is the correct field configuration and also at infinity in such a way that the radiation condition is still satisfied.

Application to a Gaussian Beam

As an example let us consider a circularly symmetric focused gaussian beam turned on at $t=0$. The beam propagates in the positive z direction from an aperture at $z=0$. We assume steady heating at late times with no transverse flow so that the condensation is given by (7.122). The treatment here is a more compact version of that presented originally by *Ulrich* [7.25].

The solution of (7.322) for ψ_0 subject to the appropriate boundary conditions has been given in (7.40). The function $n_1(r, t)$ may be obtained from (7.45, 122, 279, 302). We find

$$n_1 = -(n_0 - 1)\frac{(\gamma - 1)}{\gamma p_0} \frac{\alpha I_0 t}{a^2(z)/a_0^2} \exp\left[-r^2/a^2(z)\right]. \tag{7.324}$$

Before proceeding further it is convenient to let

$$F(r, t) = (-2k^2/n_0)n_1(r, t)\psi_0(r, t), \tag{7.325}$$

so that (7.323) can be written

$$\nabla_T^2\psi_1 + 2ik(\partial\psi_1/\partial z) = F. \tag{7.326}$$

From (7.325) we see that F has cylindrical symmetry. Thus we make the zeroth-order Hankel transform (cf. Appendix) of (7.326) to obtain

$$-\lambda^2\bar{\psi}_1(\lambda, z, t) + 2ik(d\bar{\psi}_1/dz) = \bar{F}(\lambda, z, t), \tag{7.327}$$

where, as before, the overbar indicates the Hankel transform of the quantity. Equation (7.327) is an ordinary differential equation for $\bar{\psi}_1$ with the parameter λ and z as the independent variable. Since we wish ψ_1 to vanish at the aperture, its transform will also vanish there. Equation (7.327) may be solved by the method of variation of parameters [Ref. 7.15, p. 20]. We obtain the result

$$\bar{\psi}_1(\lambda, z, t) = (-i/2k)\int_0^z \bar{F}(\lambda, z', t)\exp\left[\frac{i\lambda^2}{2k}(z' - z)\right]dz'. \tag{7.328}$$

If (7.328) is put into the inversion theorem for the Hankel transform we obtain

$$\psi(r, z, t) = (-i/2k)\int_0^\infty d\lambda\, \lambda J_0(\lambda r)$$

$$\cdot \int_0^z dz'\bar{F}(\lambda, z', t)\exp\left[\frac{i\lambda^2}{2k}(z' - z)\right]. \tag{7.329}$$

To obtain an analytic result, we carry out the integration over λ first and then the z' integration for the special case $r=0$ and $z=f$. The calculations are tedious but lead finally to the integral for $\psi_1(r,z,t)$, which for the parameter values given takes the form

$$\psi_1(0,f,t) = -\frac{ik^3a_0^4}{f}\frac{(n_0-1)}{n_0}\frac{\gamma-1}{\gamma p_0}\alpha I_0 t\psi_0(0,0)\int_0^\infty dp\,\frac{(p-i)(1-3ip)}{(p^2+1)(1+9p^2)}. \qquad (7.330)$$

To compute the axial intensity in this case, we require

$$I_1(0,f,t) = |\psi_0 + \psi_1|^2. \qquad (7.331)$$

From (7.40) we note

$$\psi_0(0,f) = -i\frac{ka_0^2}{f}\psi_0(0,0). \qquad (7.332)$$

Since $\psi_0(0,f)$ is pure imaginary and ψ_1 is a small perturbation, only $\mathrm{Im}(\psi_1)$ makes a contribution to (7.331). The required integral in (7.330) can be evaluated by a partial fraction decomposition. We obtain finally

$$\mathrm{Im}\,[\psi_1(0,f,t)] = \frac{\ln 9}{8}\frac{k^3a_0^4}{f}\frac{n_0-1}{n_0}\frac{\gamma-1}{\gamma p_0}\alpha I_0 t\psi_0(0,0). \qquad (7.333)$$

Combining (7.331–333) we obtain

$$I(0,f,t) = (k^2a_0^4/f^2)\psi_0^2(0,0)\left(1 - \frac{\ln 9}{4}k^2a_0^2\frac{n_0-1}{n_0}\frac{\gamma-1}{\gamma p_0}\alpha I_0 t\right). \qquad (7.334)$$

It is of interest to express the last parentheses in terms of the number of radians phase shift of the unperturbed beam. Proceeding in the same manner as suggested by (7.315) we write

$$\delta N = (k/n_0)\int_0^f n_1(0,z')dz'. \qquad (7.335)$$

From (7.324) we obtain

$$\delta N = -\frac{\pi}{2}k^2a_0^2\frac{n_0-1}{n_0}\frac{\gamma-1}{\gamma p_0}\alpha I_0 t, \qquad (7.336)$$

so that (7.334) can be written

$$I(0,f,t) = \frac{k^2a_0^4}{f^2}\psi_0^2(0,0)\left[1 + \frac{\ln 9}{2\pi}(\delta N)\right]. \qquad (7.337)$$

Once again we see that a phase shift of the order of 1 rad. can be considered as the threshold for significant blooming effects.

7.4.6 Pasted Phase Approximation

In this subsection we apply the formalism of Section 7.2.6 to a specific case. We choose an example that permits us to obtain a simple analytic result with a reasonable amount of effort.

Consider a paraxial beam which depends only on one transverse coordinate and which propagates in the positive z-direction from an aperture at $z = 0$. As in (7.98), we designate the transverse coordinate in the plane of observation by x and in the plane of the aperture by ξ. We may also use ξ as a transverse coordinate in the region between the aperture and the plane of observation, but the distinction will be clear from the context.

Let the beam be focused at a distance f from the aperture and have an amplitude distribution at the aperture given by (7.66), which we repeat here for convenience,

$$u(\xi) = u_0 (1 - \xi^2/a^2)^{1/2} \exp\left[-ik\xi^2/(2f)\right]. \tag{7.338}$$

Let us also introduce a transverse flow with velocity v_0 in the positive ξ direction. We wish to calculate the amplitude and intensity distribution in the plane $z = f$ as illustrated in Fig. 7.7. The amplitude in the focal plane can be obtained from (7.98). To use this equation we require an expression for the line integral of the refractive index disturbance between a source point ξ in the aperture and an observation point x in the plane $z = f$.

We begin by considering the intensity distribution at the aperture. From (7.338), the intensity $I(\xi, z)$ can be written

$$I(\xi, 0) = u_0^2 (1 - \xi^2/a^2) \tag{7.339}$$

for $|\xi| < a$ and zero otherwise. From (7.126) the condensation in the plane of the aperture is given by

$$s(\xi, 0) = -\frac{(\gamma - 1)}{\gamma p_0 v_0} \alpha u_0^2 \left(\frac{2a}{3} + \xi - \frac{\xi^3}{3a^2}\right). \tag{7.340}$$

To simplify matters further, we make the additional assumptions that, for the purpose of evaluating the line integral in (7.98), we may replace dl by dz, introduce the z dependence into (7.340) solely through the multiplicative factor $\exp(-\alpha z)$, and finally that we may assume α is large enough to write $\alpha f \gg 1$. These assumptions are clearly valid when most of the beam is absorbed before reaching the observation plane. The impact on our results for other conditions has not been assessed.

With these assumptions we obtain from (7.279) and (7.340)

$$\int_{\xi}^{x} (\delta n) dl = -(n_0 - 1)\frac{(\gamma - 1)}{\gamma p_0 v_0} u_0^2 \left(\frac{2a}{3} + \xi - \frac{\xi^3}{3a^2}\right). \tag{7.341}$$

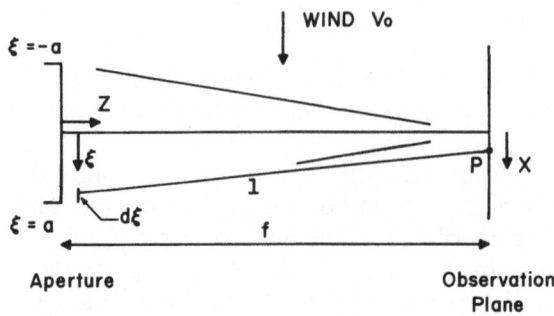

Fig. 7.7. Pasted phase approximation for a one-dimensional focused beam

Since this result is independent of x, the remaining calculations are considerably simplified. Before proceeding further it is convenient to introduce the
quantities v, p, and q defined by

$$v = \frac{\xi}{a}, \quad p = \frac{kax}{f} \quad \text{and} \quad q = ka(n_0 - 1)\frac{(\gamma - 1)}{\gamma p_0 v_0} u_0^2. \tag{7.342}$$

Substituting (7.338) and (7.341) into (7.98), setting $R_0 = f$, and making use of
(7.342) we obtain the amplitude distribution in the focal plane,

$$u(x) = \frac{(1-i)}{2}\left|\sqrt{\frac{k}{\pi f}}\exp\left[i(kf - 2q/3)\right]\right.$$

$$\cdot u_0 a \int_{-1}^{1} dv(1 - v^2)^{1/2}\exp\left[-ipv - iq(v - v^3/3)\right]. \tag{7.343}$$

Let us denote the definite integral in (7.343) by K. From the symmetry of the
integrand we note that K is a real function of p and q. We evaluate the integral by
making a Taylor expansion in the blooming parameter q. If we retain terms up
to second order in q we may write

$$K = K_0 + qK_1 + q^2K_2, \tag{7.344}$$

where

$$K_0 = \int_{-1}^{1} dv(1 - v^2)^{1/2}\cos(pv), \tag{7.345}$$

$$K_1 = -\int_{-1}^{1} dv(1 - v^2)^{1/2}\sin(pv)(v - v^3/3), \tag{7.346}$$

and

$$K_2 = -(1/2)\int_{-1}^{1} dv(1 - v^2)^{1/2}\cos(pv)(v^2 - 2v^4/3 + v^6/9). \tag{7.347}$$

From [7.7] we find

$$K_0 = \pi J_1(p)/p. \tag{7.348}$$

The integrals K_1 and K_2 may be obtained from (7.345) and (7.348) by repeated differentiation with respect to the parameter p. After some manipulation, using the recursion relations for the Bessel functions, we obtain

$$K_1 = -(\pi/p)[(2/3)J_2(p) + J_3(p)/p] \tag{7.349}$$

and

$$K_2 = \pi/(2p)\left[\frac{7}{2}\frac{J_2(p)}{p} + \frac{4}{3}J_3(p) + \left(\frac{5}{6p} - \frac{605}{3p^3}\right)J_4(p)\right]. \tag{7.350}$$

Finally, the intensity in the focal plane is given by

$$I(x) = |u(x)|^2 = \frac{u_0^2 k a^2}{2\pi f} K^2. \tag{7.351}$$

The complete analytical expression for K^2 is formidable. It is useful, however, to expand K_0, K_1, and K_2 in a Taylor series up to second order in p. We find

$$K_0 = (\pi/2)(1 - p^2/8), \tag{7.352}$$

$$K_1 = -(\pi/2)(5p/24), \tag{7.353}$$

and

$$K_2 = \frac{\pi}{2}\left(-\frac{101}{1152} + \frac{91}{4608}p^2\right). \tag{7.354}$$

If we form K^2 from (7.352–354) and include only terms up to second order in either p or q or their products, we obtain

$$I(x) = \frac{\pi}{4} u_0^2 \frac{k a^2}{f}\left[1 - \frac{1}{4}\left(p + \frac{5q}{6}\right)^2 - \frac{q^2}{576}\right]. \tag{7.355}$$

On the axis where $p = 0$, we see that the intensity decreases with respect to the vacuum value by an amount proportional to q^2. We also note by inspection of (7.355) that the peak intensity occurs when

$$p = -5q/6, \tag{7.356}$$

so that the peak value moves upstream with the onset of blooming. From (7.342) the value of x_0 for which the intensity is maximized can be written

$$x_0 = -\frac{5(\gamma-1)}{6\gamma p_0 v_0}(n_0-1)fu_0^2. \tag{7.357}$$

The additional negative term in q^2 in (7.355) means that even this peak value decreases with respect to the vacuum value for small q. The magnitude of these effects is consistent with our 1 rad criterion.

7.5 Methodology for Computer Solution of Blooming Problems

In this section we discuss some numerical techniques for solving the thermal blooming problem which apply quite generally to all of the time regimes discussed above. The ability to unify the discussion in this manner stems from the fact that the beam propagates with the speed of light, whereas the distortions of the beam follow the changes created in the medium and therefore are induced in times corresponding to either the beam size divided by speed of sound or beam size divided by convective flow speed. Thus, the assumption can be made that the medium effectively does not change in a beam propagation time and one can calculate, therefore, a series of instantaneous steady states to model the time development. Only one such calculation is needed, of course, to model the propagation of a steady-state or cw beam. It follows that the absorbing medium is taken to be frozen as if in a snapshot. The beam is propagated through the index of refraction field which has evolved up to that time, losing energy and altering the medium refractive index which controls the propagation of the beam at the next time increment. The hydrodynamics of the beam-fluid interaction can be considered to be an auxiliary calculation which uses the current intensity of the laser beam and the past space-time history of the index field to provide the required current status of the field. These calculations of index changes also will be discussed in this section, and specific computer results will be presented. The numerical approach treats the wave optics equations exactly except for truncation errors inherent in the approximation used for application of the digital computer.

The paraxial wave equation in the Fresnel approximation is given by (7.34) which we rewrite here, i.e.,

$$\nabla_T^2 \psi + 2ik\frac{\partial \psi}{\partial z} + k^2\left(\frac{n^2}{n_0^2}-1\right)\psi = 0, \tag{7.358}$$

where $(n^2/n_0^2)-1$ is dependent on the time history of $|\psi|^2$.

Since (7.358) is parabolic in form, the amplitude ψ can be found, in principle, at any plane $z=z_0+\Delta z$ given that its values are known at some other plane

$z=z_0$ [Ref. 7.26, pp. 82, 83]. We generally take for our initial conditions the field distribution across the aperture of the laser from which the beam emanates. The field is assumed to be zero far from the beam in the transverse direction. The computer solution to (7.358) then is realized in principle by replacing all derivatives by differences and all integrals by sums. Considerable refinement of this basic approach has evolved, however, since thermal blooming problems were first exposed to computer analysis. We shall survey some of the techniques which have proved useful.

7.5.1 Numerical Techniques

A numerical approximation to the partial differential equation must satisfy two conditions. It must be stable and it must produce a solution which converges to the analytic solution as sampling of the dependent variables in space and time is refined [cf. Ref. 7.27, pp. 44, 45]. The algorithms to be discussed here have all been shown to be stable and convergent. For a survey of this topic see *Ulrich* [7.2] and references therein. We shall briefly summarize the evolution of refined, efficient techniques for numerical solution of (7.358).

Equation (7.358) can be rewritten in the form

$$\frac{\partial \psi}{\partial z} = F\{\psi\}, \tag{7.359}$$

where ψ is a function of space and time and F is a functional of the dependent and independent variables. By writing the equation in this form we have isolated all the operators in the transverse direction on the right-hand side of the equation. This will lead, as we shall shortly demonstrate, to the development of an algorithm which allows the solution to proceed by marching from one plane, $z=$ constant, to another plane, $z=$ constant. In addition, one needs initial conditions at $z=0$,

$$\psi(x, y, 0) = \psi_0(x, y), \tag{7.360}$$

and boundary conditions for $z>0$,

$$\psi(x = \pm \infty, y = \pm \infty, z) = 0. \tag{7.361}$$

In this form the problem is well posed [Ref. 7.28, p. 226f.].

As an introduction to the numerical techniques used in the thermal blooming problem, we consider a simple differencing method for solution of (7.359). First we write the left-hand side in difference form,

$$\frac{\partial \psi}{\partial z} \equiv \frac{\psi(z + \Delta z) - \psi(z)}{\Delta z}. \tag{7.362}$$

We identify the functional $F\{\psi\}$ from (7.358) to be

$$F\{\psi\} \equiv \frac{i}{2k}\left[V_T^2 + k^2\left(\frac{n^2}{n_0^2} - 1\right)\right]\psi. \tag{7.363}$$

For simplicity at this point, let us set $n = n_0$ (vacuum propagation) and write the transverse Laplacian in difference form [Ref. 7.27, p. 7],

$$V_T^2\psi = \frac{\partial^2\psi}{\partial x^2} + \frac{\partial^2\psi}{\partial y^2}$$

$$= [\psi(x + \Delta x, y, z) - 2\psi(x, y, z) + \psi(x - \Delta x, y, z)]/(\Delta x)^2$$

$$+ [\psi(x, y + \Delta y, z) - 2\psi(x, y, z) + \psi(x, y - \Delta y, z)]/(\Delta y)^2, \tag{7.364}$$

where we have chosen to evaluate $V_T^2(x, y, z)$ at the plane z where we assume we have known values of the array $\psi(x, y, z)$. For vacuum propagation, (7.358) in difference form becomes

$$\psi(x, y, z + \Delta z) = \psi(x, y, z) + \frac{i}{2k}[\psi(x + \Delta x, y, z) - 2\psi(x, y, z)$$

$$+ \psi(x - \Delta x, y, z)]\frac{\Delta z}{(\Delta x)^2} + \frac{i}{2k}[\psi(x, y + \Delta y, z)$$

$$- 2\psi(x, y, z) + \psi(x, y - \Delta y, z)]\frac{\Delta z}{(\Delta y)^2}. \tag{7.365}$$

This exhibits some of the general features of the algorithms to be considered. We see that the value of the amplitude array at each point in the new plane $z = z + \Delta z$ is given by the sum of the value of the array at the corresponding transverse point in the old plane and a linear combination of adjacent values of the dependent variable. Reiteration of this algorithm provides a step-by-step approach to the focal plane which is termed "marching".

Equation (7.365) represents a relationship between adjacent planes. The complete solution is given by a sequence of such planes which are closely enough spaced so that changes in the optical phase fronts are maintained below some selected maximum value. This assures accuracy in the calculation. Indeed, for some algorithms there is an additional, generally more restrictive, criterion that step sizes in range between planes be chosen small enough to prevent the growth of numerically instability.

The numerical techniques to be discussed are a generalization of (7.365). These algorithms evolved to produce more efficient and more accurate computer solutions. They are based on marching techniques of the general form

$$\psi_n = \sum_{i=1}^{N} \alpha_i\psi_{n-i} + \sum_{i=0}^{N} \beta_i\psi_{n-i}\Delta z, \tag{7.366}$$

where i is an index specifying the planes preceding the nth plane which contribute to new data at plane n. The spacing between planes, Δz, may itself be a function of n. ψ_n is shorthand notation for $\psi(x_k, y_l, n\Delta z)$. Specific realizations of some of the algorithms used in the thermal blooming problem will be given below.

If $\beta_0 = 0$, the algorithm is termed *explicit*, since in this case ψ_n appears only on the left-hand side of the equation and the solution is obtained simply by evaluating the expression on the right-hand side of (7.366). If $\beta_0 \neq 0$, ψ_n appears on both sides of the equation. On the left it is required to be calculated at the point $(x, y, n\Delta z)$. On the right it requires knowledge of the adjacent points $(x \pm \Delta x, y \pm \Delta y, n\Delta z)$. Thus a simultaneous set of linear algebraic equations is required to be solved. In this case the algorithm is termed *implicit*. Explicit algorithms, while generally faster per range step, are usually limited by stability criteria to a maximum Δz which leads to many steps to the solution plane. Implicit algorithms are generally stable for all choices of Δz, and only consideration of accuracy dictates the size of each step. The implicit algorithm tends to take longer per step to calculate than the explicit algorithm since a matrix inversion is required to solve a set of simultaneous difference equations.

A variety of marching algorithms has been developed and we shall discuss representative examples of them below. They differ essentially only in the way the differencing is done to represent the transverse derivatives. Special techniques to represent the transverse derivatives as smoothly as possible have been used as well. Both linear and cubic splines have been successfully applied to these solution algorithms. A detailed discussion of these methods is beyond the scope of this discussion. The reader is referred to [7.29] for a detailed treatment.

In addition to differencing and spline techniques, the use of Fourier analysis has become prevalent with the advent of the fast Fourier transform (FFT) algorithm [7.30]. By choosing a suitable coordinate system and dependent variable transformations, which will be exhibited below, the ordinary differential equation that is obtained in Fourier transform space can be solved exactly to get the transformed amplitude at the new range step. Inverse transformation gives the amplitude at the new range step. The use of the fast Fourier transform algorithm [7.31] makes this method efficient enough to compete with the other methods.

The methods mentioned so far have been developed directly from the partial differential equation itself, and are therefore termed differential techniques. Integral techniques are developed from the solution of (7.358) in integral form

$$\psi(x, y, z) = -\frac{1}{4\pi} \int_{-\infty}^{\infty} \int_{-\infty}^{\infty} dx_0 dy_0 G(x, y, z; x_0, y_0, 0)\psi_0(x_0, y_0), \qquad (7.367)$$

where G is the solution for a point source at the aperture plane, i.e., the Green's function for the problem. This is a generalization of the Fresnel solution appropriate for vacuum propagation for which

$$G(x, y, z; x_0, y_0, 0) = \frac{ik}{2z} \exp \{ik[(x - x_0)^2 + (y - y_0)^2]/2z\}.$$ (7.368)

The presence of heating of the absorber by the laser beam alters the function G, and in general the integral in (7.367) must be computed numerically. In practice, an approximation to the Green's function is required [7.32], and one is led thereby to a restriction on the size of the range difference. Thus, (7.367) is applied sequentially for a number of steps to the solution plane and, therefore, has many of the features of the direct differential approach. A variant of this integral solution approach is to solve (7.367) at the solution plane using the free-space Green's function. The resultant amplitude is then used to alter the Green's function and the integration is repeated. This process is reiterated until the amplitude at some step changes negligibly from the amplitude at the previous step [7.33].

It has been found that all of these techniques compare well in generating solutions to the nonlinear blooming problem. The integral techniques seem to give more accurate answers in the near field while the differential techniques calculate the far-field patterns more efficiently. The differences between the most efficient of each kind of algorithm are not great, however, and usage of one or the other is usually a matter of taste or acquired familiarity.

Examples of these techniques applied to (7.358) will now be given. They will be treated in the order of increasing sophistication. This also happens to be the historial order in which they were developed for application to the thermal blooming problem.

Explicit Algorithm

This is the algorithm first employed in nonlinear optics propagation problems [7.34]. As pointed out by *Harmuth* [7.35] in his study of the solution of Schrödinger's equation by numerical means, equations of the type of which (7.358) is a sample are stable when a symmetric difference is used to represent the z derivative. Using the notation

$$\psi_{k,l}^m \equiv \psi(k\Delta x, l\Delta y, m\Delta z)$$ (7.369)

and defining the central difference operator to represent the second derivative, as built up from a difference of first differences,

$$\Delta_x^- \psi_{k,l}^m = (\psi_{k-1,l}^m - \psi_{k,l}^m)/\Delta x$$ (7.370)

centered at $x = k\Delta x - \frac{1}{2}\Delta x$, and

$$\Delta_x^+ \psi_{k,l}^m = (\psi_{k,l}^m - \psi_{k+1,l}^m)/\Delta x$$ (7.371)

centered at $x = k\Delta x + \frac{1}{2}\Delta x$, then one has

$$\delta_x^2 \psi_{k,l}^m = (\Delta_x^- \psi_{k,l}^m - \Delta_x^+ \psi_{k,l}^m)/\Delta x$$
$$= (\psi_{k-1,l}^m - 2\psi_{k,l}^m + \psi_{k+1,l}^m)/(\Delta x)^2 . \tag{7.372}$$

Similarly, one represents the second derivative in y by the second difference (centered at x, y) as

$$\delta_y^2 \psi_{k,l}^m = (\psi_{k,l-1}^m - 2\psi_{k,l}^m + \psi_{k,l+1}^m)/(\Delta y)^2 . \tag{7.373}$$

Now, if we apply these rules to (7.358) we have, using a symmetric differencing in z as required for stability,

$$2ik\frac{(\psi_{k,l}^{m+1} - \psi_{k,l}^{m-1})}{2\Delta z} = (\delta_x^2 + \delta_y^2)\psi_{k,l}^m + f(\psi_{k,l}^m)\psi_{k,l}^m , \tag{7.374}$$

where $f(\psi_{k,l}^m)$ represents the nonlinear term and is calculated from the change induced in the index of refraction. We transpose (7.374), putting all the already known information on the right-hand side

$$\psi_{k,l}^{m+1} = \psi_{k,l}^{m-1} + i\frac{\Delta z}{k}[(\delta_x^2 + \delta_y^2)\psi_{k,l}^m + f(\psi_{k,l}^m)\psi_{k,l}^m] . \tag{7.375}$$

We can now clearly appreciate the term "explicit" for this algorithm since as we proceed to plane $m+1$ we use only information available to us on planes $m-1$ and m.

Harmuth [7.35] has shown that this method is stable for samples $\Delta x, \Delta y, \Delta z$ chosen such that,

$$\frac{\Delta z}{k}\left[\left(\frac{1}{\Delta x}\right)^2 + \left(\frac{1}{\Delta y}\right)^2\right] < 1 . \tag{7.376}$$

Implicit Algorithm

A popular algorithm which has been successfully applied to heat conduction problems was developed by *Crank* and *Nicolson* [7.36]. Applying this algorithm to (7.358) gives

$$\psi_{k,l}^{m+1} = \psi_{k,l}^m + \left(\frac{\Delta z}{2ik}\right)(\delta_x^2 + \delta_y^2)(\psi_{k,l}^{m+1} + \psi_{k,l}^m)/2 + f(\psi_{k,l}^m)\psi_{k,l}^m . \tag{7.377}$$

Note that the field at the plane $z = m\Delta z$ has been replaced by the average of the fields at the two planes $m\Delta z$ and $(m+1)\Delta z$ in the terms containing the second differences to represent the Laplacian. This device has the effect of dampening unstable growth. Unlike the explicit method of the preceding subsection, this method is stable for all choices of $\Delta z/(\Delta x)^2$ and $\Delta z/(\Delta y)^2$. Solutions are achieved only after matrix inversion to completely specify ψ at the plane $z = (m+1)\Delta z$. Thus, the calculation per step is longer for this algorithm, but, due to its unconditional stability, it requires less steps to achieve a given range and hence competes well with the explicit algorithm for efficiency.

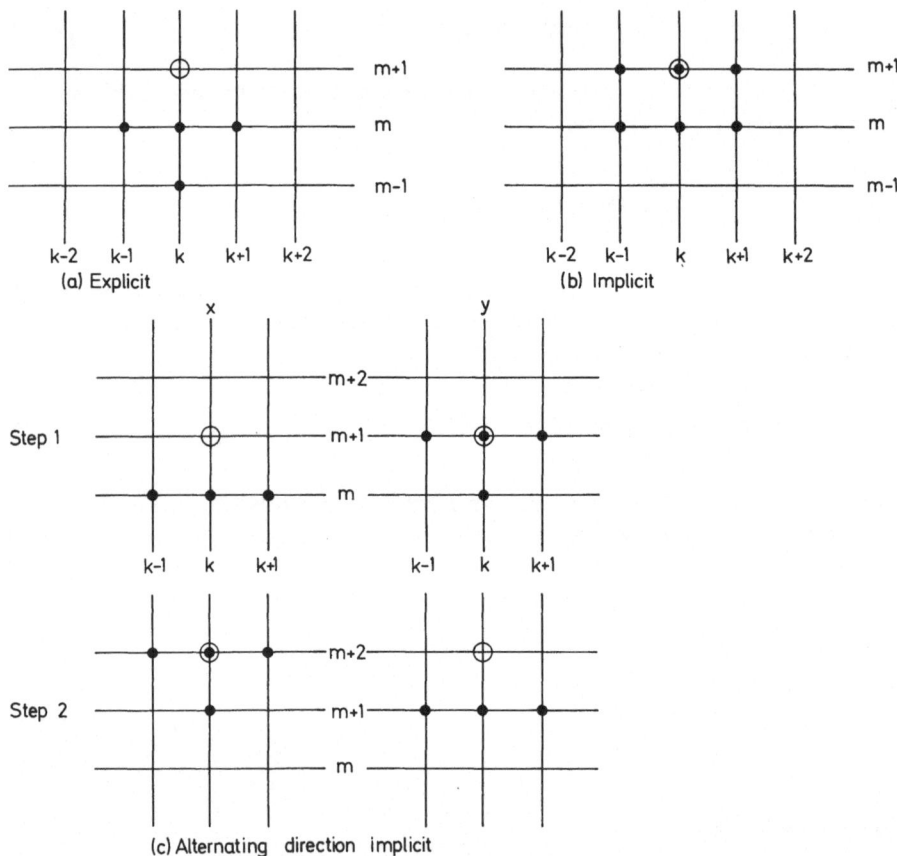

Fig. 7.8a–c. A schematic representation of the computational mesh: the intersections are called nodes and are represented by a triple of numbers (k, l, m); the y or l dimension is not shown in the figure for ease of presentation. (a) Explicit algorithm which has data at nodes on z planes at m and $m-1$ contributing to the solution (open circle) at the plane $m+1$; the contribution from nodes adjacent to (k, l, m) is due to the use of a central-difference approximation to the second derivative in x; the y dimension counterpart is similar. (b) Schematic representation of the Crank-Nicolson stable implicit method requiring knowledge of the dependent variable at the plane $m+1$ to which the solution is proceeding; thus, the full set of simultaneous difference equations must be solved by matrix inversion or by a relaxation process. (c) Schematic of the alternating direction implicit method (ADI) and the coordinate directions, x and y, take turns playing host to an implicit algorithm

Combined Implicit and Explicit

A novel two-step procedure which alternates the nature of the algorithm from explicit to implicit for the x direction and vice versa for the y direction provides an efficient stable algorithm [7.37–39].

At the first step the algorithm is explicit in x, implicit in y,

$$\psi_{k,l}^{m+1} = \psi_{k,l}^m + \left(\frac{\Delta z}{2ik}\right)(\delta_x^2 \psi_{k,l}^m + \delta_y^2 \psi_{k,l}^{m+1}) + f(\psi_{k,l}^m)\psi_{k,l}^m . \tag{7.378}$$

Note that the x-differencing involves only known field values whereas the y differencing requires knowledge of the field on the plane $z = (m+1)\Delta z$. At the next step the algorithm is implicit in x, explicit in y,

$$\psi_{k,l}^{m+2} = \psi_{k,l}^{m+1} + \left(\frac{\Delta z}{2ik}\right)(\delta_x^2 \psi_{k,l}^{m+2} + \delta_y^2 \psi_{k,l}^{m+1}) + f(\psi_{k,l}^{m+1})\psi_{k,l}^{m+1} , \tag{7.379}$$

where now the differencing is reversed with regard to operation on known or to-be-known values. The method is unconditionally stable for the two-step process as a whole, and at each substep the solution requires only the inversion of a tridiagonal matrix which is rapidly accomplished by gaussian elimination [7.40]. This method is called the alternating direction implicit, or ADI, method and it has proved extremely useful for studying nonlinear propagation problems.

Schematic representation of the above three algorithms is given in Fig. 7.8.

Fast Fourier Transform Solutions

As will be shown below [discussion following (7.390)], it is possible to transform the amplitude of the laser beam at each step so that the nonlinear term does not appear in the partial differential equation (7.358) but is incorporated, instead, in the phase of a new dependent variable $\bar{\psi}$, which satisfies the equation

$$2ik\frac{\partial \bar{\psi}}{\partial z} + \nabla_T^2 \bar{\psi} = 0 , \tag{7.380}$$

i.e., free-space propagation. A finite Fourier transform is performed on (7.380) by writing

$$\bar{\psi}(x, y, z) = \sum_{m=1}^{N_x} \sum_{n=1}^{N_y} C_{mn}(z)\exp(ip_m x + iq_n y), \tag{7.381}$$

where p_m, q_n are given by $p_m = 2\pi m/L$, $q_n = 2\pi n/L$, L being the length spanned by the computation mesh in (x, y) space, and N_x, N_y are the number of discrete

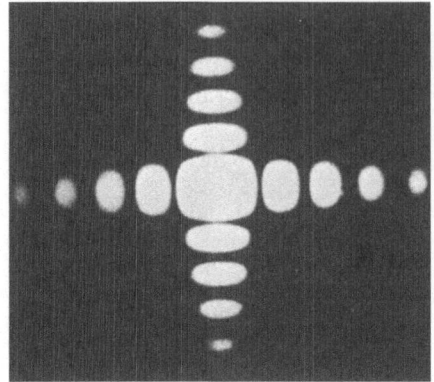

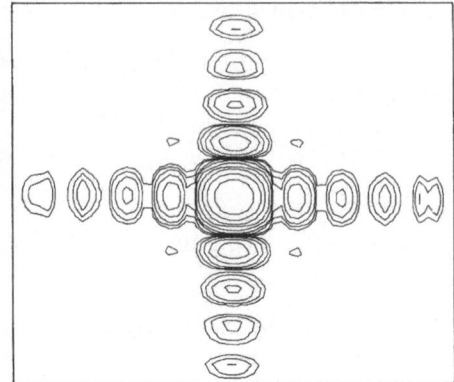

I/I_o	Experiment	Computer
Center	1	1
1st	0.0469	0.0491
2nd	0.0168	0.0169
3rd	0.0083	0.0091
4th	0.0050	0.0073

Fig. 7.9. Comparison of diffraction patterns as recorded in the laboratory (left) and as calculated by the thermal blooming codes with zero absorption; the agreement is seen to be quite excellent (photograph courtesy of Pergamon Press, reprinted from: M. Born, E. Wolf: *Principles of Optics*, 4th ed. (Pergamon Press, London 1970) p. 396

samples of $\bar{\psi}$ in x and y, respectively. Thus (7.380) becomes a simple ordinary differential equation

$$2ik\frac{dC_{mn}}{dz} = (p_m^2 + q_n^2)C_{mn}. \tag{7.382}$$

Let us assume that C_{mn} is known at a particular value of z. We can very simply obtain the solution at $z + \Delta z$ by direct integration,

$$C_{mn}(z+\Delta z) = C_{mn}(z)\exp\left[-i\frac{\Delta z}{2k}(p_m^2 + q_n^2)\right]. \tag{7.383}$$

The amplitude $\bar{\psi}(x, y, z + \Delta z)$ is obtained by inverse Fourier transformation of the new C_{mn} coefficients [7.30, 41]. The true field is then reconstructed by reintroduction of the nonlinear phase as will be discussed below. For fields, ψ, which can be expressed as a superposition of Fourier components as in (7.381), the propagation of the components is exact according to (7.383). Thus, fields which are well represented by spatial frequencies lying below the basic sampling frequency, $1/\Delta x$ or $1/\Delta y$, are accurately treated by the Fourier transform technique. This means that laser intensity patterns which are characterized by rapid intensity fluctuations such as truncated or obscured beams are handled well with this algorithm [7.42]. As an example of the accuracy with which a complicated diffraction pattern can be calculated, Fig. 7.9 shows a comparison between experimental and calculated patterns for a rectangular aperture [7.2]. The location of the zeros and maxima are accurately calculated. The values of the maxima are given in the table and compared with the theoretical values. When one considers that these maxima are in some cases almost two orders of magnitude below the central peak intensity, the agreement is seen to be extremely good.

A large contributing factor to the accuracy of the calculation of the rectangular aperture problem is the use of a square computational mesh oriented parallel to the aperture. The patterns calculated with circular or other shaped apertures are not quite as accurately done, because the aperture edges are not aligned with the mesh.

Integral Methods

When an integral solution of the form of (7.367) is taken as the starting point for creation of a computational algorithm, approximation is made to represent the Green's function as

$$G_a(x, y, z; x_1, y_1, z_1) \equiv G_0(x, y, z; x_1, y_1, z_1)$$

$$\cdot \exp\left\{\frac{i}{2k} \int_{z_1}^{z} dz' \phi[x_s(z'), y_s(z'), z']\right\}, \tag{7.384}$$

where x_s and y_s are transverse variables which follow the focusing or defocusing of the beam and G_0 is the free-space propagator. ϕ is a function of the beam intensity and represents the contribution of the nonlinearity to the index change. In order to neglect an amplitude correction which must also be applied to G_0, in general, a restriction is placed on the step size. Thus, an iterative procedure for the complete solution is required as was the case for the differential approach.

7.5.2 Improvement of Computational Efficiency

There are a number of ways in which the accuracy and speed of the computation can be improved by judicious choice of coordinate system and by appropriate changes in the dependent variable. These changes can be applied a priori to the basic equations. The coordinate systems in such cases are called "nonadaptive" systems. In a more sophisticated approach, transformation of variables can be applied as the problem is iterated toward solution, automatically adjusting the coordinate system to have the shape of the equiphase surfaces of the wave fronts. This technique leads to more accurate calculations since rapid fluctuations transverse to the beam propagation direction are absent in such a coordinates system. Such alterations are known as "adaptive" coordinate and phase changes.

Nonadaptive Coordinate System

An example of a collection of nonadaptive changes to the basic equations which has proven useful in thermal blooming calculations [7.43] will be discussed here. Other workers use similar systems [7.44]. All of these systems are essentially equivalent in their ability to enhance both the speed and the accuracy of the computation.

The basic idea and rationale for these coordinate changes is to restrict the computation to calculating only the changes induced in the vacuum amplitude rather than requiring calculation of the complete amplitude. This procedure removes the necessity for sampling the high frequency oscillation induced in the phase by focusing. The coordinate transform alters the independent variables so that the dependent variables take a different functional form but are numerically identical to the untransformed amplitude at equivalent points in space in the two coordinate systems. This altered form is very smooth and easier to calculate. It becomes less smooth as the calculation procedes and changes are induced in the amplitudes.

The particular coordinate transformations selected are based on the vacuum behavior of a focused gaussian beam. At $z=0$ this amplitude has the form

$$\psi(x_0, y_0, 0) = \frac{1}{\sqrt{\pi a^2}} \exp\left[-(x_0^2 + y_0^2)/2a^2 - ik(x_0^2 + y_0^2)/2f\right], \qquad (7.385)$$

where a is a transverse scale length which is a measure of the initial size of the beam and f is the geometric focus. With this mode at the initial plane, the solution for any range is given by performing the integrations in (7.367) and (7.368) and is,

$$\psi(\bar{x}, \bar{y}, \zeta) = \frac{1}{\sqrt{\pi a^2 D}} \exp\left(i\left[(\bar{x}^2 + \bar{y}^2)\frac{d}{d\zeta}\ln D^{\frac{1}{2}}/2\right]\right.$$
$$\left. + i\{\tan^{-1}\left[\zeta/(1 - \zeta/\hat{f})\right]\} - \left[(\bar{x}^2 + \bar{y}^2)/2D\right]\right), \qquad (7.386)$$

where $\zeta = z/ka^2$, $\hat{f} = f/ka^2$, $\bar{x} = x/a$, $\bar{y} = y/a$, and

$$D(\zeta) = \zeta^2 + (1 - \zeta/\hat{f})^2.$$

The presence of the quadratic terms in the imaginary part of the exponential gives rise to high-frequency oscillations in the amplitude for $x > a$, $y > a$. These can exceed the sampling frequency of the computational mesh. Thus, a phase change is made to remove these terms. Additionally, the amplitude is seen to grow in range as $z \to f$ due to the presence of the term $(\pi a^2 D)^{-1/2}$. Thus, it is desirable to remove this factor as well. As the beam focuses and becomes smaller the mathematical mesh should shrink at the same rate to accurately model the transverse distribution. These considerations lead to the following sequence of transformations:

1) removal of the quadratic phase and amplitude, leaving a pure real gaussian amplitude in vacuum;

$$\psi_1(\bar{x}, \bar{y}, z) \equiv \sqrt{D(\zeta)} \exp\left[-i\frac{\bar{x}^2 + \bar{y}^2}{2}\frac{d}{d\zeta}\ln D^{1/2}\right.$$

$$\left. - i\tan^{-1}\left(\frac{\zeta}{1 - \zeta/\hat{f}}\right)\right]\psi(\bar{x}, \bar{y}, \zeta),$$

2) scaling of the transverse coordinates with beam size,

$$\tilde{x} = \bar{x}/\sqrt{D}$$

$$\tilde{y} = \bar{y}/\sqrt{D}$$

$$\psi_2(\tilde{x}, \tilde{y}, \zeta) = \psi_1(\bar{x}, \bar{y}, \zeta),$$

3) alteration of step size Δz to automatically take smaller steps in the region of high beam intensity,

$$d\bar{z}/d\zeta = 1/D$$

$$\psi_3(\tilde{x}, \tilde{y}, \bar{z}) = \psi_2(\tilde{x}, \tilde{y}, \zeta).$$

Equation (7.358) then takes the form

$$2i\frac{\partial\psi_3}{\partial\bar{z}} + \tilde{V}_T^2\psi_3 + (2 - \tilde{x}^2 - \tilde{y}^2)\psi_3 + Dk^2a^2\left(\frac{n^2}{n_0^2} - 1\right)\psi_3 = 0. \tag{7.387}$$

For the case where $n^2 = n_0^2$, that is, vacuum propagation, we note from (7.386) and the indicated coordinate and phase changes that the transformed field has the form

$$\psi_3(\tilde{x}, \tilde{y}, \bar{z}) = \exp[-(\tilde{x}^2 + \tilde{y}^2)/2] \text{ (vacuum)}.$$

If we apply (7.387) to this expression we find

$$\frac{\partial\psi_3}{\partial\bar{z}} = 0 \text{ (vacuum)}.$$

Thus, the form of the vacuum amplitude is unchanged as we propagate in this coordinate system. This is rationale for its selection. When $n^2 \neq n_0^2$, changes are induced in this vacuum function. The computer calculation is able to accurately compute these changes very efficiently compared with the computation of the rapidly varying phases contained in the untransformed amplitudes. It is the presence of the last term on the left-hand side of (7.387) which gives rise to contributions to the imaginary part of Ψ. As the blooming becomes severe these contributions become difficult to compute. The surfaces of constant phase of the distorted amplitude become quite different from the coordinate surfaces. For these cases, a method for adapting the tilt and radius of curvature of the coordinate surface to match, on the average, the corresponding quantities for the distorted wave front is needed to efficiently calculate the propagation. A technique for doing this is discussed below. First, however, we indicate how a further phase change introduced at each range step during the calculation allows the equation to be rendered in quite a simple form. This then allows one to utilize simple and efficient algorithms such as the fast Fourier transform method discussed earlier.

Removal of the Nonlinear Phase

If one examines (7.387) closely one sees that there are essentially three different types of terms: 1) operation by $\partial/\partial \bar{z}$ on the wave amplitude, 2) operation by $\tilde{V}_T^2(V_T^2$ with $\tilde{x}$ and $\tilde{y})$ on the wave amplitude, and 3) simple multiplication of the wave amplitude by other variables. If one can achieve the removal of these latter terms from the equation, one has remaining an equation which has the appearance of the vacuum propagation equation in a cartesian coordinate system. It is trivial to show that this desired result is achieved by defining a new wave amplitude which is related to ψ_3 by the following phase change:

$$\psi_3(\tilde{x}, \tilde{y}, \bar{z}) = \psi_4(\tilde{x}, \tilde{y}, \bar{z})$$
$$\cdot \exp\left\{ -\frac{i}{2} \int_{z_0}^{z} \left[\tilde{x}^2 + \tilde{y}^2 - 2 - ka^2 D\left(\frac{n^2}{n_0^2} - 1\right) \right] dz' \right\}, \tag{7.388}$$

where z_0 is a constant to be specified. Equation (7.387) now becomes

$$2i\frac{\partial \psi_4}{\partial \bar{z}} + H(\bar{z})\psi_4 = 0, \tag{7.389}$$

where the operator H is defined to be

$$H(\bar{z}) = \exp\left\{ -\frac{i}{2} \int_{z_0}^{z} \left[\tilde{x}^2 + \tilde{y}^2 - 2 - k^2 a^2 D\left(\frac{n^2}{n_0^2} - 1\right) \right] dz' \right\} \tilde{V}_T^2$$
$$\cdot \exp\left\{ +\frac{i}{2} \int_{z_0}^{z} \left[\tilde{x}^2 + \tilde{y}^2 - 2 - k^2 a^2 D\left(\frac{n^2}{n_0^2} - 1\right) \right] dz' \right\}. \tag{7.390}$$

The specification of $H(\bar{z})$ for numerical work is now required. $H(\bar{z})$ is chosen [7.45] to operate midway between the computational planes at $\bar{z}$ and $\bar{z} + \Delta \bar{z}$. In addition the lower limit $\bar{z}_0$ is chosen to be at $\bar{z} + \Delta \bar{z}/2$ as well. Thus $H(\bar{z})$

becomes $H(\bar{z}) = \tilde{V}_T^2$ so that (7.389) is now much simplified and looks like a free-space propagation equation. In summary, the operations taken to propagate one step in range are: a) removal of nonlinear phase at z by (7.388), b) propagation of the new amplitude from z to $z + \Delta z$ by (7.389), and c) replacement of phase at the new plane $z + \Delta z$ by the inverse to (7.388).

The solution to (7.389) is obtained by a variety of means which have been discussed above: explicit, implicit, alternating direction, and fast Fourier transform.

Adaptive Coordinate Systems

The nonadaptive changes leading to (7.387) can be generalized to allow their application without a priori knowledge of the phase front tilt and curvature as a function of range [7.46]. At each range step the amplitude ψ is altered by removing the average tilt, curvature, and scale of the phase front. At some plane z the phase Φ of the wave amplitude measured with respect to the equiphase surface Φ_0 of the local coordinate system is

$$\Phi(x, y) = \Phi_0 + V\Phi \cdot d\mathbf{r}, \tag{7.391}$$

where $\mathbf{r}$ lies in the coordinate equiphase surface. Now, a tilt and refocusing correction is performed on the amplitude to minimize this phase gradient. Before the correction the amplitude is of the form

$$\psi = A e^{-i\phi}.$$

Correction is applied to introduce compensating tilt via linear phase in x, y and curvature via quadratic phase in x, y and defines thereby a corrected amplitude

$$\psi_c = A e^{-i\phi} e^{-i(\alpha_x x^2 + \alpha_y y^2 + \beta_x x + \beta_y y)} = A e^{-i\phi'}, \tag{7.392}$$

where α_x, α_y, β_x, β_y are determined by minimizing $V\Phi'$ *on the average* over the computational plane. The minimization is weighted by the beam intensity so that the curvature and tilt at the highest intensity portion of the beam contribute most strongly. That is, the condition,

$$\iint dx\,dy |\psi(x, y)|^2 [V\Phi'(x, y)] = \text{minimum} \tag{7.393}$$

is satisfied by taking partial derivatives of (7.393) with respect to the α's and β's to get four equations for the four unknowns. These are then used in (7.392) to reduce the linear and quadratic phases. In addition, coordinate changes in scale and a shift of the center of the coordinate system are also applied to preserve the form invariance of the basic partial differential equation. This permits the algorithm to remain unchanged throughout the calculation. The process by which we derived (7.387) can now be generalized. We want to do this because to be automatically adaptive we need a $D(\bar{z})$ determined by the calculation itself.

Equation (7.387) is now generalized to

$$
\frac{2i\partial\psi_c}{\partial\bar{z}} + \frac{1}{D_x}\frac{\partial^2\psi_c}{\partial x'^2} + \frac{1}{D_y}\frac{\partial^2\psi_c}{\partial y'^2} + [2(\gamma_x + \gamma_y)
$$
$$
- f_x(x' - x_0)^2 - f_y(y' - y_0)^2]\psi_c
$$
$$
+ \sqrt{D_x D_y}\, k^2 a^2(n^2 - 1)\psi_c = 0, \tag{7.394}
$$

where D_x, D_y are now calculated from α_x, α_y, β_x, β_y at each step, as are γ_x, γ_y, f_x, f_y. This algorithm, while requiring more calculation per step, actually is not much more expensive of computer time for a given problem than the nonadaptive system. This is because a coarser sampling both in the transverse plane and in range steps is allowed since the coordinate surfaces better follow the phase surfaces of the distorting beam.

7.5.3 Computer Results

Thus far we have described the various methods for calculating the thermally bloomed wave amplitude at a series of steps between the laser and the final range of interest. The assumption was made that the index of refraction field was properly calculated as we progressed. We now turn our attention to the subroutines which calculate the condensation and hence changes in the index of refraction as the blooming proceeds.

Recall from the discussion of Section 7.3 that it is convenient to isolate various time regimes of absorber response. We review these results here and indicate for each time regime just how the hydrodynamic response of the absorber is treated for the purpose of computation.

Early Time Regime

This regime is the period of time between the turn on of the laser and the time when some cooling mechanism such as convection or conduction creates a steady state. There is a further subdivision within this regime which differentiates between pulses which are short compared to the acoustic transit time of the beam and those which are much longer than this time.

For the purpose of deriving an analyticially tractible expression for the induced density changes in each of these regimes, recall that we made suitable approximations to the full linearized set of hydrodynamic equations leading to short pulse and long pulse expressions, (7.115) and (7.122), respectively. No such approximation is required for solution of the problem by computer. That is, we can integrate the partial differential equation (7.105) directly on the computer [7.47]. Thus, the long pulse and short pulse regimes are studied uniformly without any further approximation. Some workers have chosen to deal directly with the numerical solution of the hydrodynamic equations for conservation of mass, momentum, and energy [7.48].

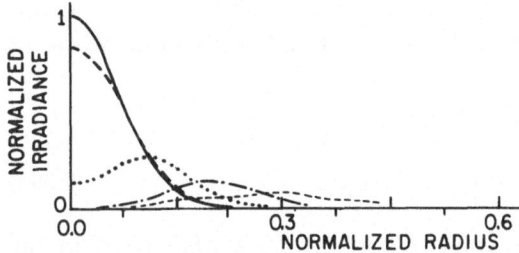

If the output of the laser is circularly symmetric, the propagation problem will have cylindrical symmetry throughout this time regime since the external nonsymmetrical influences of forced and free convection are too slow to alter the symmetry. Laser modes with this property have been used in the single-pulse studies so that the problem could be written in three variables, r, z, t, in order to save on computer run times. As we have discussed above, the pulsed blooming of a Gaussian beam is characterized by a spread of the beam away from the axis and a drop of on-axis intensity at any range. The self-consistent code confirms this behavior; typical results are given in Fig. 7.10, where the radial distributions of an initially gaussian beam at focus are given for a series of times after the arrival of the laser pulse.

Transient Regime

There is a variety of possible numerical approaches to solutions in this regime. One way is to solve the isobaric fluid relation, (7.104), for the condensation with transverse flow velocity v_0 in the x direction

$$\frac{\partial}{\partial t} s(r, t) + v_0 \frac{\partial}{\partial x} s(r, t) = \frac{-(\gamma-1)}{\gamma p_0} \alpha I(r, t) \tag{7.395}$$

by numerically integrating along characteristics. Another is to numerically evaluate the integral representing the solution of (7.395),

$$s(x, y, z, t) = -(\gamma-1)\alpha/(\gamma p_0) \int_0^t I[x - v_0(t - t'), y, z, t'] dt'. \tag{7.396}$$

Still another is to take the Fourier transform of (7.395) and use fast Fourier transform algorithms to efficiently evaluate the condensationn.

Equation (7.396) gives the condensation at time t. To create an algorithm we write the analogous expression at time $t + \Delta t$ and express it in terms of the condensation at the earlier time.

$$s(x, y, z, t + \Delta t) = -(\gamma-1)\alpha/(\gamma p_0) \int_0^{t+\Delta t} I[x - v(t + \Delta t - t'), y, z, t'] dt'$$

$$= s(x - v\Delta t, y, z, t) - (\gamma-1)\alpha/(\gamma p_0)$$

$$\int_t^{t+\Delta t} I[x - v(t + \Delta t - t'), y, z, t'] dt'. \tag{7.397}$$

We need the condensation at $t + \Delta t$ in order to calculate $I(t + \Delta t)$. But this requires previous knowledge of $I(t + \Delta t)$. This follows, for example, when we attempt to solve (7.397) by trapezoidal rule, so that

$$s(x, y, z, t + \Delta t) = s(x - v\Delta t, y, z, t) - (\gamma - 1)\alpha \Delta t/(2\gamma p_0)$$
$$\cdot [I(x, y, z, t + \Delta t) + I(x - v\Delta t, y, z, t)]. \tag{7.398}$$

Note, however, that we do not know $I(x, y, z, t + \Delta t)$ at this point in the computation. In fact, we shall use $s(x, y, z, t + \Delta t)$ to calculate it. Thus we *estimate* $I(x, y, z, t + \Delta t)$ from the time derivative in difference form,

$$I(x, y, z, t + \Delta t) \cong I(x, y, z, t) + \Delta t \frac{\partial I}{\partial t}(x, y, z, t) + \dots$$
$$\cong I(x, y, z, t) + \tfrac{1}{2}[I(x, y, z, t + \Delta t)$$
$$- I(x, y, z, t - \Delta t)]. \tag{7.399}$$

This differencing procedure for the representation of the time derivative is essentially a "predictor" which can prove to be unstable. If the intensity is rapidly changing in time, care must be exercised to sample the time development sufficiently often to assure stability.

Solving (7.399) for the required intensity, we have

$$I(x, y, z, t + \Delta t) \cong 2I(x, y, z, t) - I(x, y, z, t - \Delta t). \tag{7.400}$$

This expression is correct to order $(\Delta t)^2$ which is the same order of accuracy as the trapezoidal rule. Thus, the algorithm used to evaluate (7.396) is

$$s(x, y, z, t + \Delta t) = s(x - v\Delta t, y, z, t) - (\gamma - 1)\alpha \Delta t/(2\gamma p_0)$$
$$\cdot [2I(x, y, z, t) - I(x, y, z, t - \Delta t)$$
$$+ I(x - v\Delta t, y, z, t)]. \tag{7.401}$$

When these dependent variables, s and I, are made discrete, it will generally turn out that $v\Delta t$ will not necessarily be an integral multiple of Δx. Thus, a minor complication is that the values of s and I needed to calculate the right-hand side of (7.401) must be linearly interpolated from data at mesh points.

The Fourier method for computer solution of (7.396) proceeds as follows. Both s and I are represented by finite Fourier series

$$s = \sum_{n=1}^{N_x} \sum_{m=1}^{N_y} s_{nm}(t) \exp(2\pi i n x/L + 2\pi i m y/L), \tag{7.402}$$

$$I = \frac{-\gamma p_0}{\alpha(\gamma - 1)} \sum_{n=1}^{N_x} \sum_{m=1}^{N_y} f_{nm}(t) \exp(2\pi i n x/L + 2\pi i m y/L). \tag{7.403}$$

For simplicity, we omit writing the variables y and z and assume a convective flow of absorber in the x direction. Equations (7.402) and (7.403) when substituted into (7.395) yield, on account of the orthogonality of the functions $\exp(2\pi inx/L)$ over the interval $(0, L)$,

$$\frac{\partial s_n}{\partial t} + \frac{2\pi inv}{L} s_n = f_n. \tag{7.404}$$

This has solution at $t + \Delta t$,

$$s_n(t + \Delta t) = s_n \exp(-2\pi inv\Delta t/L)$$
$$+ \int_t^{t+\Delta t} f_n(t') \exp[2\pi inv(t + \Delta t - t')/L] dt'. \tag{7.405}$$

This, of course, is just the finite Fourier transform of (7.397). The exponential $\exp(2\pi inv\Delta t/L)$ when multiplying a finite Fourier transform will create an inverse finite Fourier transform which is translated along the x axis according to $x \to x + v\Delta t$, provided that

$$v\Delta t = n\Delta x, \quad n = 0, 1 ... N_x - 1. \tag{7.406}$$

For $v\Delta t$ not an integral multiple of Δx, the exponentials in (7.405) will cause the transformed function $s(x, t)$ to ring in an undesirable manner. This situation can be remedied in two ways. First, $v\Delta t/\Delta x$ can be rounded to the nearest integer value before evaluating the exponentials in (7.405). A more exact procedure [7.42] is to represent the exponential by a weighted average

$$\exp(2\pi inv\Delta t/L) \cong q \exp(2\pi inn_1/L)$$
$$+ (1 - q) \exp[2\pi in(n_1 + 1)/L], \tag{7.407}$$

where n_1 is the greatest integer less than $v\Delta t/\Delta x$ and q is given by $q = (v\Delta t - n_1\Delta x)/\Delta x$. This procedure is analogous to the linear interpolation mentioned above in the direct time integral solution [7.49]. Both (7.401) and (7.405) are in a form suitable for self-consistent calculation of the condensation included in the absorbing medium by the laser beam of intensity $I(x, y, z, t)$. Two values of the intensity representing times t and $t - \Delta t$ are stored for each point in the transverse plane. These locations are written over as the solution marches to the next range step. The values of intensity used are always the current updated values so that the condensation is correctly calculated. Similarly, the propagation of the laser beam amplitudes to the next calculational plane is always done with the updated values of the density and refractive index arrays so that the problem is solved self-consistently. Results of a calculation using these techniques are presented in Fig. 7.11. Shown is a plot of a ratio of peak intensity at any time to the peak intensity at zero time vs time for a laser beam

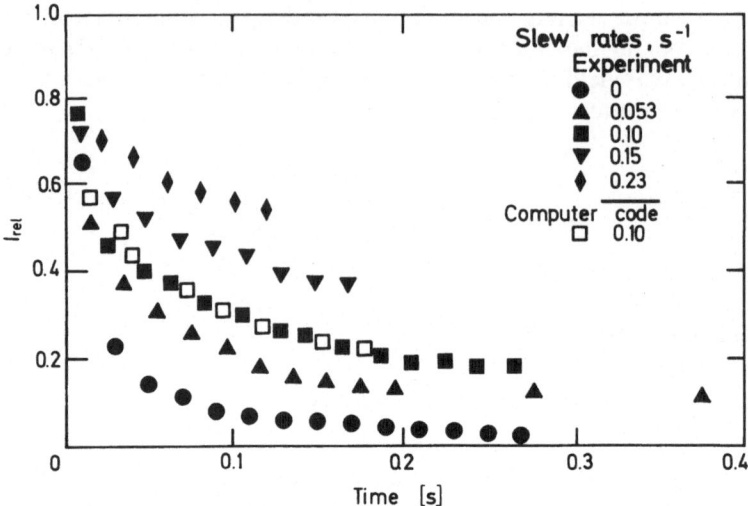

Fig. 7.11. Results of an experiment which shows the transient behavior of laser beam intensity when it is passed through a region of stationary absorber at different slewing rates; shown superimposed are results of a computer calculation of the experimental arrangement for one of the slewing rates, $\omega = 0.10$; the agreement is excellent

propagating through an absorbing cell which is being made to rotate in a simulation of a slewing beam. Good agreement is seen between the experimental points and a calculation based on (7.401) for the slew rate of $0.10 \, \text{rad s}^{-1}$.

Multiple Pulses

A special case of laser waveform is created when the laser emits a train of short pulses. If the pulse length t_p of the pulses is very much shorter than the interpulse spacing $t_n - t_{n-1}$, then (7.395) becomes

$$\frac{\partial s}{\partial t} + v\frac{\partial s}{\partial x} = -\frac{(\gamma - 1)\alpha}{\gamma p_0} \sum_{n=1}^{N} t_p I_n(x, y)\delta(t - t_n), \qquad (7.408)$$

where $t_n = n\Delta t$, $t_p I_n$ represents the pulse fluence (energy/area), and $\delta(t)$ is the Dirac delta function representing a vanishingly short pulse.

The solution to (7.408) is

$$s(x, y, z, n\Delta t) = -\frac{(\gamma - 1)\alpha t_p}{\gamma p_0} \sum_{j=0}^{n-1} I_j[x - v(n - j)\Delta t, y, z], \qquad (7.409)$$

where we have stopped the sum at $n-1$ since it has been assumed that the pulses are so short that they make negligible contribution to the density change until the pulse has long since passed. The energy absorbed leaves a residual density distribution, however, which, on the time scale of beam diameter over

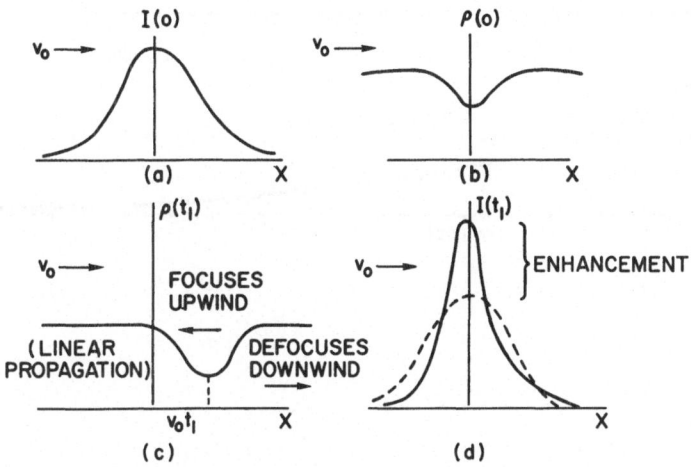

Fig. 7.12a–d. Schematic of the overlap enhancement phenomenon which can occur for multiple pulse laser beams in an absorbing fluid; see the text for an explanation of the sequence shown here

sound speed and larger, contributes to the total accumulated density change. Each pulse is refracted by the index field which has evolved up to the instant the pulse is delivered. The absorbed energy of the pulse is stored in the medium, translated by the convective flow, and added to the accumulated index field changes for the passage of the next pulse in the train.

An interesting effect is noted when pulses are propagated through heated air that has not been completely swept away since the preceding pulse. Reference to Fig. 7.12 will help provide a qualitative explanation of the phenomenon. At $t=0$, the gaussian irradiance shown in Fig. 7.12a produces the index of refraction field shown in Fig. 7.12b after pressure relaxation. If the next pulse is delivered at time $t=r/v_0$ where r is the "radius" of the gaussian beam and v_0 is the velocity of the flow convecting the heated medium away from the beam path, the upstream half of the new pulse will propagate in essentially unheated medium and will suffer little or no perturbation. The downstream half of the beam will be focused back upstream due to the negative gradient of the index field as indicated in Fig. 7.12c. Some of the low intensity portions of the new beam will be bent downstream and transverse to the stream velocity. The dominant effect, however, is the focusing into the upstream portion to add to the unperturbed intensity, there producing an enhancement of peak intensity over that of a non blooming beam, as shown in Fig. 7.12d. This enhancement is crucially dependent on stream velocity and beam shape. It is instructive to plot the peak intensity of the beam vs the pulse repetition rate (PRR) for the case where the average power is constant, so that, as PRR increases, the energy per pulse decreases in inverse proportion (see Fig. 7.13). When the PRR is near zero, the time average peak irradiance approaches the low power limit since, for fixed stream velocity, low PRR implies complete sweep-out of distorted medium between pulses. In the other extreme of high PRR, the time average

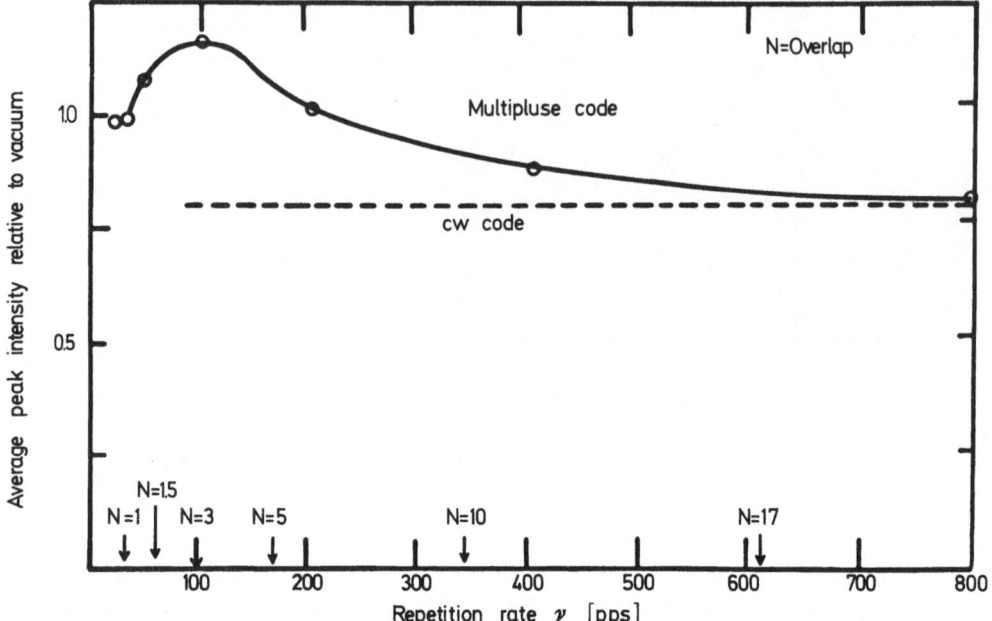

Fig. 7.13. Average peak relative intensity of a multiple pulse laser beam as a function of repetition rate: for each repetition rate the beam energy per pulse is chosen to keep the average total power constant; the enhancement phenomena discussed in the text and explained by the previous figure is clearly seen here at overlap numbers of 1.5-5

peak irradiance approaches the steady-state result for the same launched power. For intermediate PRR's, where about 50 % of the beam passes through heated medium, the enhancement phenomenon is seen to take place.

Steady-State Regime

In the limit of times long compared with stream transit time of the largest portion of the beam, a steady state evolves and (7.395) becomes

$$v\frac{\partial s}{\partial x} = -\frac{(\gamma-1)\alpha I}{\gamma p_0}. \tag{7.410}$$

If we assume that $I(x, y, z)=0$ at $x= -\infty$, $(v>0)$, then the solution to (7.410) is

$$s(x, y, z) = -\frac{(\gamma-1)\alpha}{\gamma p_0} \int_{-\infty}^{x} I(x', y, z)dx'. \tag{7.411}$$

This condensation is computed by a simple algorithm to evaluate the integral using the beam intensities at z which have been propagated to this z plane by a series of finite steps, $n\Delta z$. For most beam shapes with a single on-axis intensity maximum the index field produced by (7.411) is higher upstream than it is

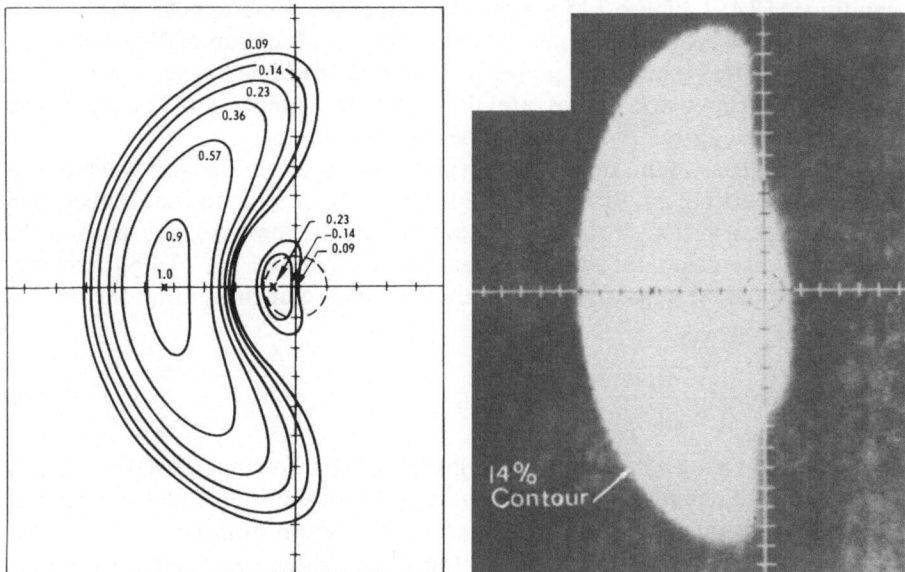

Fig. 7.14. Comparison of steady-state cw thermal blooming profiles with experimental results (burn pattern) for a 10.6 μm laser beam

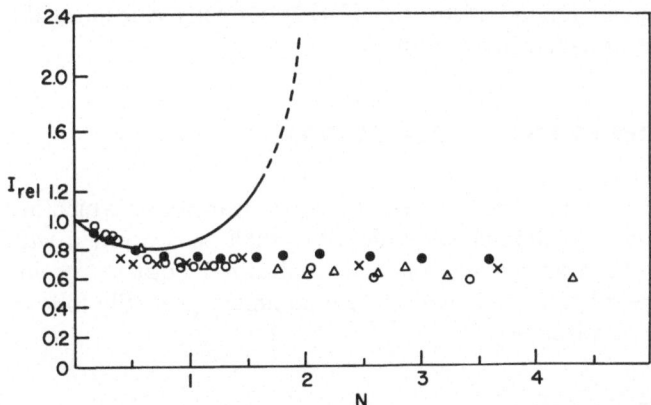

Fig. 7.15. Ratio of intensity with and without absorption as a function of a distortion number N, which is a quantity proportional to beam power: the solid-dashed line is a geometric optics calculation, the solid points are computer wave optics calculations, and the other points are experimental results; clearly, a self-consistent wave optics calculation was required to model these experiments

downstream so that light rays are bent *into* the stream, i.e., into regions of higher index. In addition, rays are bent away from the beam in the positive and negative y directions. The result of this refraction is to produce a crescent shape such as shown in Fig. 7.14, which is a computer calculation of the isoirradiance contours for a typical case compared with a laboratory result of similar conditions [7.45]. Figure 7.15 is a comparison of the computer calculation for a steady-state cw beam with experimental results as a function of N, a distortion parameter related to the laser power. The solid-dashed line is a geometric optics calculation. The black circles are the computed points for a wave optics treatment. The agreement is seen to be excellent.

A special case of the steady-state regime occurs when the laser waveform is a series of short pulses. After the pulse train has been on for a time longer than the stream transit time, a steady state evolves for the same reasons given in the preceding discussion for the cw beam. In this case I_j in (7.409) becomes independent of the subscript j except through its argument, so that the density change is given by,

$$s(x, y, z) = -\frac{(\gamma - 1)\alpha t_p}{\gamma p_0} \sum_{j=0}^{\infty} I(x - vj\Delta t, y, z). \tag{7.412}$$

The density change in this case is given by the stream-translated density changes due to an infinite series of identically distorted beams. This steady state is rapidly calculated since the propagation calculation from aperture to range of interest need only be done once for the stream-shifted sum to be performed. Thus, computer run times for this case are comparable to the cw laser steady-state case of the preceding section. Comparison of results with transient multiple pulse cases in the limit of many more pulses propagated than the number per stream transit time confirm that (7.412) correctly calculates the asymptotic state of the time-dependent problem.

Appendix: Properties of Integral Transforms

In this section we review selected properties of integral transforms which are required for our treatment of thermal blooming. We shall be concerned with both Fourier and Hankel transforms and their application both to specific functions and to the complete wave equation. Our summary generally follows the conventions given by *Sneddon* [7.50].

A.1 Fourier Transforms

We define the Fourier transform $F(\alpha)$ of a function $f(x)$ by the relation

$$F(\alpha) = \frac{1}{\sqrt{2\pi}} \int_{-\infty}^{\infty} f(x) \exp(i\alpha x) dx. \tag{A.1}$$

Then under certain conditions, the function $f(x)$ can be recovered by taking the inverse transform

$$f(x) = \frac{1}{\sqrt{2\pi}} \int_{-\infty}^{\infty} F(\alpha) \exp(-i\alpha x) d\alpha. \tag{A.2}$$

The sufficient conditions required to ensure the validity of (A.2) are that

$$\int_{-\infty}^{\infty} |f(x)|dx$$

be convergent and that $f(x)$ satisfy the Dirichlet conditions, namely:

1) $f(x)$ has only a finite number of maxima and minima
2) $f(x)$ has only a finite number of finite discontinuities and no infinite discontinuities.

These conditions are generally satisfied for most physical problems although there are important exceptions which require special handling. It is also true that under some circumstances one can find weaker conditions for the validity of the inversion theorem (A.2).

We frequently encounter the Fourier transform of the derivative of a function. Thus we may write for the Fourier transform of the rth derivative

$$F^{(r)}(\alpha) = \frac{1}{\sqrt{2\pi}} \int_{-\infty}^{\infty} \left[\frac{d^r}{dx^r} f(x)\right] \exp(i\alpha x) dx. \tag{A.3}$$

If we integrate by parts we obtain the expression

$$F^{(r)}(\alpha) = \frac{1}{\sqrt{2\pi}} \left[\exp(i\alpha x) \frac{d^{r-1}f}{dx^{r-1}}\right]_{x=-\infty}^{\infty}$$

$$- \frac{(i\alpha)}{\sqrt{2\pi}} \int_{-\infty}^{\infty} \exp(i\alpha x) \frac{d^{r-1}f}{dx^{r-1}} dx. \tag{A.4}$$

If the $(r-1)$st derivative vanishes as $|x|$ tends to infinity the first term in (A.4) may be dropped. The process of partial integration may be repeated so that we find eventually

$$F^{(r)}(\alpha) = (-i\alpha)^r F(\alpha), \tag{A.5}$$

provided that the first $(r-1)$ derivatives vanish as $|x|$ goes to infinity.

A.2 Hankel Transforms

It is most useful for our purposes to define the Hankel transform $F_n(\lambda)$ of a function $f(r)$ by the relation

$$F_n(\lambda) = \int_0^{\infty} rf(r)J_n(\lambda r)dr, \tag{A.6}$$

where $J_n(\lambda r)$ is the nth order Bessel function of the first kind. Under conditions similar to those given above for Fourier transforms the inversion theorem can be written

$$f(r) = \int_0^\infty \lambda F_n(\lambda) J_n(\lambda r) d\lambda. \tag{A.7}$$

We can also evaluate the Hankel transforms of derivatives of a function by repeated partial integration in the manner given above. However the process of differentiation introduces Bessel functions of other orders which generally makes the result appear more complicated than the simple expression (A.5). There is one special case which is of particular importance to us that does yield a simple result. If we take the zeroth-order Hankel transform of the Laplacian of a function $f(r)$ we obtain after two partial integrations

$$\int_0^\infty r\left(\frac{d^2 f}{dr^2} + \frac{1}{r}\frac{df}{dr}\right) J_0(\lambda r) dr = -\lambda^2 F(\lambda), \tag{A.8}$$

where $F(\lambda)$ is the zeroth-order Hankel transform of $f(r)$ and we have assumed that both rdf/dr and $rf(r)$ vanish as r tends to either zero or infinity.

The required conditions are generally satisfied for problems of physical interest. There are important exceptions, however, such as the radiation of cylindrical waves from an extended harmonic cylindrical source where the required solution is not absolutely integrable. These cases require special handling. The summary presented above will be adequate for our needs in the analysis presented in this chapter.

References

7.1   R.C.C.Leite, R.S.Moore, J.R.Whinnery: Appl. Phys. Lett. **5**, 141 (1964)
7.2   P.B.Ulrich: *"Numerical Methods in High Power Laser Propagation"*, AGARD Conference Proceedings No. 183, Optical Propagation in the Atmosphere, Paper No. 31. 27–31 October 1975 (National Technical Information Service, Springfield, Va. 22151)
7.3   A.C.Melissinos, F.Lobkowicz: *Physics for Scientists and Engineers*, Vol. 1 (W.B.Saunders Company, Philadelphia 1975) Chaps. 14–16
7.4   J.D.Jackson: *Classical Electrodynamics*, (Wiley and Sons, New York, London, Sydney 1962) p. 618
7.5   H.Kogelnik, T.Li: Proc. IEEE **54**, 1312 (1966); Appl. Opt. **5**, 1550 (1966)
7.6   M.Born, E.Wolf: *Principles of Optics*, 4th ed. (Pergamon Press, New York 1970)
7.7   A.Erdelyi, W.Magnus, F.Oberhettinger, F.G.Tricomi: *Tables of Integral Transforms*, Vol. I (McGraw-Hill, New York 1954) p. 11
7.8   P.M.Morse, H.Feshbach: *Methods of Theoretical Physics*, Part II (McGraw-Hill, New York 1953) p. 1092
7.9   B.B.Baker, E.T.Copson: *The Mathematical Theory of Huygens Principle*, 2nd ed. (Clarendon Press, Oxford 1950) pp. 23–26

7.10 Yu, A.Kravtsov, Z.I.Feizulin: Izv. VUZ, Radiofiz. **12**, 886 (1969); [English transl. Radiophys. Quant. Electron. **12**, 706 (1969)]
7.11 R.F.Lutomirski, H.T.Yura: Appl. Opt. **10**, 1652 (1971)
7.12 L.D.Landau, E.M.Lifshitz: *Fluid Mechanics*, Engl. transl. (Pergamon Press, London 1959) Chap. I
7.13 J.Wallace, M.Camac: J. Opt. Soc. Am. **60**, 1587 (1970)
7.14 J.N.Hayes: *Thermal Blooming of Laser Beams in Gases*, Naval Res. Lab. Rept. 7213, AD 720508 (Feb. 11, 1971)
7.15 E.L.Ince: *Ordinary Differential Equations* (Dover Publ., New York 1956)
7.16 G.N.Watson: *A Treatise on the Theory of Bessel Functions* (Cambridge University Press 1966) p. 394
7.17 M.Abramowitz, I.A.Stegun: *Handbook of Mathematical Functions* (U.S. Government Printing Office, Washington, DC 1964) pp. 297, 298, 319
7.18 I.N.Sneddon: *Fourier Transforms* (McGraw-Hill, New York, Toronto, London 1951)
7.19 A.D.Wood, M.Camac, E.T.Gerry: Appl. Opt. **10**, 1877 (1971)
7.20 J.B.Keller: "*Proceedings of Symposia in Applied Mathematics*", Vol. XIII (American Mathematical Society, Providence 1962) pp. 230, 231
7.21 P.V.Avizonis, C.B.Hogge, R.R.Butts, J.R.Kenemuth: Appl. Opt. **11**, 554 (1972)
7.22 P.G.Bergmann: Phys. Rev. **70**, 486 (1946)
7.23 M.Kline: Comm. Pure Appl. Math. **14**, 473 (1961)
7.24 A.H.Aitken, J.N.Hayes, P.B.Ulrich: Appl. Opt. **12**, 193 (1973)
7.25 P.B.Ulrich: "A Wave Optics Calculation of Pulsed Laser Propagation in Gases", Naval Res. Lab. Rept. 7413, NTIS No. AD 745739 (June 7, 1972)
7.26 A.Sommerfeld: *Partial Differential Equations in Physics* (Academic Press, New York 1964)
7.27 R.D.Richtmyer, K.W.Morton: *Difference Methods for Initial-Value Problems*, 2nd ed. (Wiley-Interscience, New York 1967)
7.28 R.Courant, D.Hilbert: *Methods of Mathematical Physics*, Vol. II (Wiley-Interscience, New York 1962)
7.29 T.N.E.Greville: *Theory and Applications of Spline Functions* (Academic Press, New York, London 1969)
7.30 J.Wallace,Jr., J.Q.Lilly: J. Opt. Soc. Am. **64**, 1651 (1974)
7.31 W.T.Cochran, J.W.Cooley, D.L.Favin, H.D.Helms, R.A.Kaenel, W.W.Lang, G.C.Maling, D.E.Nelson, C.M.Rades, P.D.Welch: "What is the Fast Fourier Transform?", G-AE Subcommittee on Measurement Concepts, IEEE Trans. **AE-15** (2), 45 (1967)
7.32 P.J.Lynch, D.L.Bullock: "An Integral-Equation Formulation of Thermal Blooming"; TRW Rept., TRW Systems Group, 1 Space Park, Redondo Beach, CA, 90278 (14. August 1974)
7.33 R.F.Lutomirski, W.L.Woodie, A.R.Hines, M.A.Dore: "Maritime Aerosol Effects on High-Energy Laser Propagation"; Pacific-Sierra Res. Corp. Rept. 510-S (September 1975) p. 54
7.34 P.L.Kelley: Phys. Rev. Lett. **15**, 1005 (1965)
7.35 H.F.Harmuth: J. Math. Phys. **36**, 269 (1957)
7.36 J.Crank, P.Nicolson: Comb. Phil. Soc. Proc. **43**, 50 (1947)
7.37 D.W.Peaceman, H.H.Rockford: J. Soc. Indust. Appl. Math. **3**, 28 (1955)
7.38 J.Douglas,Jr.: J. Soc. Indust. Appl. Math. **3**, 42 (1955)
7.39 J.Douglas,Jr., H.H.Rockford: Trans. Am. Math. Soc. **82**, 421 (1956)
7.40 G.E.Forsyth, W.R.Wasow: *Finite Difference Methods for Partial Differential Equations* (Wiley and Sons, New York, London 1967)
7.41 H.J.Breaux: "An Analysis of Mathematical Transformations and a Comparison of Numerical Techniques for Computation of High Energy cw Laser Propagation in an Inhomogeneous Medium", U.S. Army Ballistic Res. Lab. Rept. 1723 (June 1974)
7.42 J.A.Fleck,Jr., J.R.Morris,M.D.Feit: "Time Dependent Propagation of High Energy Laser Beams through the Atmosphere", Lawrence Livermore Lab. Rept. UCRL-51826 (June 2, 1975); Appl. Phys. **10**, 129 (1976); **14**, 99 (1977)
7.43 P.B.Ulrich: "PROP-I: An Efficient Implicit Algorithm for Calculating Nonlinear Scalar Wave Propagation in the Fresnel Approximation"; Naval Res. Lab. Rept. 7706 (May 29, 1974)

7.44 L.C.Bradley, J.Herrmann: J. Opt. Soc. Am. **61**, 668 (1971)

7.45 L.C.Bradley, J.Herrmann: "Numerical Calculation of Light Propagation", MIT Lincoln Lab. Rept. LTP-10 (July 12, 1971)

7.46 L.C.Bradley, J.Herrmann: "Notes on the Lincoln Laboratory Propagation Code" (1974) (unpublished notes)

7.47 P.B.Ulrich, J.Wallace,Jr.: J. Opt. Soc. Am. **63**, 8 (1973)

7.48 P.B.Ulrich: "A Numerical Calculation of Thermal Blooming of Pulsed, Focused Laser Beams". Naval Res. Lab. Rept. 7382 (December 30, 1971)

7.49 P.J.Berger, P.B.Ulrich, J.T.Ulrich, F.G.Gebhardt: Appl. Opt. **16**, 345 (1977)

7.50 I.N.Sneddon: *Fourier Transforms* (McGraw-Hill, New York 1951)

Subject Index

Applied Physics

A monthly journal

Board of Editors
S. Amelinckx, Mol. **V.P. Chebotayev**, Novosibirsk
R. Gomer, Chicago, IL., **H. Ibach**, Jülich
V.S. Letokhov, Moskau, **H.K.V. Lotsch**, Heidelberg
H.J. Queisser, Stuttgart, **F.P. Schäfer**, Göttingen
A. Seeger, Stuttgart, **K. Shimoda**, Tokyo
T. Tamir, Brooklyn, NY, **W.T. Welford**, London
H.P.J. Wijn, Eindhoven

Coverage
application-oriented experimental and theoretical
physics:

Solid-State Physics   *Quantum Electronics*
Surface Sciences   *Laser Spectroscopy*
Solar Energy Physics   *Photophysical Chemistry*
Microwave Acoustics   *Optical Physics*
Electrophysics   *Integrated Optics*

Special Features
rapid publication (3-4 months)
no page charge for **concise** reports
prepublication of titles and abstracts
microfiche edition available as well

Languàges
mostly English

Articles
original reports, and short communications review
and/or tutorial papers

Manuscripts
to Springer-Verlag (Attn. H. Lotsch), P.O. Box 105 280
D-6900 Heidelberg 1, F.R. Germany

Springer-Verlag
Berlin
Heidelberg
New York

Place North-American orders with:
Springer-Verlag New York Inc., 175 Fifth Avenue,
New York, N.Y. 100 10, USA

R.H. Kingston

Detection of Optical and Infrared Radiation

1978. 39 figures. Approx. 160 pages
(Springer Series in Optical Sciences,
Volume 10)
ISBN 3-540-08617-X

Contents:
Thermal Radiation and Electromagnetic
Modes. – The Ideal Photon Detector. –
Coherent or Heterodyne Detection. –
Amplifier Noise and Its Effect on Detector Performance. – Vacuum Photodetectors. – Noise and Efficiency of Semiconductor Devices. – Thermal Detection. –
Laser Preamplification. – The Effects of
Atmospheric Turbulence. – Detection
Statistics. – Selected Applications.

Laser Monitoring of the Atmosphere

Editor: **E.D. Hinkley**

1976. 84 figures. XV, 380 pages
(Topics in Applied Physics, Volume 14)
ISBN 3-540-07743-X

Contents:
E.D. Hinkley: Introduction. –
S.H. Melfi: Remote Sensing for Air
Quality Management. –
V.E. Zuev: Laser-Light Transmission
through the Atmosphere. –
R.T.H. Collis, P.B. Russell: Lidar
Measurement of Particles and Gases by
Elastic Backscattering and Differential
Absorption. –
H. Inaba: Detection of Atoms and Molecules by Raman Scattering and Resonance Fluorescence. –
E.D. Hinkley, R.T. Ku, P.L. Kelley: Techniques for Detection of Molecular Pollutants by Absorption of Laser Radiation. –
R.T. Menzies: Laser Heterodyne Detection Techniques.

Optical and Infrared Detectors

Editor: **R. J. Keyes**

1977. 115 figures, 13 tables. XI, 305 pages
(Topics in Applied Physics, Volume 19)
ISBN 3-540-08209-3

Contents:
R.J. Keyes: Introduction. –
P.W. Kruse: The Photon Detection
Process. –
E.H. Putley: Thermal Detectors. –
D. Long: Photovoltaic and Photoconductive Infrared Detectors. –
H.R. Zwicker: Photoemissive Detectors. –
A.F. Milton: Charge Transfer Devices for
Infrared Imaging. –
M.C. Teich: Nonlinear Heterodyne
Detection.

B. Saleh

Photoelectron Statistics

With Applications to Spectroscopy and
Optical Communication

1978. 85 figures, 8 tables. XV, 441 pages
(Springer Series in Optical Sciences,
Volume 6)
ISBN 3-540-08295-6

Contents:
Tools from Mathematical Statistics:
Statistical Description of Random
Variables and Stochastic Processes. Point
Processes. – Theory: The Optical Field:
A Stochastic Vector Field or, Classical
Theory of Optical Coherence. Photoelectron Events: A Doubly Stochastic
Poisson Process. – Applications: Applications to Optical Communication.
Applications to Spectroscopy.

Springer-Verlag
Berlin Heidelberg New York